Surface Preparation and Microscopy of Materials

Surface Preparation and Microscopy of Materials

Brian Bousfield

Buehler Europe Ltd, Coventry, UK

JOHN WILEY & SONS
Chichester · New York · Brisbane · Toronto · Singapore

Copyright ©1992 by John Wiley & Sons Ltd.
Baffins Lane, Chichester
West Sussex PO19 1UD, England

All rights reserved.

No part of this book may be reproduced by any means,
or transmitted, or translated into a machine language
without the written permission of the publisher.

Other Wiley Editorial Offices

John Wiley & Sons, Inc., 605 Third Avenue,
New York, NY 10158-0012, USA

Jacaranda Wiley Ltd, G.P.O. Box 859, Brisbane,
Queensland 4001, Australia

John Wiley & Sons (Canada) Ltd, 22 Worcester Road,
Rexdale, Ontario M9W 1L1, Canada

John Wiley & Sons (SEA) Pte Ltd, 37 Jalan Pemimpin #05-04,
Block B, Union Industrial Building, Singapore 2057

Library of Congress Cataloging-in-Publication Data
Bousfield, Brian.
   Surface preparation and microscopy of materials / Brian Bousfield.
     p.   cm.
   Includes bibliographical references and index.
   ISBN 0 471 93181 0
   1. Metallographic specimens. 2. Metallography. I. Title.
TN690.7.B68   1992   91-32125
621.382—dc20   CIP

British Library Cataloguing in Publication Data
A catalogue record for this book is available from the British Library.

ISBN 0 471 93181 0

Typeset in 10/11 pt Palatino by Dobbie Typesetting, Limited, Tavistock, Devon
Printed and bound in Great Britain by Courier International, East Kilbride, Scotland

To Mavis

Contents

Book Conventions   xiii

PART I: SPECIMEN PREPARATION

1. Introduction   3

    1.1 What is Metallography?   3
    1.2 Damage Depth   4
    1.3 Abrasive Combination   4
    1.4 Z Axis Curves   4

2. Sectioning   7

    2.1 Resultant Structural Damage   7
    2.2 Characteristics Affecting Structural Damage   10
    2.3 Common Faults and Remedies in Abrasive Cutting   15

3. Mounting   16

    3.1 Considerations and Features   16
    3.2 Results and Reasons   18
    3.3 Common Mounting Faults and Remedies   23
    3.4 Sample/Specimen Definition   24

4. Single-point Tools   25

    4.1 Stress Applicators   25
    4.2 Characteristics Affecting Structural Damage   27

5. The New Concept   30

    5.1 Reasons for a Systematic Approach   30
    5.2 The Application of Logic   32
    5.3 Abrasives and Their Functions   34
    5.4 Abrasive Function Definitions   37

5.5	Platen Surface Functions	38
5.6	Efficiency of Charged Platen Surfaces	42
5.7	Parameter Optimization	44
5.8	Preparation Examples and Family Classification	47

6. Grinding — 58

6.1	Traditional Metallography with the 'New Concept'	58
6.2	The Two Grinding Stages	60
6.3	Scratch Pattern Monitoring	63

7. Polishing — 64

7.1	Effects on Surface and Microstructure	64
7.2	Cloths, Abrasives and Carriers	66
7.3	Classification	67
7.4	Electrolytic Polishing	68

8. Grinding and Polishing Lubricants — 71

8.1	Selection Considerations—Grinding	71
8.2	Selection Verification—Grinding	72
8.3	Selection Considerations—Polishing	72
8.4	Selection Verification—Polishing	72

9. Towards a Metallographic Standard — 74

9.1	Introduction to Grinding Surface Classification	74
9.2	Deformation Classification	76
9.3	Dangers Associated with Preparation Time	77
9.4	Material Cutting Related to Microstructure	78
9.5	Polishing Cloth Classification	79
9.6	Databank by Z Axis Curves	79
9.7	Platen Surface Threshold	81
9.8	Stumbling Blocks to Standards	82

10. Characterization: Auditing and Traceable Standards — 84

10.1	Introduction	84
10.2	Creating a Standard Preparation Procedure	85
10.3	The Standard Procedure	87

11. Traditional Methods Only — 89

11.1	Consideration	89
11.2	Preparation Examples	91

12. Preparation of Spray Coatings — 97

12.1	Background	97
12.2	How to Identify a Faithful Reproduction	99
12.3	Aids to Metallography	101
12.4	Microscope Techniques	104
12.5	Plasma-sprayed Preparation Examples	108

13. Preparation of Composites — 116

13.1 Background — 116
13.2 Identifying Bad Techniques — 117
13.3 Preparation Examples — 117
13.4 Biomaterials — 137
13.5 Material Investigations — 138

14. Preparation of Minerals — 143

14.1 Thick Sections for Reflected Light Microscopy — 143
14.2 Thin Sections for Transmitted Light Microscopy — 145
14.3 High Integrity Thin Sections — 146
14.4 Thin Sections for Reflected Light Microscopy — 151
14.5 Preparation Examples — 153

15. Preparation of PCBs and Electronic Components — 164

15.1 Specimen Characteristics — 164
15.2 Preparation Techniques — 167
15.3 Preparation Examples — 169

16. Thin Film Measurement — 175

16.1 Preparation Technique — 175
16.2 Measuring Techniques — 176
16.3 Graticules Measuring — 177

17. Preparation of Soft Materials — 178

17.1 Introduction: The Problems — 178
17.2 Introduction: The Solution — 179
17.3 Preparation Examples — 182

18. Preparation of Ceramics — 191

18.1 Background — 191
18.2 Preparation Techniques for Reflected Light Observation — 192
18.3 Establishing a Procedure — 194
18.4 Preparation Examples — 196

19. Hardness — 203

19.1 Hardness by Indentation — 203
19.2 Microhardness (Micro Loads) — 206
19.3 Possible Operating Errors — 207
19.4 Vickers Hardness Under Load — 208
19.5 Certified Test Blocks — 209

20. Training in Metallography — 210

20.1 Introduction — 210
20.2 Elements of Competence — 210
20.3 Conclusion — 212

21. Supplementary Materials, Techniques and Methods — 213
21.1 A Summary — 213
21.2 Preparation Examples — 213

List of Preparation Examples — 225

PART II: APPLIED MICROSCOPY

22. The Microscope—A Résumé — 229

23. Microscope Types and Nomenclature — 232
23.1 Upright Compound Microscope — 232
23.2 Inverted Compound Microscope — 236
23.3 Stereoscopic Microscope — 237
23.4 Macro Microscope — 238

24. Creating the Microscope Image — 240
24.1 Lens Functions—Objectives — 240
24.2 The Compound Microscope — 242

25. Objective Aberrations — 245
25.1 Spherical Aberration — 245
25.2 Chromatic Aberration (Axial) — 246
25.3 Types of Objectives — 246
25.4 Chromatic Difference of Magnification (Transverse) — 247
25.5 Curvature, Coma, Astigmatism and Distortion — 249
25.6 Contrast — 250

26. Improving the Image — 252
26.1 Numerical Aperture (NA) — 252
26.2 Refractive Index — 252
26.3 Resolution and Magnification — 253
26.4 Depth of Field (df) — 256
26.5 Tube Length Correction — 257
26.6 Intensity of Image — 257

27. Measurements — 259
27.1 Measuring Systems — 259
27.2 Measurement X-Y — 260
27.3 Depth Measurements — 262

28. Illumination Systems — 263
28.1 Illumination Types — 263
28.2 Critical Illumination — 263
28.3 Köhler Illumination — 264
28.4 Oblique Illumination — 265
28.5 Diaphragm Functions — 265
28.6 Procedure for Setting up the Microscope for Köhler Illumination — 266
28.7 Reflected Light Ghost Images — 268

29. Eyepieces and Condensers — 269

29.1 Eyepieces — 269
29.2 Substage Condensers — 271

30. Introduction to Interference — 273

30.1 Diffraction Theory — 273
30.2 Creating the Image — 273
30.3 Phase Contrast — 274
30.4 The Becke Line — 276
30.5 Dispersion Staining — 276

31. Surface Finish Interference — 278

31.1 Quantitative Z Measurements — 278
31.2 Interference Principles — 278
31.3 Equipment Types — 280

32. Contrast Interference — 282

32.1 Differential Interference Contrast (Nomarski) — 282
32.2 Dark Ground — 285

33. Video Imaging and Archiving — 287

33.1 Closed Circuit Television (CCTV) — 287
33.2 Methods of Archiving — 289

34. Polarizing Light Microscopy — 291

34.1 The Equipment — 291
34.2 Practical Objectives — 292
34.3 Techniques — 294
34.4 Crystallography — 297
34.5 Fibre Identification — 299
34.6 Conditioning Samples for Polarized Light Applications — 300
34.7 Practical Examples — 301

35. Fluorescence, Reflectance and Con-focal Microscopy — 303

35.1 Fluorescence Microscopy — 303
35.2 Reflectance Microscopy — 307
35.3 Con-focal Scanning Optical Microscope (SOM) — 308

36. Photomicrography — 310

36.1 Introduction — 310
36.2 Technical Considerations — 310
36.3 Microscope Photographic Systems — 313
36.4 Filters — 316
36.5 Photomicrographs—Check-list for Success — 319

37. Inverted Techniques — 326

37.1 Explanation — 326
37.2 Bright Field — 326

37.3	Dark Ground	326
37.4	DIC	327

38. Photomicrography in Practice — 328

38.1	Background	328
38.2	Equipment	329
38.3	Choice of Film	329
38.4	Black and White Micrographs	330
38.5	Micrographs in Colour	331
38.6	Optical Technique	331
38.7	Magnification Selection	332
38.8	Constituent or Structural Identification	332
38.9	Looking Within	333
38.10	Combined Techniques	333
38.11	Selective Extinction of Colours	334
38.12	Capitalizing on Reciprocity Failure	334
38.13	Constructing the Picture	334
38.14	Novel Photomicrographs	335
38.15	Poster Presentation	336
38.16	Facing up to the Competition	336
38.17	That Once in a Lifetime Shot	336

Bibliography — 338

Acknowledgements — 339

Index — 340

Book Conventions

All surface preparation procedures give complete information with the exception of machine speeds; these must be optimized for specific pieces of equipment. From the procedure format, all information to the right-hand side of the thick black vertical line is for auditing purposes. Each step must be progressed to its best position, noting the Z axis, reflectivity value and scratch pattern.

Micrograph Magnification

Throughout the book, magnification quoted will be unity magnification, achieved by multiplying objective $\times 10$.

Resolution

Each magnification will have a resolving power based on the objective lens numerical aperture. For example:

$$\times 25 = 0.075$$
$$\times 50 = 0.10$$
$$\times 100 = 0.25$$
$$\times 200 = 0.40$$
$$\times 400 = 0.65$$
$$\text{Dry} \times 1000 = 0.90$$

Objectives are available with increased resolution relative to a given magnification. To use a $\times 100$ oil objective, for example, would increase the numerical aperture to 1.25.

Chromatic Corrections

All the micrographs have been taken using achromatic objectives. To improve chromatic corrections one would have to use fluorite or apochromatic lenses.

Scale

To avoid the use of a bar scale with every micrograph, the following comparison bar scale will apply throughout the book:
$\longmapsto = 1.0$ cm in length.
Bar scale:

Unity magnification	Actual size (μm)
$\times 25$	400
$\times 50$	200
$\times 100$	100
$\times 200$	50
$\times 400$	25
$\times 1000$	10

The reproduced micrographs in this book have, in general, been reduced from their original 5 in $\times$ 4 in format to 2½ in $\times$ 2 in. Therefore the 1.0 cm bar scale represents:

Unity magnification	Actual size (μm)
×25	800
×50	400
×100	200
×200	100
×400	50
×1000	20

NOTE: Where the reproduction is different a recalculated μm size is required.

PART I

SPECIMEN PREPARATION

The preparation of surfaces for microstructural analysis has, since the days of Sorby, been the practice of subjective empirical methods and recipes. It is fair to assume that it would have continued in that way had it not been for the introduction of new engineering materials where incorrect surface preparation techniques not only yield erroneous results but have drastic implications in the event of component failure. The author, being closely involved in the innovative aspects of metallography, soon realized that it was a deeply controversial subject where the *art* of the metallographer cast doubt on any technically derived approach. The answer of industry was to develop more automatic equipment in order to achieve repeatability, the inevitable outcome being consistent repeatable, possibly erroneous, results. Clearly the situation was unsatisfactory and what was called the 'round-robin' approach was adopted where various companies, organizations or individuals grouped together to agree on the best method and recipe! This usually resulted in a bias towards a particular manufacturer's machine or consumable product, the method often being an agreed compromise and the result of empirical practice.

A technically derived systematic approach that is totally noncommercial was first introduced in 1987 and has since been developed to its present day format which is the subject of this book. It is now possible, from a database, to select a technically sound procedure. This procedure will match the needs and facilities of each individual user but above all they will be methods capable of achieving a true and faithful surface preparation. In order to monitor the progress of the preparation, statistical and visual information will be used at each stage. Auditing to traceable standards is also possible.

Education plays an important role in any form of technology transfer. It is to this end that a level of technician standard (Level II and Level IV) has been developed in conjunction with the Engineering Training Authority.

There is something about the word *metallography*. This word adequately describes that area of science where metals are prepared and observed for micro- and macrostructural analysis; the person carrying out this function is called a *metallurgist*. This book, however, is not exclusively about metals; it is directed at materials and hence the person carrying out the function of specimen preparation and analysis will not necessarily be a metallurgist.

The name 'materials analyst' has been suggested but this is not acceptable since analysis does not necessarily include the preparation. In the book I considered using the word 'materials' technologist or scientist, but this equally is an inadequate expression in describing the *materialographer*. Finally, the views expressed in this book are based on the author's experience and are neither representative or influenced by any commercial manufacturer.

1

Introduction

1.1 What is Metallography?

Metallography relates to the structures of metals. The existence of grain composing structures was, in the Middle Ages, becoming apparent from the fractured granular appearance of metals and the dendritic tree-like shapes evident in cast materials. This belief remained unsubstantiated until the work of Henry Sorby in the nineteenth century when he was able to show the grains and constituent elements in iron and steel. Metallography therefore not only relates to the structure of metals but also embraces the science necessary in the preliminary surface preparation. Once it was possible to prepare the material surface in such a manner as to faithfully reveal the true structure it was possible to relate the material characteristics such as tensile strength, ductility, hardness, plasticity, etc. to actual structural appearance. This led to the development of new and better materials; it also signalled the development of the metallurgical microscope. The original microscopes were transmitted light biological instruments fitted with, at first, an oblique illumination followed by a semi-reflecting mirror situated between the objective and eyepiece.

1.1.1 The microstructure

One of the principles of material science is that the properties of a material are a direct consequence of the microstructural features of that material. The preparation of materials to faithfully reveal all microstructural features is therefore paramount.

Microstructural sample preparation involves a process consisting of a number of preparation steps ultimately resulting in a faithful reproduction of the material under investigation. The example in Figure 1.1 (see Plate 1) is used to illustrate the types of damage and the depth of deformation one can expect to find when a surface has been mechanically worked, such as is the case when grinding with silicon carbide paper. The depth of deformation varies with the size and condition of the abrasive being used, larger sized abrasives giving greater deformation than smaller ones, blunt abrasives being more severe than sharp ones.

It is not always obvious, when optically observing the prepared surface, to what degree deformation or structural damage remains. To analyse such a sample would lead to erroneous information. To illustrate this condition the 0.37% carbon steel, shown in Figure 1.1, was first ground on the upper surface with 80 grit silicon carbide paper and then carefully sectioned to reveal the deformed layer.

Three types of damage occur when grinding, two of which are clearly illustrated in the figure:

(i) Fragmented grains. This is the top layer and is a result of severe stress, probably attributed to the use of *blunt* abrasives, i.e. abrasives with negative rake angles and those without sharp points.
(ii) Structural changes visible on the micrograph resulting from directional shear.
(iii) Non-visual structural changes beyond those observed due to residual stress.

Another type of damage occurs when polishing (as opposed to grinding); this is the flowed surface due to surface rubbing. This is more prevalent when soft materials have been prepared with one or all of the following:

(i) High speed
(ii) High pressure
(iii) High nap cloth

Non-ductile materials such as ceramics and minerals react differently when having material removed via sharp-pointed tools (stress applicator). Nevertheless, they are still affected by structural damage in the same way as ductile materials manifest deformation.

Whenever a material is cold worked a degree of residual damage is inevitable; it is the degree or depth of damage with which the materialographer is concerned. The total deformation shown in Figure 1.1 is 120 μm. This figure is derived from the visible deformation plus an additional amount for residual stress.

1.2 Damage Depth

Before any work is performed on materials it would be an advantage to know the depth of existing damage in order to assess the size, type and combination of abrasives required for subsequent preparation. This depth can be derived as shown in Figure 1.1. It would, however, be more advantageous if these figures could be available for all mechanically worked surfaces relative to different materials. Take, for example, four identical samples submitted for microstructural analysis: the first without previous mechanical working, the second cut using a hacksaw, the third sectioned with an abrasive cutter on a high-speed machine and finally the fourth sectioned on a low deformation saw at low speed. All four samples will exhibit different degrees of deformation. Without this knowledge a suitable procedure is difficult. On this occasion take the structural damage as being 120 μm deep.

1.3 Abrasive Combination

Traditional metallography depends upon the use of silicon carbide for grinding steel samples. With the knowledge that the sample has a deformation depth of 120 μm we can from the database chart shown in Figure 1.2 make the initial abrasive size selection. Each horizontal line on the chart indicates the position of least damage or deformation that can be achieved with that type and size of abrasive (for 0.37 steel). If we look at the position for 180 grit silicon carbide we can see that the least damage will leave 110 μm still remaining. This would be too close to the residual damage figure of 120 μm. The Z axis values in Figure 1.2 (least damage) are arbitrary and are used for comparison only. The best choice would be to start with 240 silicon carbide. The next steps can be a combination of many different surfaces. The object of the following chapters is to define a systematic approach ultimately resulting in a faithfully reproduced sample of integrity.

1.4 Z Axis Curves

Instead of looking upon a particular abrasive size as being capable of achieving a given surface finish, we must relate abrasive functions to what are called 'Z axis curves'. These give us an indication of the rate of progress towards integrity of the sample and the limiting curve for each abrasive size and surface. With this information we can make a selection of Z axis curves to suit the requirement of the specimen. Some abrasives become blunt during use and actually induce specimen damage; this causes the Z axis curve to rise upwards. These abrasives are said to be degrading (when abrasives do not degrade the lower curve remains horizontal). Figure 1.3 is

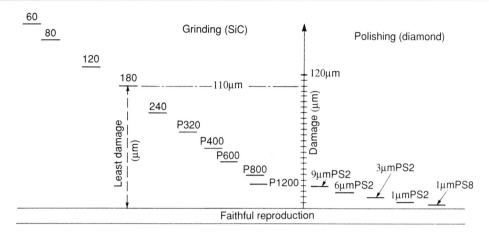

Figure 1.2 Database chart for 0.37 steel (for comparison only)

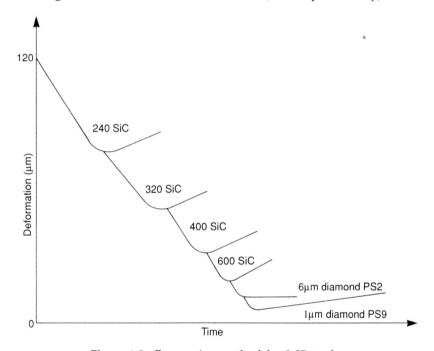

Figure 1.3 Preparation method for 0.37 steel

used to illustrate one such combination using degrading abrasives (SiC papers) and diamond abrasives on two different cloths. Polishing surface (PS) No. 2 with 6 μm diamond is not degrading, yet PS No. 9 with 1 μm is gradually degrading.

The reason why two polishing cloths differ in function is due *not* to the abrasive type but to the polishing cloth surface. This introduces another reason for the need to evaluate by Z axis and not surface finish.

Figure 1.4 illustrates the integrity curves for two different preparation methods, one using three silicon carbide steps and one polishing step, the other three platen surfaces (to be explained later) and one polishing step. The two large dots indicate a preparation completion point for both methods. It is clear from the illustration that the three-platen system gives reduced time, better integrity and is less sensitive to time.

The purpose of this graph is to illustrate how

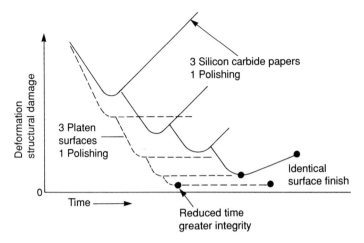

Figure 1.4 Integrity chart for material MMC

different preparation methods can be compared when displayed by the Z axis curve and also to dispel, if possible, the misconception of relating surface finish to sample integrity. All four large dots could display identical surface finishes yet all four exhibit different degrees of sample integrity.

1.4.1 Setting the ground rules

The theme for future considerations, be it cutting (sectioning), grinding or polishing, is not what is the surface finish but what is the least damage we can expect from each operation, will that surface degrade or can we make continuous use of the abrasive, and with what platen and polishing surface combination? Ultimately a whole series of Z axis curves will enable materialographers to make not just the correct choice but also the choice that most favours their needs.

1.4.2 Traceability

Having established a preparation route it is essential for quality control auditing to have some reference in order to relate the technique and the consequential analysis. To this end it is desirable to have a traceable procedure and reference material along with the custodian of such material. This then is the format for the subsequent chapters.

2

Sectioning

2.1 Resultant Structural Damage

Sectioning is not a prerequisite of subsequent preparation but it can have a dramatic effect on subsequent preparation techniques. Every mechanical working procedure (sectioning, grinding and polishing) has a relative destructive action on the parent material and it is the degree to which this takes place that influences future work. On some occasions this damage is so great that it is impossible to reveal the true structure by grinding and polishing. One such example would be a plasma coating where the sectioning technique displaced the coating in a manner indicative of delamination. Another example is one where the structure of the material is changed; this is usually associated with heat and can be readily illustrated on a hardened steel component where the heat generated in sectioning has not been dissipated. The rate of traverse can also influence the material structure. Take, for example,

Table 2.1 Deformation sectioning chart for 0.37% carbon steel

Wheel type	Rate of traverse	Macro observation	Optical observation (×1000)	Depth of damage (μm)
Soft bond Al_2O_3	Slow	Little wheel wear, good surface finish	Minor deformation	10
Soft bond Al_2O_3	Fast	Slightly more wheel wear, good surface finish	Deep deformation	45
Hacksaw	Normal by hand	Coarse surface finish	Deformation and uneven surface	70 + 200
Hard bond Al_2O_3	Slow	Nil wheel wear, good surface finish	Slight structure change	20
Hard bond Al_2O_3	Fast	Little wheel wear, sample burned	Gross structure change	900

8 Surface Preparation and Microscopy of Materials

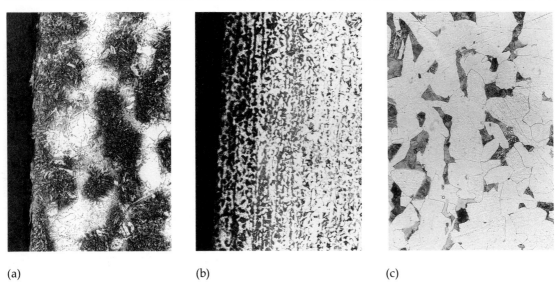

(a) (b) (c)

Figure 2.1 Deformation and burning. (a) Fast traverse (×400), (b) normal traverse (×100) and (c) normal structure (×400)

materials that work-harden; when using the correct wheel at a slow traverse the material can be sectioned, but to increase this rate of traverse can create insurmountable problems, i.e. it will not cut.

The deformation sectioning chart for 0.37% steel (Table 2.1) is used to illustrate the effect that wheel selection and rate of traverse has on the resultant damage. Using a 25 mm diameter rod of 0.37% carbon steel, the material was sectioned using two different density-bonded aluminium oxide wheels and a hacksaw. An abrasive cut-off machine was used with speed and lubricating conditions identical for all tests, the only variable being the rate of traverse of the wheel through the material. The cut specimen was carefully sectioned and prepared on a plane normal to the original cut in order to identify the depth of damage. From the introductory chapter the figure of 120 μm was used as a Z axis depth of damage measurement. From this figure a suitable preparation recipe was constructed. From Table 2.1, we can see how a soft bond (low density resin) wheel accompanied by a slow rate of traverse resulted in a depth of damage of 10 μm. With this sample we could omit all the silicon carbide paper stages and have a *one-step* preparation procedure using, say, 3 μm diamond on a hard cloth (PS 2). When traversing at a higher rate with the same wheel as the second example in Table 2.1, the measured damage was 45 μm with a higher wheel wear. This sample would require some silicon carbide grinding prior to polishing; the question is what grade and how many?

The traditional route for preparing this material would be to grind the sample (after sectioning) on a series of silicon carbide papers. These papers would initially induce damage some 20 times greater than existed. This brings into question the validity of traditional metallography techniques which will be investigated more fully under 'Grinding' in Chapter 6.

An interesting feature from the deformation sectioning chart is the relatively low deformation that takes place when using a hacksaw (70 μm). This is due to the correct use of positive rake angles (see Chapter 4).

The last two examples show the effect that traverse has on the sectioned specimen when using a hard bond (high density resin) wheel. This type of wheel is very economical and is the wheel favoured by many materialographers. Unfortunately the fast traverse can, and often does, destroy the structure beyond recovery by traditional four-step silicon carbide methods. It could be wrongly concluded from the 20 μm damage when using a slow traverse that this would be a suitable wheel; unfortu-

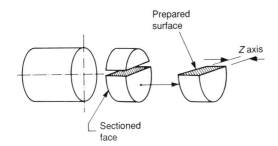

Figure 2.2 Depth of sectioning damage

nately the wheel would soon glaze and the 20 μm would soon become 200 μm.

Damage in any cutting, grinding or polishing operation can be caused by heat, which can completely change the structure of the materials. From Figure 2.1(a), the extent of structural change when using the fast traverse hard bond wheel can be seen. Figure 2.1(b) shows the stuctural change resulting from a slower traverse compared to the normal structure illustrated in Figure 2.1(c).

The experienced metallographer would recognize this condition when analysing the structure but at that stage it would be impossible to say whether the structural change was as a result of the preparation, i.e. induced, or whether it was from the original condition. Many materials that change their structure through cold working and/or exposure to high temperature do so in a manner not necessarily obvious at the analysis stage.

The object of sectioning is to extract the specimen to be prepared from the parent metal with the minimum amount of damage to the material adjacent to the cut. Although there are many ways of achieving this (hand saw, shearing, flame cutting, fracturing, chemical or spark cutting), abrasive cutting is the favoured route and will be investigated further.

Sample integrity is influenced from the very first operation, so we need to look into the various forms of abrasive wheels and make a commercial judgement. This last point is added since sample integrity could be better on, say, a low speed saw taking 30 minutes to cut, as opposed to a high speed wheel taking 30 seconds.

The type of sectioning chosen or the choice of abrasive wheel will give varying degrees of structural damage. When a standard procedure is being formulated it is essential to define the depth of damage since subsequent operations depend upon this information. Take, for example, a standard preparation procedure that has been issued for inter-company and subcontract quality control adherence. If stock removal during the preparation procedure is 200 μm and the initial sectioning operation is carried out incorrectly or in an inefficient manner, resulting in a damaged depth of 250 μm, then good components will inevitably be rejected. Whenever a sectioning technique is proposed the sectioned specimen should be resectioned and prepared (see Figure 2.2) to show the depth of sectioning damage (Z).

This operation must be carefully carried out without inducing further preparation damage; if electrolytic polishing is available (or suitable) this should be favoured. The Z axis dimension will be estimated from the etched surface using a microscope with maximum objective resolution relative to the field of view, i.e. too low magnifications will obscure any change within a small field of view. Having estimated the damaged depth this must be increased by 50% in creating a realistic working Z axis dimension.

The introductory chapter indicated the importance of knowing the least damage one can expect from each operation; with this information the most efficient preparation recipe can be formulated. If, for example, the specimen is sectioned on a low deformation saw resulting in a deformed layer of 30 μm, one would be ill advised in progressing to 180 grit silicon carbide paper if the least damage achievable is 45 μm using this grit size.

Table 2.2 is used to illustrate the influence different types of wheels, speeds, load and lubricant have on the cutting time and resultant Z axis damage. It will be noticed how the least time gave the least deformation and Z axis damage. This was the abradable wheel rotating at 2000 rev/min. These conditions apply only to the aluminium in question and could be totally reversed if, for example, a plasma-coated alumina were to be sectioned. It will also be noted from Table 2.2 how an increase in speed can result in a loss of cutting efficiency and how different conditions affect the resultant Z axis structural damage.

Table 2.2 Cutting aluminium

Variable speed saw (rev/min)	Wheel	Lubricant	Load	Z axis (μm)	Time (min)
100	Diamond	Soluble oil	Low	10	20
100	Diamond	Mineral oil	Low	9	18
100	CBN	Mineral oil	Low	8	15
1000	CBN	Mineral oil	Low	50	150
1000	CBN	Mineral oil	High	48	120
1000	SiC abradable	Soluble oil	Low	9	2
2000	SiC abradable	Soluble oil	Low	8	1
2000	SiC abradable	Mineral oil	Low	7	1

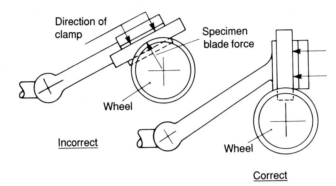

Figure 2.3 Reduce contact, reduce time

2.1.1 Low speed, low deformation, sectioning

It is more important to maximize the operating parameters with low speed cutting since incorrect use can adversely affect the resultant sample deformation. The specimen-holding chucks must give maximum support yet they must not be overclamped; it is common practice to have a large selection of different clamping arrangements available including the flat-faced chuck where the sample has to be stuck with cement, resin or double adhesive tape. Special purpose vacuum chucks can also be advantageous.

The thin sectioning blades are best used with optimized load and speeds, specific for each different material. New diamond and cubic boron carbide blades could require dressing before use in order to reveal the cutting abrasive. It will be necessary with the non-abradable wheels to periodically dress when cutting ductile materials.

Although cutting time may not be the most critical factor in low speed sectioning, it is, however, desirable to achieve the shortest time; this can be achieved, as shown in Figure 2.3, by reducing the blade contact. The incorrect method also illustrates bad clamping practices; the direction of clamp should where possible complement the blade force and not oppose it.

It is often required to produce thin parallel sections from low speed saws, thus requiring the sample to rotate against the direction of the wheel. When the thickness of a sample is important it is necessary to take into account what is called *kerf* loss; this is the slot thickness produced by the slitting wheel. All cut-off wheels cut larger than their actual thickness and are influenced by material, wheel thickness and any eccentricity of the wheel arbor.

2.2 Characteristics Affecting Structural Damage

When optimizing the cutting conditions we need to consider lubrication, wheel speed, abrasive type, wheel bond, wheel size and mechanical factors affecting the cut.

2.2.1 Lubrication

Lubrication can be used for one or more of the following reasons:

To lubricate the abrasive, ensuring the swarf does not adhere to the cutting surface. Abrasive cutting is a multi single-point tool operation and is influenced in the same way as, say, turning on a lathe or using a drill. When we turn or drill aluminium, for example, we must ensure the cutting surface is adequately lubricated otherwise the tool becomes blunt (through chip and swarf adherence). This ultimately results in poor or nil stock removal. One advantage over turning and drilling is that, as the abrasive becomes blunt, it is broken away introducing new abrasives. This, however, is not the case when sectioning with a diamond or cubic boron nitride (CBN) wheel, since expense demands a more efficient use. When using expensive abrasives great care must be taken in selecting the correct lubricant and this ultimately has an effect on the operating speed. In general it is advised with ductile materials to operate at low speeds when using diamond or CBN abrasives.

To dissipate the generated heat. Heat will be generated with all types of materials when operating at high speeds, but in general it is the ductile materials where particular care has to be taken (metals being better conductors than minerals and ceramics). The best medium for dissipating heat is water; this, however, is not suitable since it corrodes the cutting machine and also creates a steam barrier when water comes into contact with hot metal. These problems are overcome by using a soluble oil (for lubrication) mixed with water (for heat dissipation) having a wetting agent (to avoid the steam barrier) and copious supplies. In some circumstances it is necessary to totally immerse the sample in order to dissipate the heat. This is particularly important with materials susceptible to thermal shock cracking or thin sections where there is little material mass to dissipate the heat.

To wash away swarf from the cutting area and/or keep down the dust level. When cutting hard brittle materials it is common practice to use a metal bonded fine diamond wheel, this will cut dry, the only problem being trapped swarf between the blade cutting surface and the material being cut. Under these conditions the cutting rate can be improved by a factor of 10 by washing away swarf.

2.2.2 Fixed abrasive wheels

Sectioning can be likened to any grinding operation where the multi single-point tools move across the workpiece (sample) removing material. These single-point tools must, at the point of cutting, be retained or bonded, in this case to the cut-off wheel. Fixed abrasive (single-point tool) wheels are those where chosen abrasives will not readily fracture during use and, due to the high cost, will not break away from its bond. Such abrasives are diamond and cubic boron nitride. Because the abrasive will not break away many constraints are placed on its use. For example:

(i) Fixed abrasives must not be favoured when sectioning ductile materials unless the speed is very low. Aluminium, for example, could be cut using a 125 mm diameter wheel rotating at 35 rev/min using an oil lubricant, but would not cut at all if the speed of the wheel were to be increased to, say, 2500 rev/min. Not only would it not cut but the abrasive would be damaged and in some cases the cut-off wheel could be destroyed. You could say, 'why use a fixed abrasive with ductile materials when this can be readily cut using an abradable wheel?' The answer is that low speed fixed abrasive cutting, although taking a long time, could exhibit a lower deformation than high speed abradable wheel sectioning, particularly with delicate or aluminium ceramic composite material.

(ii) Fixed abrasive wheels using diamond or CBN are very expensive. The abrasive is usually metal or resin bonded onto the periphery of a thin blade, which makes them very prone to breakage if subjected to shock load (in particular any shear stress).

(iii) The cutting efficiency of fixed abrasives

is greatly influenced by the correct choice of lubricant. Low speed cutting uses an oil lubricant while high speed cutting would use a water-soluble oil. Examples have been shown where mineral oil compared to soluble oil can reduce the cutting time by as much as 50% on fixed abrasive low speed cutting.

2.2.3 Abradable wheels

Abradable wheels make use of sharp inexpensive abrasives and can therefore tolerate the cost of allowing the abrasives to break away. They are used for all types of ductile materials and hard materials such as hardened steel VHN 600. The only limitation with abradable wheels is the hardness of the material being cut; tungsten carbide, for example, would require a resin-bonded diamond (fixed abrasive) wheel.

Abradable wheels have the abrasive bonded such that, as it becomes blunt, the increasing cutting force breaks the abrasive clear. Due to the high operating speed, lubricant is required to:

(a) dissipate the generated heat and
(b) wash away the broken abrasives.

Heat must be quickly dissipated; this is best accomplished when lubricant is applied to both sides of the cutting wheel just above the cutting area or when submerged as described earlier. Coolant is only necessary in extreme circumstances such as when lengthy cutting cycles are encountered with particularly heat-sensitive materials.

2.2.4 Wheel speeds

There is an optimum wheel speed, relative to the cutting load, and this in turn affects the resultant structural damage. To change wheel speeds without consideration of the wheel feed or load can be counter-productive. This can best be illustrated using the low speed diamond slitting saw. A ceramic component can be cut using a set load and speed in, say, 10 minutes; to increase the speed from this optimum condition without changing the load could, for example, take 20 minutes to cut. Conversely, to keep the same speed yet increase the load could increase the structural damage.

With abradable wheels the effect of increasing the feed rate (or load) would reduce the wheel life or break the wheel. To increase the peripheral speed could result in wheel burst or glazing.

The most efficient rim speed for diamond wheels is generally lower than that for abradable wheels. Probably a good general optimum speed of 25 metres per second is to be recommended, as opposed to 42 metres per second for abradable wheels. With fixed-speed cutters this can be achievable by using a smaller diameter wheel for diamond blades.

2.2.5 Abrasive types

Diamond

This is used in a metal bond for minerals, ceramics and hard metals. The abrasive concentration affects the cutting, but it does not always follow that the greater the concentration, the greater the cutting efficiency. Take, for example, the cutting of a hard 'burnt flint' material. A high concentration diamond wheel with normal load and speed will take 3 times as long as a low concentration wheel with high load and lower speed.

This book makes many references to the shock-absorbing characteristics relating to the abrasive bonding agent. To absorb more shock from the cutting abrasive is to reduce the deformation structural damage likely to reside in the specimen as a result of that cutting action. This can be used to advantage by using resin-bonded diamond wheels instead of metal-bonded wheels when high structural damage is experienced.

Diamond wheels are expensive but are very hard wearing. The abrasive is only used around the rim of the wheel. These wheels must not be used at anything but extremely low speeds with ductile materials otherwise they will become 'clogged' and break away the diamond.

Cubic boron nitride

This is used for cutting hard materials—again metal or resin bonded. Due to the abrasive shape this abrasive tested at low speed cuts ductile materials quicker than diamond. Due to its cutting efficiency and also its ability to dissipate heat CBN will make a cooler cut.

Like diamond CBN is expensive and must not be used at high speed for ductile materials. It should be the first choice with alloy tool steels when the conventional aluminium oxide wheels have failed.

Silicon carbide

This is used in bonded self-dressing expendable wheels and is a friable abrasive used mainly on low tensile and/or non-ferrous and non-metallic materials. Typical materials on which this abrasive is used are:

Carbon
Soft ceramics
Titanium
Non-ferrous metals

It can be used for ferrous materials but tends to have a reduced life.

Silicon carbide is a sharp, hard, brittle abrasive ideally suited for abrasive cut-off wheels. The abrasive size tends to be the same for all grades of wheels but some suppliers mix a finer sized abrasive to improve wheel life or surface finish. Because these abrasives are hard and sharp they *will cut* where alumina *will not*; this is particularly important with ferrous materials. The reason for emphasizing this point is because suppliers of cut-off wheels only recommend alumina as the abrasive for ferrous materials, the reason being economics, not performance. You could find, for example, that a plasma coating on a steel substrate when cut with the recommended alumina wheel induced damage; the same grade of silicon carbide wheel cuts without damage.

Aluminium oxide

This is mainly used in the 95% pure state (brown) to give a tough abrasive for use on most ferrous materials from grey iron and mild steel up to nickel alloys. It is also found in a purer form as white abrasive (99.5% pure) which is a more friable and cooler cutting abrasive. Brown abrasives are more robust than brittle silicon carbide and therefore have a longer life.

The deformation caused by the abrasive is related to its size, the larger abrasive causing the greatest damage. The abrasive size also has an effect on the cutting temperature. Although most wheels consist of similar sized abrasives, special mixed abrasive wheels are available, optimizing the cutting conditions with minimum deformation. Wheels combining the mixed qualities of resin and rubber can also result in lowering deformation.

2.2.6 Wheel bond

Abradable wheels are made from a mixture of abrasive grain, bonding agent, fillers and other additives to assist in the bonding or performance of the wheel. For cut-off wheels, three main bonding agents are available:

(i) Resin is the most common and least expensive. Its disadvantages are that it may be prone to distortion when the diameter–thickness ratio is greater than 120:1 and it cannot generally be made thinner than 0.8 mm at any diameter.
(ii) Rubber may be natural or synthetic. Its advantages are that it has an excellent surface finish and can be made very thin commensurate with diameter. It can also be made to very close tolerances. Its disadvantages are high price and environmental problems due to excessive fumes and smell.
(iii) Shellac. Its advantage is a cool cutting wheel. Its disadvantages are poor life (only available normally in two grades, soft and very soft) and it is expensive, usually twice the price of a resin wheel.

Wheel grade or hardness is not dependent on the abrasive but on the bond that retains the abrasive. The bond should release the abrasive grain just as it starts to become blunt. Apparent hardness is achieved by varying the amount of bond in the wheel. The bond itself consists of the main bonding agent, resin, rubber or shellac mixed with a variety of fillers (wetting agents) which provide varying degrees of tenacity, thus altering the apparent hardness of the wheel.

Some wheels are supplied with what looks like a paper backing on both sides of the wheel, the abrasives exposed only on the peripheral face of the wheel. These wheels are usually more expensive but do offer advantages in cooler cutting and could be used for heat-sensitive materials before resorting to submerged cutting.

A high proportion of total bond will give a hard wheel; a low proportion will give a soft wheel.

2.2.7 Wheel selection

Wheels are graded not by the abrasive volume but by the bonding volume. A series of wheels for cutting ferrous materials will be given numbers from what is called a *soft wheel* with less bond to the other extremity, a *hard wheel* with more bond.

A *soft wheel* would be chosen for a *hard material* since the abrasives will blunt more quickly and a *hard wheel* for the *softer materials* since the abrasive can be used longer. Two questions arise:

(i) Can you use a soft wheel on a soft material?
   Yes, it's just not economical.
(ii) Can you use a hard wheel on a hard material?
   No, it will create heat and structural damage.

A thin or small section can be cut with a harder wheel than a large solid section. When working near the size limit of the machine, with a soft material of large section it could be necessary to use a soft wheel. In addition to the material hardness, the arc of contact between the work and the wheel should also be considered, since there is a direct relationship between maximum machine power, area of wheel in contact with the material and strength of the bonding agent (wheel hardness).

If the cut-off machine has a reciprocating type action on the forward cutting stroke then the wheel contact remains constant and no change of wheel grade is necessary.

2.2.8 Wheel size

For efficient and safe use the wheel must conform to the safe working surface speed which is influenced by its size. Of greater microstructural importance is the wheel thickness since this governs the depth of deformation during the cut, i.e. thin wheels create less deformation than thick wheels.

In general the thickness will increase with the wheel diameter from 1 to 3 mm, but special purpose thin wheels are made and are to be encouraged for delicate or minimum deformation applications.

2.2.9 Mechanical factors

When components that are internally stressed are cut, movement often occurs in the area of the cut slot. This can result in the cut-off wheel being pinched and can in extreme cases cause breakage. To avoid this the component should be stress relieved before cutting; however, if this is not practical, the abrasive wheel must be constantly reciprocated or incremental cuts applied.

Samples that are of irregular shape can be distorted when clamped on both sides of the cut. If there is any doubt relating to sample parallelism then only one vice should be fully clamped, the other acting to hold the cut-off portion only.

Some manufacturers supply cut-off machines with only one vice. This is acceptable providing the sample retrieved does not include a burn flash. Many low speed cutters offer just a single vice and this again is acceptable with non-ductile materials. In general all machines should be supplied with two vices: one vice for clamping, the other very lightly clamped to just retain the eventual specimen. One good example supporting this point is the sectioning of plasma-sprayed coatings where the eventual sample, if clamped, can cause delamination and coating degradation.

There are occasions when it is advantageous to rotate the sample as the rotating wheel advances. This ensures a uniform area of contact irrespective of the sample shape, viz. cutting a washer or coin in half. Sectioning samples that require to be flat and parallel are best cut with a rotating sample and wheel configuration.

2.2.10 Safety

Machines and wheels must be marked with operating speeds, and all persons concerned with the mounting of wheels (but not operation of the machines) should be certificated in compliance with the appropriate Abrasive Wheel Regulations. Under no circumstances should the machine be operated without the safety hood or screen in position.

2.3 Common Faults and Remedies in Abrasive Cutting

Fault	Cause	Remedy
Wheel does not cut or cutting ceases after a short time	Incorrect abrasive Wheel abrasive has become blunted or glazed	Use alternative abrasive or softer grade of wheel
Wheel wears rapidly	Wheel is too soft	Use harder grade of wheel
Wheel breaks	Excessive cutting force	Reduce cutting pressure
	Sample moved during cutting	Clamp sample more securely
	Wheel not clamped securely	Tighten wheel flanges more securely but do not overtighten
	Uneven wear on wheel rim causing deflection of cut	Coolant flow not equal on both sides of wheel—check for blockages
	Wheel pinched in cut by sample	(a) Sample clamping arrangement causing pinching action on cut when partially complete—reduce pressure on one vice (b) Internal stresses in sample causing pinching—take step or intermittent cuts
	Wheel vibration	Check wear on cutting shaft bearings
Overheating and burning of cut surface	Abrasive blunted causing excessive friction	Use softer grade of wheel
	Partial or complete failure of coolant supply	Check coolant flow to wheel and sample
	Cutting rate too high	Reduce cutting pressure and/or feed rate
	Stressed component pinching in on wheel sides	Use softer wheel or use a reciprocating wheel action
Excessive wheel vibration during cutting	Worn cutting shaft bearings	Replace bearings
	Air in hydraulic system (where fitted)	Bleed air from hydraulic system
	Mechanical damper too loose (where fitted)	Tighten mechanical damper
	Wheel too hard	Use softer wheel
Coolant froths	Antifrothing agents in coolant have deteriorated	Clean out coolant system and refill with fresh coolant
	If coolant is new and frothing occurs	Wrong type of coolant used—replace with recommended grade
Cutting motor overheats and/or lacks power	Excessive cutting pressure causing overloading	Reduce cutting pressure
	One or more of phase voltage low	Check voltages on each phase of three phase supply to machine
	Failure of one of motor windings	Check motor windings for each phase
	Wheel too hard	Use softer wheel

3

Mounting

3.1 Considerations and Features

Mounting is an encapsulating process which takes place mainly after sectioning (with loosely bonded subjects, it can be advantageous to mount prior to sectioning). The object is to facilitate further processing.

There are two important physical features associated with encapsulation: they are the rate of contraction and abrasion resistance characteristics. One very important omission from this list is hardness, since this contributes little to the matching of sample-to-mount contraction and resistance to abrasion. If we are to achieve a planar sample after grinding and polishing the abrasion rates must be matched as close as possible, even if it means the hardness does not match.

When a sample is being prepared on an abrasive surface where the abrasive backing allows the abrasive to rise and fall, then the difference in sample-to-resin hardness *will* influence preferential material removal. Take, for example, silicon carbide paper where the weight of backing material can be the 'A' grade which is thin or the 'C' grade which is thicker (other grades are available). 'A' grade will create the least material removal difference, 'C' grade the least structural damage.

Encapsulation can take place using a hot mounting press where the resin and specimen are subjected to heat and pressure. The thermosetting resins will set with appropriate temperature and pressure and can therefore be ejected hot. The thermoplastic resins are also subjected to heat and pressure but must be cooled to set prior to ejection. The alternative to using a hot mounting press is to utilize a cold mounting resin, which sets when mixed with the appropriate hardener, thus avoiding the high temperatures and pressures so necessary in hot mounting.

The selection of the method and mounting material will depend upon many factors, one of these being the lack of any cleavage (fissure) between the mount and the specimen interface. The reasons for keeping cleavage down to a minimum are:

(i) Abrasive particles become trapped in the gap.
(ii) Soft specimen materials will flow (smear) into the gap giving false results (plated layers).
(iii) Friable materials will be unsupported and break away during grinding.

One way to eliminate cleavage is to ensure that the mounting material physically adheres to the metal specimen. Unfortunately only the epoxy resins comply with this requirement.

The component shape can also have an effect on the quality of the mounting. Square shapes in a round mount, for example, result in unequal contraction, with the greatest stress at the corners. It is therefore wise to avoid sharp corners; where this is not possible make the mount as large as possible in relation to the component and/or use a resin with a lower coefficient of thermal expansion.

3.1.1 Edge retention and relief

A mounted specimen is usually critical to one, if not both, of these qualities. Relief between the mount and sample can be reduced by balancing the different abrasion rates, reducing the nap on the polishing stage and replacing the final grinding paper stage with a charged or fixed platen surface. Edge retention is the gap seen at the mount/sample interface, in many cases wrongly attributed to contraction. When a gap is present on the outside diameter of a hot mounted sample it is often a result of poor mounting conditions, in particular with phenolic resins. The chance of poor granular adhesion with a mounting resin is greater with the thermosetting material than is the case with thermoplastics. The grain size also influences the intergranular adhesion. When present with a compression-mounted phenolic resin gaps can be attributed to the grains being too coarse or incorrect pressure or temperature parameters.

Good edge retention, when not satisfactory by matching abrasion and contraction, can be overcome by using some extra coating technique. One such technique is the use of electroless coating where an electroless deposition of metallic ion on the specimen surface is carried out prior to mounting. Various solutions or mixtures are available and are suitable for both metallic and non-metallic materials.

3.1.2 Cold mounting

If the requirements of cold-setting resins are considered then one of the objectives is to use mounting materials with the least contraction when setting. These figures are often difficult to acquire and vary with different mixes. It is wise, however, to exercise caution with some quick-setting resins since contraction can increase with reduced setting times. Setting

Table 3.1 Relative coefficient of thermal expansion

Material	Coefficient of thermal expansion
Phenolic	$3–4.5 \times 10^{-5}$
Acrylic	$5–9.0 \times 10^{-5}$
Epoxy	$4–7.0 \times 10^{-5}$
Metals	$1–3.0 \times 10^{-5}$

can be induced by heating; this can also result in increased contraction.

Having established the best combination of abrasion and shrinkage with a particular resin only to find the contraction is still too high, this can then be reduced by the introduction of small particles such as alumina in the resin mix.

3.1.3 Hot mounting

When considering mounting materials that require heat and pressure to activate the setting, as is the case with hot mounting materials, then the coefficient of thermal expansion has to be taken into account. For those who recall the shrinking of steel rims by the blacksmith on to wooden wheels, this is not unlike our expectations of hot mounting specimens. It follows that the best shrinkage would come from those mounting resins exhibiting the highest coefficient of thermal expansion; these values are illustrated in Table 3.1.

From Table 3.1 it can be seen that the acrylic offers the highest coefficient of thermal expansion. This is a thermoplastic material whose abrasion rates, as will be illustrated later, are compatible with phenolics. It does, however, become plastic with heat and 'clogs' the abrasive papers when grinding. To say that high shrinkage is most desirable is *only* relevant when the mounting resin is on the outside of the specimen. Consider a washer or tube-shaped sample; ideally we would like a resin with a high coefficient of thermal expansion for the outside with a low contraction on the inside. Contraction can be reduced with both hot or cold mounting resins by mixing small particles of, say, alumina in the moulding resin.

Figure 3.1 illustrates how reduced contraction can be achieved by the use of inserts. In those cases where it is essential to have nil

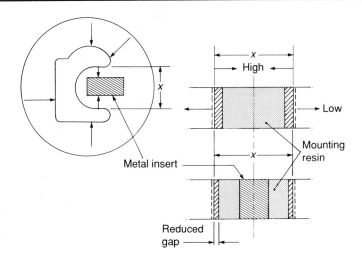

Figure 3.1 Reducing contraction

contraction it is better to 'dip' the component into an epoxy resin to create a hard skin prior to encapsulating in a hot mount. Electroless specimen coating is another favoured option.

3.1.4 Abrasion rate

Suitably mounted the sample will then be subjected to a series of grinding steps followed by polishing. The rate at which mounting material is removed in relation to the specimen material is important; ideally they should be similar.

Table 3.2 Abrasion rates

Class	Material	Abrasion constant (μm/min)
Hot mounting	Phenolic	560
Preload for sensitive materials	Diallyl phthalate	440
	Epoxy resin	130
	Acrylic	520
Cold mounting	Epoxy	870
	Acrylic	1040
	Polyester	1100
	Acrylic + A_2O_3	920
Metal	Copper alloy	35
	Steel (0.37)	12.5
	Aluminium alloy	142

Table 3.2 illustrates the abrasion rates for many of the commercially available mounting resins and some common materials. The abrasion rate was achieved using 25 mm samples on new discs of P600 grit silicon carbide paper for 1 minute with a force of 35 kPa, the 12 in diameter platen rotating at 150 rev/min and the specimen in contra motion at 66 rev/min. The measurement mode is via a hardness indent reduction, measuring the diagonals and multiplying by 0.202 to give the Z axis material removal. An instant observation from this table shows the vast difference between abrasion rates of mounting mediums compared to metals. Also note how cold-mounting resins abrade at a much higher rate than the hot-mounted resins.

Mounting materials of the thermosetting type (which require heat to set after which they will remain set) have a higher abrasion rate than the thermoplastic materials (can be made plastic with heat after setting). The fact that thermosetting materials have an abrasion rate much higher than metals explains the reason for the all too familiar interface relief.

3.2 Results and Reasons

Traditional metallography very often results in what is called relief polishing and is usually associated with soft, high nap polishing cloths; this is illustrated in Figure 3.2(c) where the hard elements stand proud. Unfortunately,

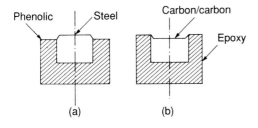

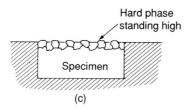

Figure 3.2 Relief grinding and polishing. (a) Positive relief, (b) negative relief and (c) polishing relief

this is not the only form of relief; it is very often derived from a mismatch between resin and sample abrasion rates and is created at the grinding stage. Figure 3.2(a) illustrates the condition of positive specimen relief. This grinding relief is further exaggerated when using grinding papers such as silicon carbide, where the abrasive is allowed to preferentially grind the soft resin.

Mounting materials are often related to hardness rather than their abrasion rate characteristic. Selection of mounting material type should firstly be judged on a balance between the specimen abrasion rate and the nearest equivalent mounting medium abrasion. When this combination results in a cold mounting medium with a high contraction rate, alumina particles (silicas or carbides) of particle size between 3 and 0.05 μm should be mixed with the mounting resin to reduce the specimen/mount interfacial cleavage.

The reverse of positive relief, i.e. negative relief, is shown in Figure 3.2(b), where the specimen abrades at a greater rate than does the mounting medium. When a grinding operation creates negative relief one will find the 1 μm polishing stage to be totally ineffective (it polishes the mount only).

When a *standard* preparation technique is being developed the mounting medium must be designated and conformed to. A recent 'standard' issued for a carbon–carbon material did not define the type of mounting resin. This resulted in total rejection when in fact it was negative relief that was restricting the preparation.

Another example of what at first appeared to be a faulty 'standard' preparation technique was the density of the resin–sample ratio. The preparation technique for composite materials can vary with the differing composite ratio. On this occasion it was an iron oxide (100 μm diameter) epoxy mixed composite. A recipe was created for the research material but when it went into production the iron oxide ratio was halved and the preparation technique failed. Because the technique was a 'standard' it carried the Z axis dimension at each stage and it was the non-achievement of this dimension that told the operator something had gone wrong.

When considering the abrasion rate at the polishing stage metal having a much higher rate than both thermoplastic and thermosetting usually results in good edge retention, assuming that both are ground plane. The materialographer likes to see polished mounts accompanying a polished specimen. This is relatively easy to achieve with phenolics and thermosetting resins but can be extremely difficult with thermoplastic material.

From Table 3.3 it is easy to see how samples mounted in cold setting acrylics can be well polished yet surrounded by a badly scratched resin. This table also includes Vickers hardness values, which must be somewhat subjective in the 20–25 region. They are, however, constructive values for comparison purposes.

3.2.1 Reasons for mounting

(i) The specimen is too small or of awkward shape for ease of handling in subsequent stages of the preparation process.
(ii) To support the outermost edge of the specimen's surface to prevent damage or rounding during the subsequent grinding or polishing operations.
(iii) The specimen is of a delicate or friable nature.
(iv) To provide a uniform size facilitating further processing in the standard sizes used in specimen holders of automatic preparation equipment.

Table 3.3 Combined abrasion/polishing/hardness chart

Class	Material	Abrasion constant (μm/min)	Polishing rate	Hardness (Hv)	Characteristics
Hot mounting Preload for sensitive materials	Phenolic Diallyl phthalate Epoxy resin Acrylic	560 440 130 520	High Medium Medium to low Low	46 60 71 20	Thermoset Thermoset Thermoset Thermoplastic
Cold mounting	Epoxy Acrylic Polyester Acrylic + A_2O_3	870 1040 1100 920	Low Very low Low	25 23 24 30	Thermoset Thermoplastic Thermoset
Metal	Copper alloy Steel (0.37) Aluminium alloy	35 12.5 142	High High High	265 350 52	

When mounting for automatic preparation the process can be accomplished by two distinctly separate methods:

(i) Samples with a regular shape can be clamped without prior mounting in a purpose-made specimen holder.
(ii) Others can be moulded by:
    (a) casting cold in moulding cups,
    (a) moulding hot in a mechanical press,
    (c) moulding *in situ* within the specimen holder (see Figure 3.3),
    (d) being stuck to the plain specimen holder by means of adhesive.

3.2.2 General comments

There is a fault and remedy table at the end of this chapter in Section 3.3. General cleanliness is important in prolonging the life of the rams and cylinders, and regular cleaning and coating with a mould release agent is important. With the epoxy resins there is a tendency for the mould to stick to the upper and lower rams; this can be reduced by *polishing* the flat surfaces of the rams to produce a smooth scratch-free surface.

Care must be exercised when compression mounting those materials with poor shear and/or compression characteristics. Many of these materials are compression mounted; the resulting analysis must always be subjective. Such materials are plasma-coated aluminas, ceramics, multilayer printed circuit boards, sintered materials with high Vf porosity levels, etc. If it is essential to compression-mount, then the preload facility must be employed. Acrylic resins could be preferred, since they can be used at lower pressures. Those materials withstanding compression, yet poor in shear, could be compression mounted provided the shape of the sample was round, rather than square, and a low coefficient of thermal expansion resin was used. When cooling, and therefore contracting, the resin will exert unequal forces on the sample if the geometry is not uniform. For this reason, it is also unwise to subject soft materials to these conditions if, for example, an analysis of slip planes were to be undertaken. The release of this stress is evident when the specimen shows cracks in the resin at the corners of the sample/resin interface.

Mounting using cold mounting compounds

The use of cold pouring resins to mount specimens has a number of attractions. They are used:

(a) where too few specimens are prepared, making investment in a press uneconomical;
(b) where so many specimens are prepared that compression moulding does not provide sufficient capacity;
(c) where specimens are required to be mounted as an integral part of the specimen holder;

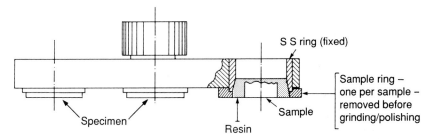

Figure 3.3 *In situ* specimen holder

(d) where turn-around time prohibits the use of long-cycle hot mounting;
(e) where the specimen will not stand any increase of temperature and/or pressure during moulding;
(f) for impregnation of porous specimens.

The resins are based on readily available polyester, acrylic or epoxy materials. Two or three components are mixed together immediately before mounting and the mixture is poured into moulds containing the specimens.

Although this process is often referred to as cold mounting this is a misnomer since there is heat evolved in the polymerization process which produces the solid plastic. If fast-setting resins are used then this heat cannot escape and causes a significant heating of the mount, which can be as high as 100°C. It is possible by using very slow-setting resins, which take over 10 hours before the mount is hard enough to process further, for the temperature rise after casting to be only a few degrees. There are also some acrylics that incorporate some already polymerized materials that have a negligible exotherm and are quite quick setting.

Usually the materials provided for casting are pure resins without fillers so that the mounts will be transparent. Shrinkage, however, can be a problem which is only readily overcome by the incorporation of fillers in the resin, resulting in:

(a) loss of transparency,
(b) less shrinkage,
(c) improvement in abrasion resistance.

A problem with some cold resins is the limited shelf life. This is particularly true for the polyester which should be stored below 15°C.

When mounting very porous or friable materials it is desirable to have the resin permeate all voids in the specimen. This can be accomplished by vacuum impregnation when all the air that may be trapped in the specimen by the casting liquid is drawn off by vacuum. The vacuum removes the atmospheric pressure and forces the liquid into the unfilled areas.

Epoxy resins actually stick to most samples; this can be used to great advantage with friable or porous materials but can also benefit from the 'drip-dry' technique. To put a protective coat around the sample prior to any mounting or grinding operation will benefit many materials. Take, for example, submicrometre plated surfaces; dipping in epoxy prior to subsequent mounting will ensure that contraction will not take place round the plated surface. Drip-drying with epoxy resin also allows subsequent hot mounting of friable materials. If samples can be in the form of a coupon this can often be initially epoxy mounted within the commercially available ceramic tile spacers; they are then split up for normal mounting. This technique can be carried out under vacuum when required (plasma coatings) and is illustrated in Figure 3.4.

Epoxy resins are often used in mineralogy where there is an advantage in knowing the *refractive index* when mounting thin sections. The refractive index (1.54) can be an aid in mineral identification and can also contribute to the 'Becke line' technique (explained in the Microscopy Part).

Bond strength can also be important. One of the reasons slides are 'frosted' for mounting sections is to improve the bond strength. This step could be avoided with high bond strength epoxy resins.

Pot life varies, with some resins being useful for up to 5 days after mixing, others for less than 1 hour. Production departments have resisted the use of epoxy resins, in part due to the continuous need to mix and the care needed in ensuring exact ratios between the

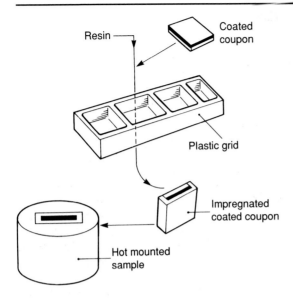

Figure 3.4 Coating coupons

hardener and resin. With the introduction of 5 day pot life resins the 'drip-dry' technique could be used without these problems and all alumina plasma coating (for example) could utilize the tile spacer technique. Another advantage of the long pot life resin is that it can be set in minutes by the introduction of heat.

The *cure time* can vary from 3 minutes to 12 hours. The time taken for any of the cold mounting resins to set (epoxy, polyester and acrylic) can be reduced by induced heat. Care must be exercised when heating the sample since a reduced curing time is often at the expense of increased contraction.

It is vitally important that the moulded sample should be fully set before any grinding operation takes place. Failure to do this will result in:

(a) silicon carbide paper 'grabbing' when grinding, causing gross specimen deformation,
(b) negative relief,
(c) impressing abrasives into the resin (see Chapter 5).

It is the practice of some laboratories to subject the mount, after setting, to a temperature of 150°C for 15 min, in particular when subsequent grinding operations are to be carried out with a charged (loose) abrasive.

There are many materials such as plasma coatings where it is important to impregnate the sample with resin penetrating all the internal pores. Special *low viscosity* resins are available for such occasions. Viscosity is rated in centipoise, high numbers on the scale indicating the lowest viscosity. If the epoxy is claimed to be of low viscosity this number must be quoted (or available).

Hot mounting

Before any compression hot mounting takes place, two essential points have to be addressed:

(i) Will the sample tolerate a temperature of between 140 and 200°C without structural change?
(ii) Compression mounting involves using a pressure of between 21 and 29 MPa. This therefore excludes the use of ceramics and brittle minerals and extreme care must be taken with any sintered or plasma-coated material. In an attempt to overcome this problem manufacturers have produced equipment where it is possible to preheat the resin prior to applying the pressure.

These resins fall into two major groups, those based on thermosetting plastics and others that remain soft at elevated temperatures, i.e. thermoplastic materials. The thermosetting compounds require heat and pressure to ensure correct curing. Reheating this mount will have little or no effect. Thermoplastic compounds have to be cooled prior to ejection in order that they set to produce a hard mould.

Ideally the mount material should have minimum shrinkage and have a similar abrasion rate to the specimen material. Specimen abrasion rates vary depending upon its composition. Therefore manufacturers of mounting compounds offer a range of products to meet these requirements.

Hot mounting involves the pouring of mounting powder into a moulding chamber having first placed the specimen in position. Dust arising from this procedure can be an irritant and on such occasions premoulds should be used. Premoulds are lightly compressed resins (phenolics) that take up the size and shape of the desired mount. They are factory prepared under controlled health and safety conditions and are not intended to be used with friable or easily stressed components.

Another important property that is desirable in the finished mount is its resistance to solvents or chemicals used in subsequent procedures. For example, acrylics are not very resistant to acetone.

Types of hot and cold mounting materials

Hot Mounting

(i) *Phenolic*. This is a woodflour-filled compound and is the most common thermosetting mounting material. It is often available in different colours such as red, green or black and also as premoulds.
(ii) *Diallyl phthalate*. A glass fibre or mineral filler resin makes this an ideal choice when lower abrasion and less shrinkage is required.
(iii) *Epoxy*. Good for bonding to the sample, it has a very low abrasion rate which makes it suitable for all but the soft or friable materials. Ensure that the mounting press ram faces are well polished to resist epoxy adhesion.
(iv) *Acrylic*. This is the only thermoplastic choice if the mounting medium must be transparent. It has a similar abrasion rate to the phenolics yet it is much softer. Its high coefficient of thermal expansion offers many attractions. Samples can be mounted at a lower pressure than thermosetting resins.

Cold mounting

(i) *Epoxy*. This exhibits the lowest cold setting abrasion rate and is the hardest of the air curing resins. Its low viscosity makes it essential for vacuum impregnation. One of its best features is its low contraction (nil) and very low exotherm (30°C).
(ii) *Polyester*. This has the highest abrasion rate, can exhibit the highest contraction and is usually the least expensive but can have a high exotherm (130°C). The high exotherm can be reduced by increasing the setting time. If a hot air blower is used at the exotherm temperature it will induce setting without increasing from the exotherm temperature. The mount temperature as a result of the resin exotherm is related to the ability of the specimen to dissipate the generated heat. To reduce the mount temperature, increase the ratio of the mount to resin.
(iii) *Acrylic*. This is a thermoplastic resin with a very low polishing rate. Some versions are available with an extremely fast setting rate (5 min) without the accompanying contraction normally associated with quick setting resins. Its abrasion rate is good, making it a good general purpose cold mounting resin.

3.3 Common Mounting Faults and Remedies

These are with particular reference to hot compression moulding.

Fault	Cause	Remedy
Radial cracking	Sample too large for mould size	Use large mould or reduce specimen size
	Sharp corners on specimen	Round corners before mounting or use larger mould
Transverse cracking	Evolution of gases from hot inner regions of mount	Do not release pressure until mount is cold or degas
	Steam pressure from damp moulding powder	Keep moulding powder container closed when not in use
Distortion of mount	Mount ejected above heat distortion temperature of plastic	Maintain moulding pressure until mount is colder
	Insufficient curing of thermosetting materials	Increase curing time or temperature

(continued)

24 Surface Preparation and Microscopy of Materials

Fault	Cause	Remedy
Cotton ball in centre of mount	Can be seen in transparent acrylic mounts due to gas evolution when centre is hot	Maintain pressure until mount is cold; reduce cooling rate (stop down input water valve)
Specimen is damaged	Specimen being crushed	Use thermoplastic powder and do not apply pressure until powder is fluid
	Insufficient moulding powder allowing specimen to contact ram	Use more powder in mould
Friable areas	Insufficient moulding temperature or time	Increase moulding time and/or temperature
	Mounting plastic is overfilled if additional filler is used	Reduce added amount of filler
Gap at junction of specimen and mount	High shrinkage or expansion differential	(a) Use mounting material with lower shrinkage (b) Increase cooling period (c) Use a more highly filled mounting material
	Contaminated specimen	Thoroughly clean specimen before mounting
Mount sticking to mould components	Insufficient lubricants in moulding compound	(a) Use alternative powder (b) Use mould release fluid
	Moulding cylinder or rams damaged or scored	Replace defective components

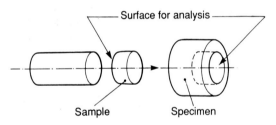

Figure 3.5 The specimen

3.4 Sample/Specimen Definition

The words 'sample' and 'specimen' are used in the preparation of materials and are intended to describe the object under investigation. Confusion arises when it is required to describe the object *to be investigated* in its initial state, such as the 'as cut' condition, with that of the object *being investigated*, such as the 'mounted' condition. Do we, for example, vacuum impregnate the specimen or the sample? Since this has been a cause of confusion and differing interpretation, Figure 3.5 is intended to clarify, viz.:

Sample. This is the piece of material or liquid to be investigated. If it is to be investigated without modification it becomes the specimen in describing the analysis.

Specimen. This is the material or liquid in its analysis state. This would include any mounting, staining, impregnation, coating or any other modification necessary in its preparation for analysis.

4

Single-point Tools

4.1 Stress Applicators

The introduction of what looks like a production engineering subject may at first sight seem strange, but it has proved to be very applicable to the removal of material for subsequent microstructural analysis. The only difference between single-point tool applications and our requirements is that we use multi single-point tools. All the rules for efficient stock removal by mechanical means apply equally to multi single-point tools just as they do for single-point tools. Take, for example, the different flutes used on traditional drill bits; they have a purpose relative to the material ductility. Notice the different rotational speeds available on the engineering lathe; they also are used to maximize cutting efficiency. The cutting tools will have different rake angles, the cutting tool materials will vary and choice is again determined by the material being worked. Machines operate using different types of lubricant; they all have a purpose.

Although this technology is not new it has, however, remained the domain of the production engineer and has not been part of the materials technologist curriculum. Without a complete understanding of the tools that are used by the materials technologists we are unable to identify the most appropriate and nor can we influence improvements.

It could also come as a shock to learn that these single-point tools, although essentially having a sharp point, do not cut the sample. Material is removed not by cutting, not by tearing, but by the application of stress. This stress will at a specific point cause the material under stress to separate; the point at which this occurs changes as the specimen structure and composition changes. Any load below this point will not remove material; any load above this point will speed up the operation.

Tool sharpness has a major effect on the required stress necessary in creating dislocation, the blunt tool changing the stress position and requiring greater loads to remove material. This has an effect upon the induced deformation and also the scratch pattern. In order to avoid confusion when describing the scratch pattern the words *cutting* will be used for efficient material removal (optimum to critical rake angle), *ploughing* (critical to negative angle) as damage occurs during removal and *rubbing* (blunt tool) as the surface is burnished.

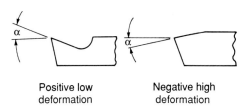

Positive low deformation   Negative high deformation

Figure 4.1 Optimum rake angle for ductile materials (optimum rake—least deformation)

The tool (abrasive) rake angle is extremely important in the degree to which it imparts deformation, as can be seen from Figure 4.1. Rake angles for ductile materials vary from the negative which creates a greater degree of deformation to the positive with lower deformation. The optimum rake angle varies for different materials: pure aluminium, for example, would require a high positive angle, steel a low positive angle and a ceramic could be zero to positive.

Another important consideration is that induced deformation increases as the preparation material ductility increases. It is interesting to apply the deformation/rake angle principle to silicon carbide papers. These papers can be considered as multiple-point cutting tools with a variable distribution of rake angles.

Efficient cutting takes place from the optimum rake angle to a 'critical angle'. When using an abrasive with an angle beyond the critical angle the material is removed by what is called 'ploughing'. Ploughing causes high deformation and is therefore very undesirable.

Steel, for example, has an abrasive critical angle of 90° (as illustrated in Figure 4.2), beyond which the material is simply displaced. Figure 4.3 shows the silicon carbide grains on the surface of the paper with optimum, critical and negative angles relative to the specimen which is steel. When considering more ductile materials the critical angle would be much less than 90° (50–60°), in which case the example shown in Figure 4.3 could be devoid of any optimum rake angle abrasives and only one critical angle, the remainder causing ploughing.

New sheets of silicon carbide paper can often have less than 25% of the abrasives with optimum rake angles. When in use this figure drops dramatically.

Figure 4.4 shows how positive grains soon become negatives in use. This necessitates disposing of the papers when in fact they will still remove material. Living in this world of reality, the operator will continue to use the papers as long as they continue to remove material. When this happens, the results achieved from soft, brittle, friable or porous materials must be questionable.

Abrasives in use shear across the plane of principal stress, resulting in negative abrasives causing a ploughed scratch. These abrasives will in a short time become blunt, causing *rubbing*.

The materials scientist must learn to observe optically the difference between a cut and a ploughed finish. Figure 4.5 illustrates the difference between clean cutting and displacing or ploughing. The ploughed surface when observed optically often displays a scaled top edge, as would be seen in a ploughed field.

The need to observe continually the surface of ground subjects optically cannot be overstressed in order to identify the different scratch patterns. When creating a new preparation technique or making comparisons with established techniques the choice of

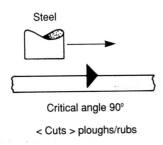

Figure 4.2 Critical angles

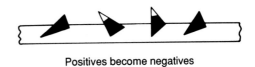

Figure 4.4 Silicon carbide paper in use

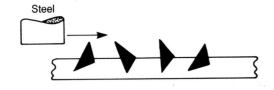

Figure 4.3 Silicon carbide paper when new

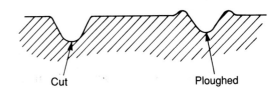

Figure 4.5 Scratch patterns

abrasives, lubricants, surfaces, etc., hinges on the combination that gives the best scratch pattern. Many so-called good preparation techniques can be ruled out on this factor alone since it is the good scratch pattern that gives the least subsurface damage or deformation. We could follow on from this last point by saying that until the specimen preparer can differentiate between grinding, ploughing and abrasive rubbing, encouragement should not be given to progress to specimen polishing, where there are even more different aberrations to encounter and overcome.

4.2 Characteristics Affecting Structural Damage

Deformation or structural damage is of major concern since we ultimately wish to reveal faithfully the parent material. A conception that the surface finish or topography is important must be avoided. What is required is efficient cutting with the least structural damage. In order to achieve this the following points need to be reviewed:

  (i) Pressure
  (ii) Lubricant
  (iii) Rake angle
  (iv) Type of abrasive
  (v) Abrasive size, abrasive shape
  (vi) Specimen material
  (vii) Speed

4.2.1 Pressure

There is a general conception that increasing pressure increases deformation; this is not entirely correct. Single-point tools act as 'stress applicators'; it is the application of this stress that causes the 'chip' to occur. In the application of this stress a point is reached where efficient stock removal occurs, causing the minimum deformation. To increase the stress value basically increases the rate at which the stock is removed. It is a fact, however, that with increased loads the stress applicator is forced further into the material and it is this which increases the deformation.

Recent work with ceramic materials has related the depth of cracking with a single-point tool, as in microhardness testing, to be related to the load applied. As faithful as this information is, it does not relate to the action of a single-point tool when grinding.

4.2.2 Lubricant

With any machining operation, lubricant is a most important factor. In general the object is one of lubrication rather than dissipation of heat (due to the small amount of metal removal and slow surface speeds). With ductile materials it would be very unwise not to use an oil-based lubricant. The water-soluble oil lubricant is suggested as a general purpose medium. Alcohol-based lubricants are also universally used; they represent a clean lubricant with good heat-dissipating qualities but their lubricating effect must be brought into question. They could nevertheless offer advantages with brittle material and ceramics.

4.2.3 Rake angle

The rake angle is very important. The question is, how can we make best use of it? The obvious area is with degrading abrasives; when efficient cutting stops, replace the surface.

Another area would be to avoid the use of blocky-type abrasives with soft/ductile materials, i.e. blocky diamonds are not the best for all materials—they have advantages with hard materials and disadvantages with soft.

New abrasives are being developed allowing rake angle comparisons to be made: man-made diamonds, for example, are shaped differently to natural diamonds, CBN can have more angular shapes, etc.

4.2.4 Type of abrasive

Silicon carbide and alumina have traditionally been used for the metallographic grinding steps. Alumina is tougher but not so sharp, silicon carbide is sharp but brittle. Alumina therefore could be useful in the coarser grades for planar grinding and also in the finer grades with soft material where impressed abrasive proves prevalent. Zirconia can also be used at the coarser stages for soft materials without incurring impressed fragmented abrasives. For silicon carbide, a general purpose abrasive, care should be exercised in its efficient use to avoid gross deformation or impressed fragments.

Diamond and cubic boron nitride are suitable with the harder materials (tungsten carbide), and are used in the fine grinding stages of most materials. These abrasives, being so hard, have wide applications and are available in the bonded/plated or loose form.

4.2.5 Abrasive size

The size of the abrasive influences the deformation and residual stress; the smaller the abrasive the lower the deformation.

Diamond grading to ensure a controlled size and restricted variation is said to be vitally important in metallography and can constitute as much as 80% of the total price. Why is this necessary? If the diamond product contains diamonds much larger than average, then so-called deep scratches occur. Take, for example, a designated 6 μm diamond paste with the occasional 9 μm sized abrasive—it has been suggested that these should be banned. If the reader thinks in terms of induced structural damage and not surface aesthetics, then these 9 μm abrasives, if present on a high shock absorbing surface, will not manifest the undesirable conditions as suggested. Those abrasives lower than the average will not be redundant, as is often suggested, provided they are employed on a soft surface, such as a polishing cloth. Where grading is important is when it is employed on a hard flat charged abrasive platen surface. This is because the largest size abrasive will restrict specimen contact with small abrasives, rendering the latter redundant.

When dealing with multiple abrasives, as opposed to single-point tools, then the abrasive concentration relative to size has to be taken into consideration. Suppliers could, for example, offer high and low concentration diamonds without any guidance for application. As the abrasive decreases in size it becomes more difficult for the chip or swarf to clear the specimen surface; this requires localized clear areas where the lubricant can wash away the debris. When the material increases in hardness the full depth of the abrasive is more readily achieved if the abrasive concentration is reduced.

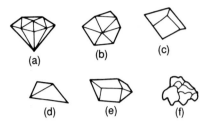

Figure 4.6 Diamond shapes

4.2.6 Abrasive shapes

The abrasive shape is, in part, connected to rake angle considerations. To the materialographer diamonds are not forever and nor are they necessarily his best friend, particularly when an incorrect shape can make the difference between good and poor results. Monocrystalline diamonds produced naturally or synthetically have a different shape or cutting action to the synthetically manufactured polycrystalline diamond. The shapes indicated in Figure 4.6 are but a few of the different abrasive shapes that are possible. A recent search (1991) for diamond shape relative to material preparation has resulted in quotes such as the following:

(i) 'Polycrystalline diamonds make it possible to develop a one-stage polishing method.' This statement is true but only for a particular range of materials, which the quotation overlooks. This type of diamond is shown in Figure 4.6(f); it is the rake angle and the number of positive cutting edges that make this shape most suitable where sample ductility is present.

(ii) 'Blocky diamonds have proved successful for most materials.' This again is true but because of the diamond shape (Figure 4.6(b) and (c)) they would be inefficient with ferrous materials and undesirable with materials exhibiting high ductility, i.e. good for hard materials. High ductile materials would be better served with abrasive shapes such as in Figure 4.6(d) and (e). Although much is documented about the importance of grading, little if any effort is put into abrasive shape. Given a choice the author would prefer an ungraded correct abrasive shape to the current graded incorrect shape.

4.2.7 Specimen material

This influences the type of abrasive to be used; in general the harder the specimen material the lower the deformation. With metals there are examples that would suggest that deformation reduces uniformly with a reduction in abrasive size. Because of the complex manner in which stock is removed from ceramic materials this relationship is not totally arithmetical at the lower micrometre sizes. There is, however, a stock removal deformation angle, relative to all materials (see Chapter 9), and this can be used systematically in transferring preparation techniques between different materials.

The problem materials in specimen preparation are not the hard ones, as would be expected, but the soft smeary materials. When preparing soft materials with degrading or loose abrasives impressed abrasives will inevitably occur. The following is a guide to overcoming these and other problems:

(i) Used fixed abrasive (if possible).
(ii) Increase surface speed.
(iii) Reduce load.
(iv) Use solid lubricants (wax or solid soap).
(v) Use strong abrasives.
(vi) Oil the abrasive surface prior to use.
(vii) Use abrasives with large rake angles.
(viii) Attack polish.
(ix) Avoid swarf rubbing (i.e. do not use finer grades of SiC > 1200 maximum).
(x) Use the largest size of abrasive possible (smaller loose abrasives impress into the sample).

4.2.8 Speed

With all material removal operations there is an optimum speed. This becomes apparent when investigating the relationship between stress and material dislocations. Machine tool technology would be greatly simplified if overnight all machine tools could operate at one single speed. This is, of course, impossible and it is used to illustrate the lack of technology behind the materials technologist who grinds at the same speed for every material or perhaps a colleague who has a variable speed machine and is not very good at guessing. Since most of the grinding/polishing machines on the market tend to have a rotary action there is an added complication to the optimum surface cutting speed due to the material removal differential (MRD). This MRD is more apparent when using automatic machines where the specimens in their holders rotate with the platen. The MRD is not only related to the physics of speed and load acting upon the sample but also to the action of the abrasive when subjected to such conditions, i.e.:

(i) Will the variable surface speed cause outer abrasives to blunt prior to the more central abrasives?
(ii) Will some charged abrasives roll on the outside as the more central abrasives grind?
(iii) Will the difference in surface speed cause uneven lubrication?
(iv) When the abrasive is fixed to a soft backing (C grade silicon carbide papers), will the outer area be squeezed less than the inner?

The best stock removal rate could, for example, be achieved at 80 rev/min, but this gives the greatest MRD. The optimization is also complicated by the specimen head speed–platen grinding speed ratio and whether operated in a complementary or contra motion. (These points will be developed later.)

Taking into consideration all the points raised above will ensure maximum efficiency when sequencing the preparation stages, giving samples of integrity (faithfully reproduced) prepared in the shortest time.

5

The New Concept

5.1 Reasons for a Systematic Approach

To talk about a 'new concept' could imply a totally new system which has been created for the preparation of materials involving newly developed equipment. This would not be entirely true since the study that took place was to investigate the possibility of developing sound technical approaches in order to systematically and scientifically derive suitable methods of specimen preparation. In the event of this study leading to the development of new equipment (which it did), this was to be considered supplementary to existing technology and not a replacement. The author, having been involved with four major manufacturers of preparation equipment along with a close association with university departments of metallurgy whose advice on preparation techniques differed, realized the necessity for such a study. On commencing this study the following terms of reference applied:

(i) Designate different abrasive functions in such a manner as to fit into existing terms such as grinding, polishing and lapping. This was essential since one man's view of grinding was another man's view of lapping, etc. It was also essential since some of the so-called lapping surfaces required the cutting abrasive not to roll. Having searched for documented references they proved contradictory.

(ii) Descriptions of procedures such as grinding, lapping and polishing should relate to the abrasive function and not the surface carrying the abrasive. This is because the abrasive can be arranged to lap or grind on a lapping surface; it can grind or polish on a lapping surface.

(iii) Abrasives used should be *fixed* as in, say, a resin-bonded wheel or *charged* as in, say, spraying diamond on a cloth.

(iv) Abrasives that chip or break in use should be called degrading abrasives, such as silicon carbide papers.

(v) Where possible all samples should be processed via a series of grinding operations to faithfully reproduce the true structure. Those samples such as minerals or brittle subjects where grinding induces severe damage should include lapping operations. In order to reduce total preparation times a combination of both grinding and lapping should be investigated.

(vi) This stage of the procedure (v) will be called the 'sample integrity stage'. The specimen is not to be progressed beyond this stage without first achieving sample integrity, i.e. faithful reproduction. Under

no circumstances is polishing to be included in the sample integrity stage.
(vii) Any cosmetic or aesthetic requirement beyond the sample integrity stage is to be called a polishing stage.
(viii) Specimen preparation procedures are to be expressed in a non-commercialized language. The implications of this point necessitate the introduction of some standard reference from which different manufacturers' trade name products can be related.
(ix) Shock-absorbing surfaces should be investigated with a view to reducing grinding impact and therefore grading reduction in terms of deformation relative to a given size abrasive.
(x) These shock-absorbing surfaces, called platen surfaces, should be designated relative to preparation-induced structure damage. This designation should be a series of numbers from a theoretical zero (nil damage) to 10.
(xi) These platen surfaces could be charged abrasive surfaces or fixed abrasive surfaces.
(xii) Polishing surfaces should also be designated from this theoretical zero (true structure) to 10.
(xiii) An optimization programme is to be implemented to identify the most efficient cutting speeds, head–platen ratio, pressure, lubricant, etc., relative to not just stock removal but also specimen-induced deformation or structural damage.
(xiv) All procedures are to be databased relative to the abrasive type and Z axis curve for each different material. (The Z axis curve is the stock removed in a given time.) This curve is positioned in the database relative to the structural damage remaining after completion of that particular step.
(xv) Devise a system that overcomes the current practice of *introducing* structural damage at each preparation stage and confines the depth of damage, where possible, to the initial preparation step.

This initial 12 month study resulted in the documentation of what is called the New Concept (1987), A Systematic Approach to Sample Preparation.

5.1.1 What is the new concept and why is it needed?

To address the last point first: it is needed because there has in the past been few (if any) specific rules for the preparation of metallographic samples. This has resulted in having empirical methods that are very much operator subjective. The following apply:

It is needed because these empirical methods differ from one laboratory to the next.
It is needed because advice given from one user or supplier to the next differs.
It is needed because users claim that only their specific methods on their specific equipment will give the desired result.
It is needed because textbooks conflict on which is the best route.
However, *it is needed* most of all because many advanced engineering materials currently in use and under development do not respond to conventional preparation methods.

Empirical development of preparation procedures is only possible if the less than optimum technique is capable of giving an adequate result (see Figure 5.1). This integrity curve is typical of the traditional four-step silicon carbide papers with two or three so-called polishing stages. This has proved to be adequate for a whole range of materials from ferrous to non-ferrous such as steels to brass. Note the angle of the curve between the arbitrary numbers 90 to 100; it is a gentle curve

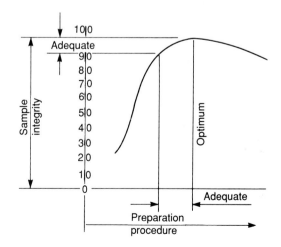

Figure 5.1 Integrity curve for 0.37% carbon steel

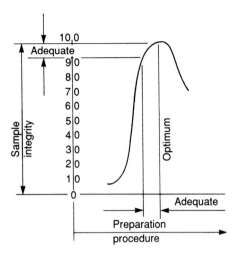

Figure 5.2 Integrity curve for alumina/yttria coating

towards the optimum and is slow to descend beyond this point. Assuming the distance between 90 and 100 to be indicative of an acceptable adequate preparation then there is a broad band on both sides of the optimum showing how relatively insensitive the technique is to 'less than optimum' preparation.

For traditional engineering materials, empirical development has been possible in many cases, due to the high tolerance of these materials to less than optimum preparation routes. Advanced engineering materials, however, have a low tolerance to 'less than optimum' preparation procedures and could be prepared in a manner that resulted in erroneous analysis (see Figure 5.2). This particular curve was generated for an alumina/yttria plasma coating. Note how steep the curve is towards the optimum and how quickly it falls off; not only is this material sensitive to the correct method, it is also sensitive as it proceeds. From this curve it can be seen how important it is to select and control the preparation procedure in order to remain within this narrow band of adequate preparation.

To summarize, one could say 'traditional materials have in general *not been too sensitive* to less than optimum procedures and it is quite rare to find one of these procedures actually destroying the targeted sample integrity'. On the other hand, when preparing advanced engineering materials they can be *very sensitive* to using the optimum procedure; the targeted integrity can readily be destroyed.

It is not intended to conclude from the above that traditional materials, because they are insensitive to less than optimum, should be treated any differently in our study than the new materials. If there is a correct procedure we should define it and use it, irrespective of the material.

5.2 The Application of Logic

The general practice in microstructural analysis is to optically observe the specimen after the polishing step. This then is a tradition one must break when auditing our preparation procedure. We are *never sure* when looking at a highly polished sample if the true integrity of the microstructure has been revealed. Without understanding the preparation procedure one can *never be sure*, for example, that the revealed porosity has not been induced nor can one say that the sample prepared is devoid of any inherent defects just because we do not see any. For these and many other reasons it was necessary to look more closely into sample preparation and define exactly what was happening.

If one considers the removal of surface material as a mechanical engineering function such as turning on a lathe or milling and drilling, then one can start to understand more deeply the action of different abrasives and surfaces. For example, when cutting a ductile material the rake angle should be greater than when cutting a more brittle material.

Lubrication is also an important factor to be considered in the removal of material. Note the different types of lubricating and cutting fluids used in the machine shop. All this information we can use together with two golden rules.

(i) Smaller sized abrasives create less surface damage.
(ii) Softer abrasive backing creates less deformation and/or structural damage.

It is this last point that has been developed with the 'new concept'. Briefly, if one uses an abrasive whose size is adequate for stock removal and utilizes a system such that the abrasive could be housed in variable shock

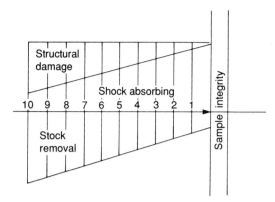

Figure 5.3 Surface classification

absorbers, then a point could be reached giving a satisfactory surface finish without structural or deformation damage, i.e. achieving *sample integrity*.

If we consider the point of sample integrity (see Figure 5.3) to be zero and we then create a series of material surfaces whose shock absorbing characteristics decrease (as stock removal increases) to, say, a given point of 10, one now has a choice of material surfaces or grinding surfaces that can be related to the sample material under investigation. These varied surfaces have been called platen surfaces.

Two types of platen surfaces were worked on:

(i) *Fixed abrasive surface*. Fixed diamonds in varying resins and nickel-plated diamond on thin brass sheets are bonded to different shock absorbing backings.
(ii) *Charged abrasive surface*. These were made from a mixture of different metals/ceramics and resins, the diamond abrasive being charged or sprayed onto the surface (not diamond paste).

It has become apparent that the fixed abrasive surfaces had limited application since most specimens exhibited ductility and were better suited to a charged abrasive. Those materials without ductility (ceramics, minerals, etc.) could also be prepared using charged abrasive surfaces but did take longer than when using fixed surfaces. It is not intended to imply that the use of fixed abrasives would not be desirable in certain circumstances; on variable shock absorbing surfaces with, say, 9 μm and less they could be very useful with ceramics and hard materials. Having established these shock absorbing surfaces on a scale of 0 to 10 on the principle of reduced deformation relative to a given abrasive size, the question to be addressed was 'What shall that abrasive size be?' Our needs are abrasives whose size and shape are adequate for stock removal that can be used in a varying shock absorbing mode to faithfully reveal the true structure. In practice this is not possible and we have to vary the abrasive size; this will be dealt with later.

If asked to prepare the sample through all the stages from 10 to 0 we would expect to achieve sample integrity. Although there is a faithfully reproduced microstructure the surface lustre or 'polish' may not be aesthetically pleasing (with softer materials in particular). This is to be expected because all the abrasive actions up to this zero point are used in a cutting mode which gives a scratched appearance (the only exception to this is when a free lapping technique is used for brittle minerals giving a matt finish).

Although reference is made to surfaces from 0 to 10, the zero position is a theoretical utopia which can never be achieved by mechanical preparation techniques. Therefore in practice the lowest numbered surface must be greater than zero.

Two points emerge from what has been said so far. One is that it is now possible to define a point where true integrity occurs. To continue with the preparation sequence into the polishing mode without achieving integrity would give erroneous results. This is something we were never sure about with conventional preparation methods. It also means one *must* optically observe the specimen prior to final polishing.

The second point is that two distinct stages have been created in our preparation procedure.

(i) The sample integrity stage
(ii) The polishing stage

With stage (ii), 'the polishing stage', one must understand that to polish the sample surface could smear the microstructure surface topography which, if prolonged, has an adverse effect on sample integrity (greater with softer materials).

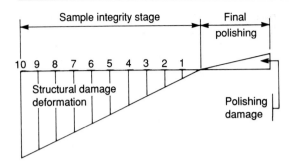

Figure 5.4 System foundation

It is also interesting to note at this point that a scratched surface that has its scratch edges rounded off actually appears polished since the compound microscope will not reveal smooth undulating surfaces without the use of interference contrast.

One of the many advantages of the 'new concept' is the reduced times needed at the polishing stage, since it is only used for aesthetic finishing, thus avoiding the traditional pitfalls. Figure 5.4 illustrates the foundation of the system; the next step is its application.

5.3 Abrasives and Their Functions

If one looks into the 'tool-box' of possible methods one now has platen surfaces. Before investigating the application of these one must look to the traditional tools in order to define their best use. It is known that fixed abrasives function in a different manner from those that are allowed to roll freely. Different types of abrasives have particular characteristics best suited to particular materials and conditions. To make the best use of these available tools, their function must be understood in order to select the right combination. (A 6.5 millimetre spanner could loosen a 6 mm nut but that does not make it the best choice nor should it be the only spanner in our tool-box).

5.3.1 Types of abrasives

A basic essential of the abrasive is that it removes stock, creates little damage and if possible dissipates the generated heat. Four main types of abrasives are in general use:

(i) Silicon carbide
(ii) Aluminium oxide
(iii) Cubic boron nitride
(iv) Diamond

A major restriction to improving our technology in material preparation is the restricted available abrasives; this situation will change as some of the new ceramics come on stream. Silicon carbide, for example, is an ideal abrasive since it is hard and is not blocky but unfortunately it is brittle. Some of the synthetic materials (man-made diamond) can exhibit totally different rake angles and these features have to be exploited if we are to make maximum efficient use of available resources. The writer is aware of situations where, for example, the end user will strongly prefer a particular supplier's diamond product in preference to others without realizing that it is coincidentally the diamond shape. This product if satisfactory for a ductile material will be unsuitable for a hard brittle material. Manufacturers themselves give no guide-lines in this direction, and they further confuse the situation by referring to their product as, say, 'heavy concentration' or 'superior quality' or even 'suitable for all materials', etc.

What is needed is a greater choice of abrasive and a more useful definition of each particular abrasive along with a standard measure of concentration. (Abrasives such as diamond, supplied by the 'pint' or 'litre' or 'heavy' or 'normal' concentration, will inevitably have to be changed to conform to some industrial standard.)

5.3.2 Silicon carbide

With ductile materials (metals) it is unwise to totally fix the abrasive, due to the 'chip' buildup on the abrasive cutting edge. To this end silicon carbide papers have traditionally been used for coarse and fine grinding of the specimen, the silicon carbide breaking away during use. A series of tests were carried out to identify the useful life of various grades of silicon carbide paper and to investigate the deformation present at each stage.

A medium carbon steel was checked at intervals for weight loss and also microstructurally prepared to measure deformation (the section was electrolytically polished at 90°

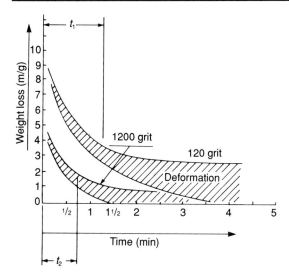

Figure 5.5 Deformation weight loss relationship

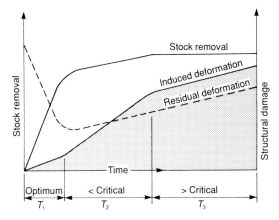

Figure 5.6 Degrading abrasive and deformation chart

to the etched surface). The results of these tests with 120 and P1200 grit papers (see Figure 5.5) illustrate microstructurally the degree to which deformation occurs.

What is emerging from the weight loss deformation chart is the relationship existing between degrading abrasive use in time and the induced damage. Another factor, that of reduced residual damage, must also be taken into account. Figure 5.6 makes an interesting relationship between residual damage, induced damage and stock removal; all three factors have a definite relationship to the rake angle of the silicon carbide abrasive. From this graph can be related the best stock removal rate, the lowest residual deformation and the least rate of induced deformation occurring during the period before the cutting abrasives exhibit a critical angle. When the abrasive rake angles become more than critical, ploughing commences; when they then become blunt, rubbing commences.

Quite clearly silicon carbide would not remain sharp long enough to remove any substantial amount of stock without creating deformation (hence the reason why many different grit sizes of paper are traditionally used). Silicon carbide paper is, however, inexpensive and therefore could be used in particular at the first preparation stage (after sectioning and mounting). Since the object at this juncture is to flatten the section or sections, this stage is called 'planar grinding', and is usually carried out with a coarse grit dependent upon the sample material condition.

To simply select a coarse paper because the specimens are not flat, however, is *not* the suggestion; careful note must be made of the residual damage from the sectioning stage and an abrasive size must be used that will not increase this Z axis damage. Take the situation where the subject, having been sectioned, exhibits a damaged Z axis of $20\,\mu m$; to use a coarse abrasive just to achieve planarity could induce $120\,\mu m$ of damage. This situation occurs very often; it is better to use two identical grades of silicon carbide paper than to incur what could be a tragedy by adopting traditional techniques (this is particularly important with brittle fibres where longitudinal damage can extend along the length of the fibre).

To continue to the 'sample integrity stage' with fixed silicon carbide papers could, as indicated, be undesirable. When silicon carbide powder was used on platen surfaces its cutting capacity was low and the grains broke down, leaving impressed abrasives in the specimen.

The discussions so far have been confined to the use of silicon carbide as a grinding agent, firstly as a fixed abrasive on a paper backing and then as a loose abrasive 'stationary at the point of cutting' on platen surfaces. Silicon carbide powder when allowed to roll on a flat surface (lapping) proves very useful in the preparation of brittle materials such as minerals and ceramics. It is used in the form of a slurry. The specimen sitting on top of the slurry is not *forced* across the surface, it is *drawn* (or dragged) across by the action of the rolling abrasive, achieved by slowly rotating the platen. The cutting force on the specimen is lower and hence the damage to both the abrasive and specimen is less, resulting in lower deformation but greatly increased preparation times. This method, though ideal for brittle subjects, is undesirable for ductile metals since rolling abrasives would 'stick' (become impressed) into the softer material.

5.3.3 Aluminium oxide

As a fixed abrasive aluminium oxide tends not to be used in materials preparation; it has, however, certain advantages over silicon carbide. It is slightly softer than silicon carbide, which is a disadvantage with hard materials; it is, however, tougher, which is an advantage over silicon carbide. It will in general leave more residual damage than silicon carbide paper but since its useful cutting life is longer it could be preferred instead of two identical sheets of silicon carbide paper.

Both silicon carbide paper and aluminium oxide paper are degrading consumable items. It is this very fact that caused the introduction of alternative surfaces for use on automatic specimen preparation equipment (platen surfaces).

Just in case history records the introduction of the so-called 'platen surfaces' as a response to the need for integrity in the preparation of new engineering materials, this is not the case. Some 25 years ago when specimen preparation had to be part of a materials quality control procedure requiring endless specimens for preparation, it was found that more time was spent actually changing papers than was taken to prepare the specimens. This was obviously not only time consuming but it could hardly be called automatic, and it is for this reason that a non-degrading charged abrasive was used on composite discs, avoiding the need to change papers. This was also the time when the three-step method was employed for ferrous materials, i.e.:

(i) Planar grind using a grinding stone (Al_2O_3)
(ii) 9 μm diamond on composite disc
(iii) 1 μm diamond (sometimes 3 μm) on hard surface polishing cloth

As with silicon carbide, alumina is used in the lapping mode for minerals and ceramics. Traditional metallography techniques call for alumina to be used in the coarse and fine polishing stages and this has been a time-honoured procedure, particularly for ferrous materials. It is still recommended today as a polishing medium, usually gamma 0.05 μm for aesthetic or cosmetic applications. Gamma alumina is losing favour to some of the activated silicas, but it is still important in particular to the coal petrologist.

5.3.4 Cubic boron nitride

Cubic boron nitride would be very expensive in the bonded abrasive paper form but has an increased life over silicon carbide and aluminium oxide. It efficiently dissipates heat and initial trials on platen surfaces with CBN powder have been encouraging, but tests are too limited as yet to be of value.

CBN is very hard (next to diamond) and should have benefits with the harder range of materials. It is unlikely in its blocky form to offer real advantages for routine ferrous materials.

5.3.5 Diamond

Diamond will cut efficiently, but again is too expensive as a bonded abrasive on a backing material. Trials using all the special platen surfaces have shown diamond as being better than any other material tested. At about 6 to 9 μm, diamond will continue to remove stock and if used with the variable shock absorbing surfaces will progress the sample towards zero deformation.

5.3.6 New abrasives

It would appear as if diamond and CBN can satisfy the requirement of hard materials. The medium range materials (general ferrous), being planar ground on SiC or Al_2O_3, do require something new for the sample integrity stage; although diamond is used, its lack of cutting efficiency in this range is well known. The softer materials (aluminium) would benefit from, say, a silicon carbide which is sharp if only it were not so brittle.

5.4 Abrasive Function Definitions

This was the first task in the terms of reference since the necessity of any dialogue is to speak the same language. The following abrasive function definitions have been published and used for the last 5 years. They clarify the many misconceptions that took place prior to such a convention.

5.4.1 Definitions

Grinding. The abrasive will be 'stationary' at the point of cutting.
  Appearance: shiny and scratched.
Lapping. The abrasive will be 'rolling' at the point of cutting.
  Appearance: dull matt finish.
Polishing. The abrasive will be allowed to 'rise and fall' within, for example, the nap of the cloth at the point of cutting.
  Appearance: shiny without scratches.

When the materials technologist requires to communicate when relating to such functions it is *essential* that these definitions be adhered to.

Grinding (see Figure 5.7)

On all grinding procedures the abrasive is fixed at the moment of cutting, the sample passing normal to the axis of the fixed abrasives. The resultant force (on the sample) is α to the horizontal. This is generally used for stock removal and gives high deformation.

Free lapping (see Figure 5.8)

With free lapping the abrasive is allowed to predominantly roll free; this is a slower process than grinding, resulting in a shallower deformation layer, less stock removal and flat surfaces. If used with ductile materials it will result in impressed abrasives and also cause compressed grains. Generally used for brittle subjects such as minerals, it is important to note the increased resultant force angle and hence its advantage with brittle subjects.

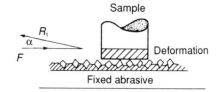

Figure 5.7 Grinding

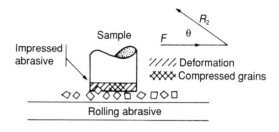

Figure 5.8 Free lapping

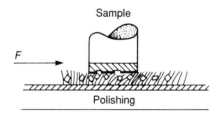

Figure 5.9 Polishing (note the deformation/plucked grains, smear and cloth nap)

Polishing (see Figure 5.9)

Polishing can be carried out with many different abrasive materials on different napped cloths. The abrasive is allowed to move up and down within the nap of the cloth. Polishing gives very low deformation, preferential polishing (depending on the nap) and can smear the top surface. The polish also varies with the type of lubricant used.

Grains can be plucked out with the nap, which is sometimes associated with damage created at a previous grinding stage.

Composite surface grinding (see Figure 5.10)

This technique is a combination of both grinding and free lapping. The degree to which both these functions interrelate offers a completely new dimension in the area of sample preparation. By varying the materials on the composite surface it is possible to have:

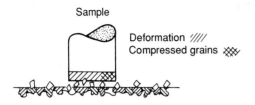

Figure 5.10 Composite surface grinding (note the impressed abrasive, porosity and rolling abrasive)

(i) abrasives that embed into the material surface (grinding);
(ii) abrasives that roll (as in free lapping) on the surface;
(iii) abrasives that are momentarily retarded from rolling by a surface interface (grinding);
(iv) abrasives that can be drawn back onto the surface from areas of controlled porosity;
(v) combinations of any of the above.

By controlling these conditions, along with newly created surfaces, it is possible to control the degree of grinding with an associated shock absorbing feature.

Surfaces that grind when charged with abrasives are not new; 50 years ago different so-called 'laps' made from wood, lead, tin, etc., were in common use. Today's composite surfaces are more stable, more efficient and more comprehensive; they are still referred to as 'laps'. This term 'lap' is obviously confusing since it conflicts with the 'abrasive function definitions'. For this reason the expression 'composite surface' or 'platen surface' is preferred when classifying.

5.5 Platen Surface Functions

This is the name given to abrasive surfaces that have some form of classification related to the ability of that surface to progress the sample towards integrity. These surfaces can be one of two types, i.e. employing fixed abrasives or loose 'charged' abrasives. Although much work has been done on the classification of 'charged' abrasives, fixed abrasives currently have no identity beyond that of abrasive size. Take, for example, two $9\,\mu m$ diamond platens, one metal bonded and the other nylon bonded; the metal-bonded platen will remove more material but will, more importantly, create the greatest structural damage. We therefore require a classification related, as with charged abrasives, to the ability of the abrasive to absorb shock when cutting. Until such classifications are available it is wise to identify the abrasive support material whenever a fixed abrasive is to be used and to use a softer backing if high induced structural damage occurs.

With charged platen surfaces there are many different types currently available, the majority being a resin matrix composite and others being inserts of metal composite within, say, a cast iron platen. Towards the lower end of the classification scale, resin and resin/fibre platens are to be found.

Surfaces that are polymer matrix composite (PMC) usually consist of small particles of metal in the resin matrix; the size of these particles can influence the surface efficiency. The action of, say, a diamond abrasive when charged onto these surfaces functions in different ways. If predominant cutting is to take place, as the abrasive is interface retarded (resin to metal) then it follows that the more this takes place the higher the cutting efficiency.

Another rather interesting concept is to have the grinding abrasive attracted to the resin and to introduce composites to the resin that are similar, but slightly softer, than the material being prepared. One such example is shown in Figure 5.11 (Plate 1) where the dark ground optical technique shows the impressed abrasives predominantly fixed into the resin or resin/metal interface. Figure 5.12 (Plate 1) is a normal bright field image before diamond dosing, showing the two metal/resin composite material also exhibiting porosity. The size and shape of the material, its type, the level of porosity, the area of resin, the type of resin etc. will all have an effect upon material removal efficiency. The original lead and tin so-called laps required the grinding abrasive to impress into the metal. This new concept very much relates to the correct resin for maximum abrasive ingress enabling the more efficient grinding to take place.

One could consider platen surfaces as being a series of platens with diminishing hardness from 10 to 0. Unfortunately, hardness, although related, is unreliable since with charged abrasives it does not take into account the action of the abrasive. Take, for example, surface 10; this is a ceramic (Al_2O_3) resin matrix composite. If it were a dense alumina surface then its hardness would be greater but the classification would have to be lower since the charged abrasives would tend to roll, achieving the *undesirable* lapping condition.

When functioning on these surfaces the abrasive can change its mode (from grinding to lapping) when a change of specimen material takes place. The choice of platen surface is based on the hardness of the specimen material, very hard materials starting with high numbered platen surfaces (10) and very soft materials low numbered platen surfaces (2). The specimen must never be softer than the platen surface. There are cases when a lower numbered surface is chosen because the specimen material is too hard and changes the abrasive cutting mode. Take, for example, a soft alumina specimen; platen surface 10 would be the prime selection. Hard ceramics such as boron nitride, for example, could dislodge the diamond abrasives, causing a greater predominance towards lapping. Experiments have shown on such occasions that it is better to use, say, surface 8 or 9 with the same sized abrasive than to use a smaller sized abrasive on surface 10.

The following is an explanation of differed charged platen surfaces and when best to use them.

5.5.1 Platen surface 9/10/11

In view of the changing mode (from grinding to lapping) of the abrasive it is necessary to group these three together; surface 11 acknowledges the possibility of a surface more aggressive than 10, which is the alumina composite. Surface 9 could be a metal/alumina composite for very hard ceramics.

This group is the hardest of the series and if charged with 35 to 45 μm diamond will function predominantly in a grinding mode (with correct selection) and is therefore ideal for use in a planar grinding mode for *hard materials*. The reader would be correct in thinking that a fixed diamond or CBN abrasive would carry out the same function more quickly. This would be true *but* the structural damage would be higher. This raises a very interesting point and one that must be stressed. These platen surfaces are intended to supplement existing technology and not to replace it; therefore if a fixed diamond or CBN wheel is available and in removing material does not induce a high level of damage then use it. A similar situation occurs when creating a procedure for some materials where a choice between platen surface or silicon carbide paper is made; always go for the quickest, least expensive unless the method has to be totally automated, in which case platen surfaces should be favoured.

Platen surfaces are not restricted to the use of 30 to 45 μm diamond. Occasions arise when their use in the sample integrity stage demands 9 or 6 μm diamond.

The abrasive can operate with this group of materials in either grinding or lapping mode without any adverse effect, the only disadvantage being that lapping can increase the preparation time by a factor of 10 and a lapping abrasive wears out the platen surface as well as removing stock from the specimen.

Obviously if this range of charged platen surfaces were to be used with, say, aluminium then the grinding abrasive would impress into the specimen. There are times, however, with, for example silicon carbide particles in an aluminium matrix composite (MMC) when it is unwise to use a fixed diamond wheel due to swarf buildup and degrading surfaces such as silicon carbide papers result in gross sample particulate damage. On such occasions platen surface 10 will give a much improved result providing the abrasive does not impress into the aluminium matrix. If the abrasive does impress, it must be removed by fixed diamond grinding or resectioning before the specimen can be progressed. To avoid impressed abrasives ensure the diamond does not roll on the integrity surface (as in lapping). Rolling abrasives are more prone to impress than lodged abrasives. Reducing the abrasive size and/or the relative surface speed reduces the tendency for lapping. Increasing or decreasing the pressure from the optimum can have an adverse effect, i.e. increase the pressure and impress the abrasive—decrease the pressure and increase the lapping.

Diamond abrasive is used in an appropriate slurry or carrier dosed onto the surface.

The type of slurry used with platen surface 10 can vary: water-based slurry for the harder materials and oil-based for the composites with a soft matrix.

Occasions arise when platen surface 10 with 30 to 45 μm diamond will not satisfactorily cut the specimen (the abrasive predominantly rolls). One such material is the glass matrix composites, reducing the abrasive size can help to alleviate this problem.

Specimens such as MMC particulate or tungsten carbide/cobalt, which ideally should be prepared with platen surface 10, can be satisfactorily planar ground using, say, a degrading abrasive such as silicon carbide paper providing the matrix is soft and the abrasive size is larger than the particulate (preferably a 3:1 ratio or greater).

5.5.2 Platen surface 6/7/8

Polymer composites of iron, copper or both usually make up this group, but note should be made of the metal insert type and the steel mesh cloths also available. This group of surfaces satisfies the preparation needs of ferrous and copper alloys. There is a school of thought that suggests that this is the area where rationalization could take place and that only surface 7 should be produced. If this were to take place then it would have to be accompanied by the design of surfaces 5 and 3, which are currently not available.

The abrasive size for this group in general would be no smaller than 9 μm, the upper limit depending on the previous step. With large ferrous sections that have been planar ground using a high speed alumina stone sizes as high as 45 μm might be necessary prior to the 9 μm stage. When prolonged time is required at this stage it usually signifies an out-of-flat surface or that a supplementary stage such as silicon carbide paper should be included.

The action of the abrasive particle when used on these platen surfaces is different to the action when the same abrasive is used on platen surface 10. Here it will tend to skid, be retarded and to a lesser degree impress into the surface. Aerosol diamond does work on these surfaces but the use of water- or oil-based diamond slurries avoids the need to supplement the lubrication. Oil-based diamond slurries have been noted to give a reduced cutting efficiency although still producing satisfactory results. With some materials the choice of oil-based slurry may be preferred.

5.5.3 Platen surface 4/5

As was mentioned above, surface 5 is not currently available, both 5 and 4 being primary and secondary mode surfaces. This means that they can be used as a primary selection for the softer materials such as pure copper and also as a secondary choice after surfaces of higher value. These surfaces usually have particles of, say, tin, lead and copper which allow the charged abrasive to impress into the surface thereby giving an improved shock absorbing function. In the secondary mode selection it can be seen how they can, by absorbing the shock, progress the sample towards integrity. What is equally important is the ability of these surfaces to retain the structure of friable elements in the specimen composition. The best example illustrating this point is the preparation of spheroidal graphite cast iron; if the friable graphite were to be studied then surface 4 would be used but if just the graphite morphology were required, surface 7 would be adequate.

Many new materials include elements that are brittle and it is the shock absorbing features of this group that reduce the shock waves when cutting and in consequence reduce the Z axis structural damage. When optically observing the progress towards integrity after each preparation stage, a dense alumina will 'come to life' on completion of this stage. There is one major disadvantage using these surfaces in their secondary mode application and that is platen surface pick-up. When the specimen is very hard and porous the progress to integrity is good, but each pore edge acts like a small cutting tool and deposits the soft platen surface material within. On such occasions it would be wise to either increase the lubricant thickness, increase the abrasive size or use a harder surface (the latter is not possible since surface 5 is not commercially available).

It is always wise to use an oil-based lubricant on these surfaces to avoid where possible any specimen-to-platen surface rubbing. This has the disadvantage that with a sample of large area the cutting abrasives are not sufficiently impressed into the integrity surface. The above condition manifests itself as poor stock removal, which is usually overcome by a reduction in platen speed (from 120 rev/min down to as low as 10 rev/min).

There is what is called a nominally 'stationary mode' where the platen is rotated just in order to maintain planarity. The specimens are rotated at between 30 and 120 rev/min in a contra motion; this mode is used to retain features that are extremely friable or loose.

Abrasive sizes for the group would be 3 to 15 μm; the writer uses and prefers 9 μm for most, if not all, applications.

5.5.4 Platen surface 2/3

In this group we have surfaces of a totally different construction. On the one hand we have a resin/metal composition (aluminium/lead) (3) and on the other a non-metallic resin construction (2). Although their functions overlap as secondary mode surfaces, they do differ when requirements are for primary mode. Surface 3 is a natural choice after using surface 5; it also has applications when extremely brittle or friable materials require improved integrity beyond that which can be achieved with surface 4. A recent experiment, using boron nitride as the sample, showed that with each decreasing surface starting at surface 8 the sample integrity improved.

This confirmed the importance of surface 3 with hard/brittle/friable materials when used in the secondary mode. When preparing soft materials this surface has many limitations: one is the problem of rolling abrasives, the other metal-to-metal surface affinity. The abrasive on surface 3 must be forced into the surface by the specimen as it passes over the surface; due to the soft specimen the abrasive is equally attracted to the sample and causes impressed or lapping abrasives. Surface affinity occurs predominantly with softer materials and causes a breakdown of the lubricating membrane between the platen and specimen, resulting in surface rubbing and binding. To satisfy the requirements of a primary surface for soft materials surface 2 is used; being a non-metal surface this overcomes the problems associated with surface affinity. To overcome the problem of impressed abrasives with soft materials, it may be necessary to use 15 to 30 μm diamond. The surface is also perforated, ensuring swarf clearance and maintenance of a lubricant membrane. Diamond sizes much smaller than this would be used when this surface is used as a secondary mode surface (3 to 9 μm). Both platen surfaces 3 and 2 are extremely important surfaces. It has already been explained how surface 3 improved integrity with the hard brittle subjects. Surface 2 can be used on nearly all materials from soft metals to hard ceramics, but to quote one specific example—an aluminium matrix/silicon carbide particulate—it will grind and progress the particulate without causing impressed abrasives in the alumina. The surface also responds to an activated slurry since it can be retained within the perforations of the platen surface.

When it is necessary to use a resin wheel for mineralogy and ceramic applications it is not necessary to have any perforations and such surfaces are commercially available. Their use, however, for soft metals is not recommended.

5.5.5 Platen surface 1

This surface (1) will have many contenders that the reader may consider to be traditional polishing cloths. These surfaces (cloths), due to being hard and napless when used at high pressures, perform as abrasive holders, ensuring that the abrasive grinds and does not polish (as described earlier). One example to illustrate this point is the old-fashioned method of using brown paper with 1 μm diamond impregnated as the final step in the preparation of tungsten carbide; this would be platen surface 1. When preparing hard materials, platen surface 1 would as mentioned be a hard surface such as a chemotextile material or nylon and can be used with diamond abrasives down to 1 μm (as low as 0.25 μm diamond for ceramics). When selecting platen surface 1 for softer materials two important points have to be considered. The first point is that the surface material must be much softer than that acceptable for preparation of hard materials (wool or cotton); the second point is that the abrasive size should not be less than 3 μm.

The abrasive used on this surface can be aerosol, slurry and (on this occasion only) paste. The abrasive must be caused to act in a grinding mode. With hard materials (ceramics) this would require an increased load—as much as twice that of previous grinding steps. This would produce a specimen of integrity requiring no polishing stage. The lubricant, if water based, would be most efficient.

Impressed abrasives can still manifest themselves even at this stage. Oil-based lubricants can help alleviate but in many cases it is wise to abandon the traditional 1 μm diamond route and change to a high pH slurry. Diamonds impress less from a used cloth than from a new cloth and also benefit from using a dummy specimen before the actual specimen after each diamond recharge.

When faced with, for example, an MMC where the matrix is soft and the fibre or particulate is hard then:

(i) slurry polishing would result in relief (high pH) and
(ii) diamond polishing would result in impressed abrasives.

The answer is to combine both (i) and (ii) together.

5.6 Efficiency of Charged Platen Surfaces

Lapping platens rarely need flattening since the abrasive action is uniform across the platen surface and removes material at a uniform rate from the lapping surface or platen. Charged platen surfaces act in a different manner in that the driven specimens (they are not dragged as in lapping) traverse the platen surface at unequal surface speeds and in consequence exhibit a different wear pattern. If the operating velocity is uniform, i.e. the specimens and the platen rotate at the same speed in a complementary direction, then uniform wear should take place and little if any uneven wear will occur providing the abrasives on the surface act in an identical manner.

Surface speed and direction will be covered more deeply under the surface optimization section but for the moment consider the specimens and platen to operate in contra motion. The peripheral speed being greater than any other position can cause those abrasives on the outer diameters to roll (lap), causing unequal surface wear.

The Questions to be addressed are (i) how much wear can be tolerated and (ii) how is it corrected?

(i) Firstly let it be clear that if planar corrections are required at frequent intervals then inefficient abrasive action is responsible and must be investigated. Efficient use of platen surfaces will have a self-levelling effect; some platens never need to be corrected for planarity even after thousands of samples have been progressed. The efficiency of platen surfaces does, nevertheless, depend upon surfaces remaining flat to less than 25 μm (0.001 in) for all but the highest numbers (8 to 10). When operating at identical surface speeds (unity specimens to platen in a complementary direction) errors similar to the abrasive size being used will automatically correct themselves. Unfortunately when surface speeds are dissimilar the error will be amplified. Sometimes it is necessary to compromise optimum stock removal with platen planarity.

Checking for out-of-flat is carried out using a straight edge accompanied by good back illumination. This could be done at regular intervals or on occasions when a loss of efficiency is recorded.

(ii) Levelling of surfaces is carried out using a flat plane sample holder with a coarse (P120) silicon carbide paper attached, both sample holder and platen rotating at 50 rev/min in contra motion.

5.6.1 Making the 'right choice'

The procedure is simply as follows:

(a) Select a planar grinding surface.
(b) Select the platen surface dictated by the hardness of the sample. If a further stage is required select a softer platen surface.
(c) Select a final polishing surface and abrasive.

From Figure 5.13 there is a guide to surface selection after the primary mode. This can be

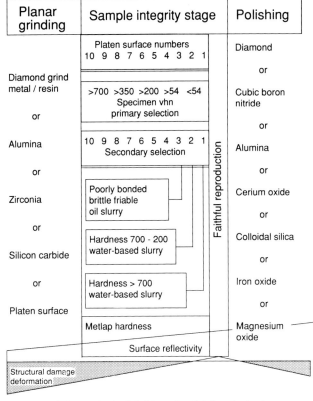

Figure 5.13 Making the 'right choice'

Table 5.1 Examples of machine conventions

Stage	Surface	12 in rev/min	8 in rev/min	Direction	Force per sample (lb/in² or kPa)		Abrasive
Planar grinding	SIC paper	150	250	Complementary	5	(35)	
	Resin bonded diamond wheel	120	200	Complementary	5	(35)	
	Surface 10	120	200	Complementary	5	(35)	Diamond
Sample integrity	Surface 10 8 6 4[a]	120	200	Complementary	5	(35)	Diamond
	Surface 3	25	35	Complementary	5	(35)	Diamond
	Surface 2	240	300	Complementary	5	(35)	Diamond/colloidal silica
	Surface 1	240	300	Complementary	5	(35)[b]	Diamond/colloidal silica/alumina
Polishing	PS 3	240	300[c]	Contra	2½	(17)	Diamond/colloidal silica/alumina
	PS 8	240	300	Contra[d]	2½	(17)	Diamond/colloidal silica/alumina
	PS 7	80	120	Contra	2½	(17)	Alumina/colloidal silica

General comments: when using a silica or alumina slurry reduce speed to 80–120 rev/min.
[a]Friable subjects reduce speed to 25–35 rev/min.
[b]Hard ceramics increase pressure to 10 lb/in² (70 kPa).
[c]Theoretical calculation dictates speed at 360 rev/min.
[d]Soft materials complementary.

any lower surface number; however, if the specimen should exhibit any form of porosity, the grains be poorly bonded or the subject be in any way friable, it will be necessary to select a surface in the region of 4 or lower. With extremely friable materials it could be advisable to use surface 4 as the primary platen surface.

The following are useful combinations:

Platen surface 10/6/1 or 10/4/1 or 10/2
Platen surface 8/2
Platen surface 6/2
Platen surface 4/1
Platen surface 10/2

The basis for making such combinations is the choice of platen surfaces available to the writer. Combinations such as 10/5/1, 7/3/1, 10/3/1, etc., could prove to be even more advantageous.

Briefly, the secondary mode selection number will be less than the primary selection number and will utilize the least number of surfaces in targeting for integrity. The extremes are to use every surface progressing to zero or to use surface 2 or 1 to prolong duration; neither example is commercially acceptable.

5.6.2 Operating conventions

Having made 'the right choice' there are still many variables in speed, time and pressure to be considered, all having an effect on sample integrity. The built-in variable to the operating conventions is time; this will vary between different machines and different diameter platens. To reduce these variables a list of conventions should be prepared and should be adhered to.

Since these conventions will vary with different manufacturers' equipment it is necessary to carry out an optimization programme to establish these conventions. The result of one such exercise on equipment having a specimen head speed of 66 rev/min and a variable platen speed is shown in Table 5.1.

From our understanding of material removal by stress dislocation it could be argued that a series of machine conventions should be available for every different material. One knows, for example, that when turning on a lathe one would change the surface speed for different materials. This information can be translated to the machine conventions without overcomplicating by observing the rules shown in Table 5.2.

Table 5.2 Machine conventions for different materials

Material	Speed	Pressure
Generous ferrous	No change	No change
Soft metals	Reduce	Same or less
Brittle hard materials	Reduce	Same
Hard materials	Increase	Increase

When preparing extremely hard materials, diamond in a fixed bond will be used as the abrasive. Greatly increased surface speeds will be necessary to remove material without causing high grinding wheel wear.

5.7 Parameter Optimization

It is the responsibility of all manufacturers and suppliers of equipment to offer guide-lines to the best use of that product. If this were already the case this section would not be necessary. An investigation into the possible fixed machine speeds available for silicon carbide paper use ranged from 150 to 1500 rev/min without any reference to a particular material. The general advice seemed to imply the best speed was 250 to 300 rev/min irrespective of material and platen diameter. All engineers know that the optimum cutting speed for steel is not the same as it is for aluminium or cast iron yet we are supplied with fixed or dual speeds for all materials (even worse, a variable speed platen with the advice 'use it at 300 rev/min'). The optimization programme is therefore best carried out using a specific class of material compatible with your specimen throughput. It must also be independently carried out for the different types of surfaces to be used, i.e. silicon carbide paper, platen surfaces, polishing surfaces. The measure of efficiency is obviously the progressive reduction in structural damage in the least time; this is to be measured in terms of stock removal (Z axis) relative to time.

5.7.1 Silicon carbide paper

Procedure

Using a fixed speed sample head/variable speed platen:

(i) Fill the sample holder with the *minimum* number of specimens in order to balance the head (3 or 4).
(ii) With P120 grit silicon carbide paper planar grind the samples in the head (using manufacturers' parameters if available).
(iii) When planar, using a micrometre fiducial micrometer, measure the thickness of each specimen allocating an identity number to each.
(iv) At set intervals of 30 seconds, using a pressure of 35 kPa and changing the silicon carbide paper at each stage note the Z axis stock removal. (Always measure from the centre of the sample.)

Graph conclusions (Figure 5.14)

(i) Contra motion gives the highest material removal rate.
(ii) The material removal rate in contra motion is very sensitive to platen speed.
(iii) Complementary motion gives the most consistent removal rate irrespective of platen speed.

Figure 5.14 shows a typical graph indicating contra motion at the higher speeds to be the most efficient. If we were to use traditional metallography techniques (i.e. four stages of SiC paper followed by two polishing stages) then these would be the conditions to adopt. Because of the unidirectional forces that occur in contra motion it would be undesirable with ductile materials, where preferential forces can cause unidirectional deformation, complementary direction being required. Figure 5.14

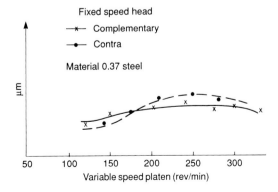

Figure 5.14 Material removal (SiC). (Reproduced by permission of Trevor Bousfield)

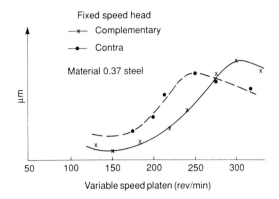

Figure 5.15 MRD (SiC). (Reproduced by permission of Trevor Bousfield)

indicates a much broader platen speed. If, however, it is intended to subsequently use charged platen surfaces, then another factor has to be considered, that of the 'material removal differential' (MRD).

5.7.2 Material removal differential

Procedure

The material removal differential occurs due to varying surface speed taking place when specimens are rotated on a rotating platen. When dealing with degrading abrasives such as silicon carbide the element of differential wear also affects the calculation. Had it not been for this last consideration it would have been possible to prove mathematically that unity speed in a complementary direction is the best choice in minimizing MRD. Figure 5.15 shows the position of the minimum MRD, and it can also be seen that optimum stock removal and minimum DMR are incompatible. There is, however, a best compromise position and this often has to be used.

In order to accurately measure the MRD it is necessary to make a VHN indent across the specimen surface after the silicon carbide stage. The specimen is then flattened using the appropriate charged platen surface until all indents have been contacted. The out-of-flat is taken to be the difference between the longest and shortest diagonal ×0.202.

Before this test can be adequately carried out it is important to optimize the parameters for platen surfaces.

Graph conclusions

(i) MRD is extremely high at high platen speeds irrespective of head direction.
(ii) Medium speed complementary head direction gives the least MRD.
(iii) The ideal compromise between planarity and stock removal (from the two graphs) would be to use medium speed in a complementary direction.

5.7.3 Head–platen ratio

Procedure

So far the considerations have been confined to a fixed head speed with variable platen speeds. We do, however, have a different set of conditions when the situation is reversed and the head is varied (see Figure 5.16).

Graph conclusions

(i) The greater the head–platen ratio the greater the material removal. (This is probably accompanied by an increase in MRD.)
(ii) The contra motion removal rate is higher than complementary motion.

5.7.4 Platen surfaces

Procedure

Material removal differential is not often associated with platen surfaces since the

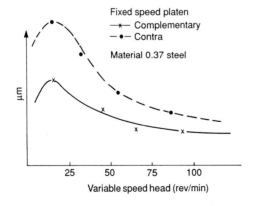

Figure 5.16 Head–platen ratio (SiC). (Reproduced by permission of Trevor Bousfield)

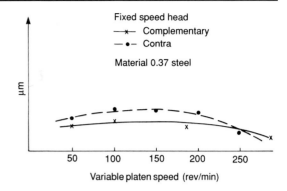

Figure 5.17 Material removal (platen surfaces). (Reproduced by permission of Trevor Bousfield)

surface must be plane in order to function efficiently. If the platen surface should not be plane then the lower area or valley instead of grinding will cause the abrasives in that area to roll. This results in a loss of efficiency and increases the out-of-flatness of the platen.

Figure 5.17 shows the best positions for complementary and contra motion, the latter giving the best condition, but this could result in a shorter surface life efficiency requiring a more frequent planar correction.

Graph conclusions

(i) Contra motion removes more material.
(ii) Contra motion speed is sensitive, particularly as speed increases.
(iii) Complementary motion is the most consistent.

These graphs are shown as illustrations only and must not be taken as representative of various specimen–platen ratios. Ideally the specimen head speed should be balanced to suit the optimum platen speed; currently the available equipment does not exhibit this feature. What is required is therefore a variable specimen head speed. It would not be satisfactory for suppliers to offer this facility without carrying out optimization programmes and making the results applicable to specific materials or composites.

5.7.5 Specimen head–platen ratio

Procedure

(i) Ensure planarity between specimens in

The New Concept

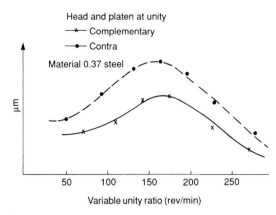

Figure 5.18 Material removal (unity ratio). (Reproduced by permission of Trevor Bousfield)

the specimen holder (use a charged platen surface).
(ii) Make indents in the centre of each sample.
(iii) Progress the sample at unity speed, head to platen, noting stock removal.
(iv) Carry out similar tests at unity speed in set increments.

Graph conclusions

(i) There is an optimum ratio for this material.
(ii) Contra motion is the more aggressive.
(iii) Both contra and complementary motion peak in a similar ratio position.

The object of all these tests is to ultimately relate machine functions to a material removal optimization in order to reduce the number of variables currently blocking the progress towards not just a metallography standard but also its procedure. The writer is very much aware of the tremendous effort that takes place in identifying what is called a repeatable preparation operations method, but this is of little value if it does not work on other equipment.

5.7.6 Induced structural damage

Traditional metallography makes use of the thinking that all damage from previous steps must be removed at the current step before proceeding to the next. Although this is satisfactory it is time consuming and unnecessary; it also presents problems with materials that adversely react to mechanical working. The target is to remove just the deformed or damaged layer created at the first preparation stage. It has been mentioned how important it is to understand this Z axis dimension. It is therefore equally important not to increase the depth of damage.

The procedure for surface and abrasive selection will be covered later. At this stage let it suffice to say 'the summation of all the Z axis dimensions at each preparation stage will equal or be slightly greater than the total structural Z axis damage from the first preparation step'.

5.8 Preparation Examples and Family Classification

Throughout the book examples of material preparation will be illustrated and will follow the following format:

- *The material:* a description of its use
- *The technique:* an insight into the material's response to mechanical working related to its microstructure
- *The method:* a suitable recipe for revealing the true structure (the abrasive quoted in methods will be diamond unless otherwise stated)
- *The micrograph:* an illustration of the expected result from the given method

The classification of materials by their preparation characteristics is a very useful exercise since it allows comparison of similar preparation techniques by information retrieval. The mistake often made is to classify materials by material class instead of preparation class. This restricts information retrieval of similar methods, therefore restricting procedural progress.

A suggested family classification is included below and is adhered to with all the examples given in the book.

5.8.1 Family classification

Index number

- BB1 Ferrous materials (except cast iron)
- BB2 Cast iron
- BB3 Aluminium, magnesium and zinc alloys
- BB4 Copper and alloys
- BB5 Nickel and cobalt super alloys
- BB6 Titanium, tantalum and alloys
- BB7 Refractory metals
- BB8 Composite matrix and electronic components
- BB9 Soft and noble metals
- BB10 Ceramics
- BB11 Metal matrix composite (MMC)
- BB12 Polymer matrix composite (PMC)
- BB13 Glass and ceramic matrix composite (GMC and CMC)
- BB14 Sintered carbides
- BB15 Rocks and minerals
- BB16 Spray coatings
- BB17 Thin platings including PCBs
- BB18 Corrosion products
- BB19 Paint
- BB20 Polymers
- BB21 Magnetic materials and superconductors
- BB22 Composites
- BB23 Refractories
- BB24 Glass
- BB25 Bone and implants (bio material)
- BB26 Bearing materials
- BB27 Solder, brazed and welded components
- BB28 Thick coatings

5.8.2 Examples

Example 5.1

Family classification BB4

The material: copper phosphorus

Bronze, being the name given to an alloy of copper with tin, can be modified by a solid solution of phosphorus. Phosphor bronze is used in corrosion-resistant springs. The eutectoid for phosphorus is 0.3%; beyond this figure the hardness of the material increases with the hypereutectoid hard constituent. These materials have been used in heavy duty bearings.

The technique

These materials offer few problems in surface preparation and can be prepared most satisfactorily using traditional silicon carbide paper techniques followed by polishing cloths. The choice of oxide polishing can be varied with equally good results. The chosen method is a short four step procedure suitable for automatic or hand preparation.

The method

AUDITABLE PREPARATION PROCEDURE   No. 1

CLASSIFICATIONS: BB4			MATERIAL: COPPER PHOSPHORUS	
SECTIONING TECHNIQUE:	EQUIPMENT TYPE	BLADE	CONCENTRATION	
MOUNTING TECHNIQUE:	METHOD		SAMPLE/MOUNT RATIO	

*Alternatively two different grades of silicon carbide paper could be used.

	SURFACE and TYPE	ABRASIVE SIZE/TYPE LUBRICANT	FORCE (kPa)	HEAD ROTATION	Z AXIS (µm)	REFLECTIVE FACTOR	MICROGRAPH YES/NO	IMAGE ANALYSIS YES/NO	REMARKS
PLANAR GRINDING STAGE	PAPER	P320 SiC WATER	35	COMP	PLANE				

	SURFACE and TYPE	ABRASIVE SIZE/TYPE LUBRICANT	FORCE (kPa)	HEAD ROTATION	Z AXIS (µm)	REFLECTIVE FACTOR	MICROGRAPH YES/NO	IMAGE ANALYSIS YES/NO	REMARKS
SAMPLE INTEGRITY STAGE	S.6*	9 µm OIL	35	COMP					
	S.1	ALUMINA 3 µm	35	COMP					

	SURFACE and TYPE	ABRASIVE SIZE/TYPE LUBRICANT	FORCE (kPa)	HEAD ROTATION	TIME	REFLECTIVE FACTOR	MICROGRAPH YES/NO	IMAGE ANALYSIS YES/NO	REMARKS
POLISHING STAGE	PS 7	COLLOIDAL SILICA	16	COMP	ONE MIN	/	YES	/	5.19 (Plate 1)

The micrograph Figure 5.19 (see Plate 1)

Example 5.2

Family classification BB3

The material: aluminium bronze

Aluminium when added to copper dramatically increases the material strength; 10% aluminium increases the basic copper strength by a factor of three. At this point aluminium bronze becomes as strong as mild steel, its inclusion (aluminium) is near the limit from which further increases would result in an extremely brittle useless material. If, however, the aluminium content were to be increased to 90% the resultant material would lose its brittle characteristics and take on a strength greater than pure aluminium. High strength aluminium bronze finds many applications where strength in corrosive environments is required. Such examples would be in marine engineering. Large ship propellers are made from this family of materials.

The technique

Very suitable material for traditional silicon carbide grinding, the choice of a suitable recipe is large; they should, however, follow the format of the enclosed.

The method

AUDITABLE PREPARATION PROCEDURE   No. 2

CLASSIFICATIONS: BB3			MATERIAL: ALUMINIUM BRONZE (CAST Cu–11.8 Al)	
SECTIONING TECHNIQUE:	EQUIPMENT TYPE	BLADE	CONCENTRATION	
MOUNTING TECHNIQUE:	METHOD		SAMPLE/MOUNT RATIO	

*Alternatively two different grades of silicon carbide paper could be used.

	SURFACE and TYPE	ABRASIVE SIZE/TYPE LUBRICANT	FORCE (kPa)	HEAD ROTATION	Z AXIS (μm)	REFLECTIVE FACTOR	MICROGRAPH YES/NO	IMAGE ANALYSIS YES/NO	REMARKS
PLANAR GRINDING STAGE	PAPER	P220 SiC WATER	35	COMP	PLANE				

	SURFACE and TYPE	ABRASIVE SIZE/TYPE LUBRICANT	FORCE (kPa)	HEAD ROTATION	Z AXIS (μm)	REFLECTIVE FACTOR	MICROGRAPH YES/NO	IMAGE ANALYSIS YES/NO	REMARKS
SAMPLE INTEGRITY STAGE	S.6*	9 μm OIL	35	COMP					
	S.1	3 μm OIL	35	COMP					

	SURFACE and TYPE	ABRASIVE SIZE/TYPE LUBRICANT	FORCE (kPa)	HEAD ROTATION	TIME	REFLECTIVE FACTOR	MICROGRAPH YES/NO	IMAGE ANALYSIS YES/NO	REMARKS
POLISHING STAGE	PS 7	ALUMINA 0.05 μm	16	COMP	ONE MIN		YES		5.20 (Plate 1)

The micrograph Figure 5.20 (see Plate 1)

Plate 1

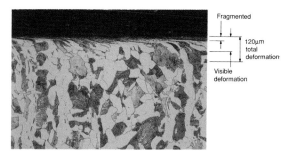

Figure 1.1 Structural damage (unity magnification ×400)

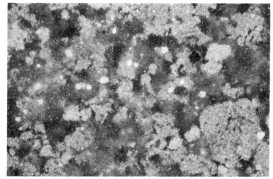

Figure 5.11 Composite surface showing impressed abrasives

Figure 5.12 Composite surface/microstructure showing porosity

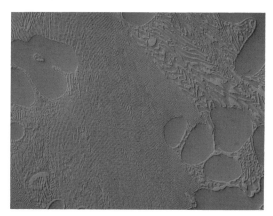

Figure 5.19 Copper phosphorus (×400)

Figure 5.20 Aluminium bronze (×200)

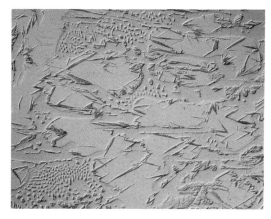

Figure 5.21 Martensitic white cast iron (×400)

Plate 2

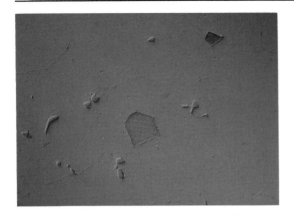

Figure 5.22 Nimonic 105 ($\times 1000$)

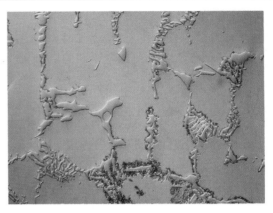

Figure 5.23 Cast alloy steel ($\times 400$)

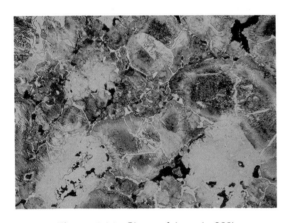

Figure 5.24 Sintered iron ($\times 200$)

Figure 5.25 Spheroidal graphite iron ($\times 1000$)

Figure 5.26 Carbon fibre ($\times 100$)

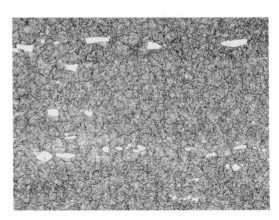

Figure 5.27 Tool steel ($\times 400$)

Plate 3

Figure 6.3 Good scratch pattern

Figure 6.4 Poor scratch pattern

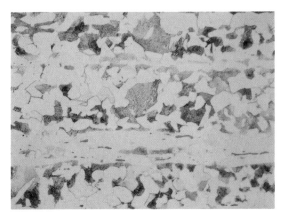
Figure 11.1 Carbon steel, 0.37% ($\times 400$)

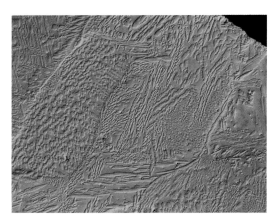
Figure 11.2 Brass ($\times 200$)

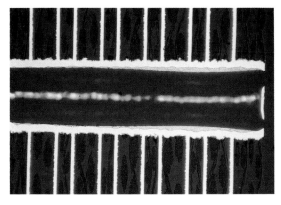
Figure 11.3 Printed circuit board ($\times 100$)

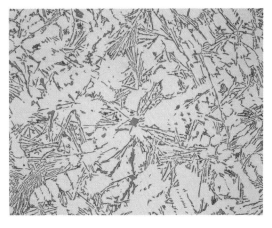
Figure 11.4 Aluminium alloy ($\times 200$)

Plate 4

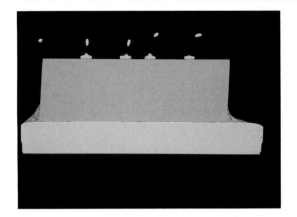

Figure 11.5 Electronic component (×50)

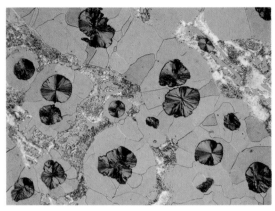

Figure 11.6 Cast iron (×200)

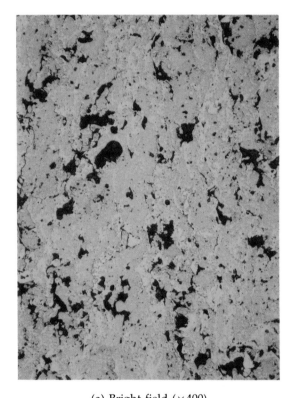

(a) Bright field (×400)

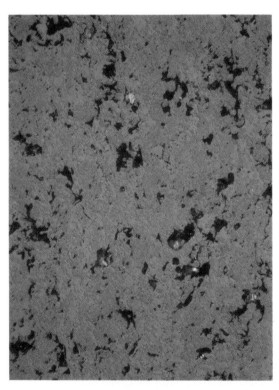

(b) First-order red (×400)

Figure 12.4 Damage identification

Plate 5

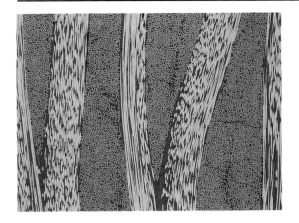

Figure 13.4 Carbon fibre PMC (×100)

Figure 13.5 Alumina fibre MMC (×400)

Figure 13.6 SiC particles MMC (large particles) (×200)

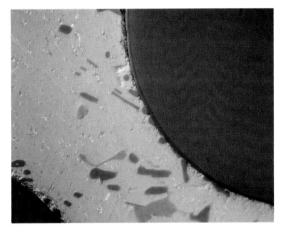

Figure 13.7 Borsic fibres MMC (×1000)

Figure 13.8 Silicon carbide fibres GMC (×200)

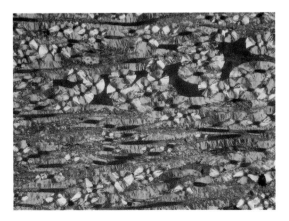

Figure 13.9 Carbon–carbon composites (×100)

Plate 6

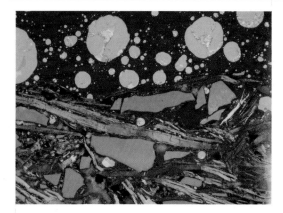

Figure 13.10 SiC/graphite/paint layer (×200)

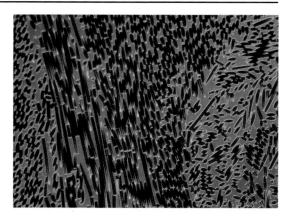

Figure 13.11 Alumina fibres in aluminium/copper matrix (×400)

Figure 13.12 SiC fibres/graphite core in aluminium matrix (×400)

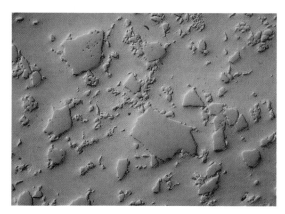

Figure 13.14 Titanium carbides in titanium matrix (×400)

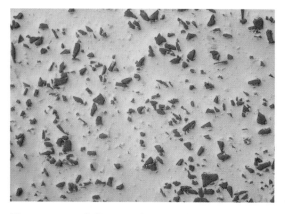

Figure 13.15 SiC particulates in aluminium/silicon matrix (small particles) (×400)

Plate 7

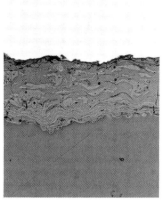  

(a) Bright field (×200) (b) DIC (×200) (c) DIC (×400)
Figure 13.16 Optical observations for hydroxyapatite

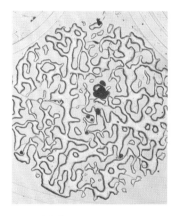 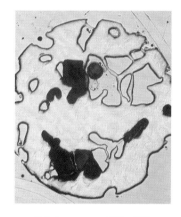 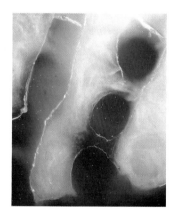

(a) Porite (BF×25) (b) Acropora (BF×25) (c) Blue resin (DF×100)
Figure 13.20 Microporosity in coral

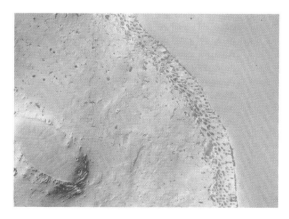

Figure 13.21 Coral structure DIC (×200)

Plate 8

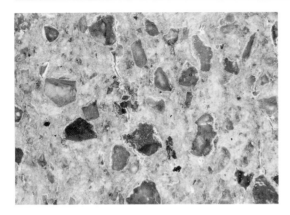
(a) Early neolithic DG

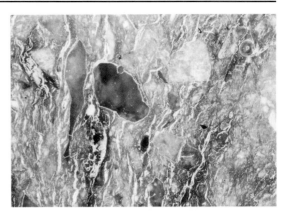
(b) Late neolithic DG

Figure 14.3 Neolithic thick sections (×50)

Figure 14.8 Chalcopyrite (×100)

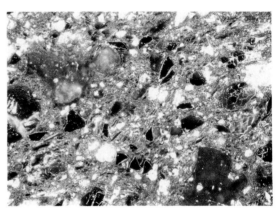
Figure 14.9 Friction material (×100)

Figure 14.10 Graphite (×400)

Figure 14.11 Iron oxide/quartz (×200)

Plate 9

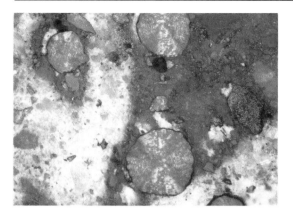

Figure 14.12 Cement clinker (×100)

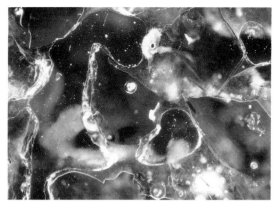

Figure 14.13 Vitrified alumina (×100)

Figure 14.14 Covellite (×50)

Figure 14.15 Hematite (×50)

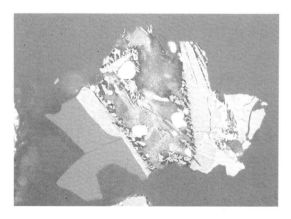

Figure 14.16 Titanium oxide (×100)

Plate 10

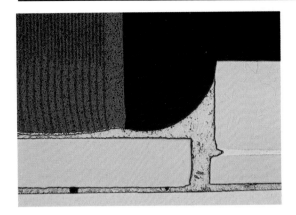

(a) Uncrossed polars (b) Dark ground (×50)

Figure 15.5 Electronic component

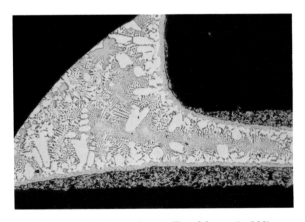

Figure 15.6 Brazed metallized layer (×200)

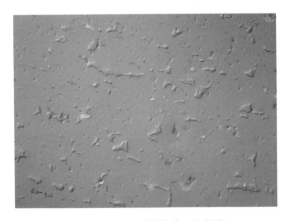

Figure 17.2 Al/4% Sn (×200)   **Figure 17.3** Lead-based alloy (×200)

Plate 11

Figure 17.4 Tin-based alloy (×100)

Figure 17.5 Al/Si overlay plate (×400)

Figure 17.6 Al/Pb plated (×100)

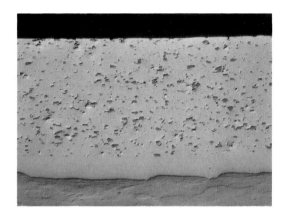

Figure 17.7 Al/Cu/Si/Sn bearing (×400)

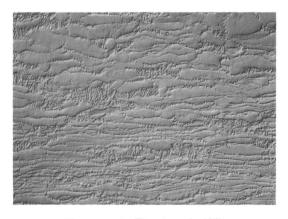

Figure 17.8 Titanium (×400)

Plate 12

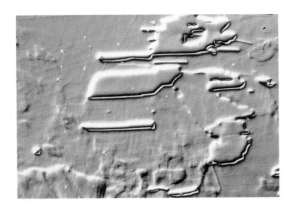

Figure 21.1 Pure copper DIC (×5)

Figure 21.2 Superconductor—uncrossed polars (×1000)

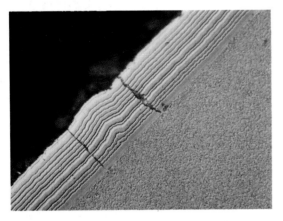

Figure 21.3 Tungsten carbide coating DIC (×1000)

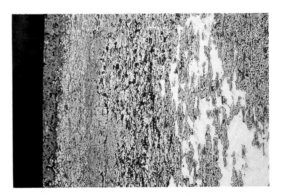

Figure 21.4 Corrosion of copper

Example 5.3

Family classification BB2

The material: martensitic white cast iron

The difference between grey cast iron and white cast iron is that grey iron manifests the carbon as free graphite and pearlite while all the carbon in white iron is present as cementite (iron carbide). This results in a very hard though brittle material, often present as a hard outer skin in a grey cast iron created by rapid cooling. The condition of the carbon varies with the rate of quenching. Large rollers used in rolling mills make full use of this condition.

The technique

Although silicon carbide paper could be and is used in preparing this extremely hard material, it is better suited to the use of diamonds or CBN as the grinding abrasive. The planar grinding stage in the example given would be equally if not better suited to 30 to 45 μm fixed or charged diamond abrasive grinding. When a hard and brittle material is to be ground, fixed abrasives are suitable and would equally be more efficient in the sample integrity stage using platen surface 8.

The method

AUDITABLE PREPARATION PROCEDURE — No. 3

CLASSIFICATIONS: BB2						MATERIAL: MARTENSITIC WHITE CAST IRON			
SECTIONING TECHNIQUE:	EQUIPMENT TYPE		BLADE			CONCENTRATION			
MOUNTING TECHNIQUE:	METHOD					SAMPLE/MOUNT RATIO			

*Alternatively two different grades of silicon carbide paper could be used.

PLANAR GRINDING STAGE	SURFACE and TYPE	ABRASIVE SIZE/TYPE LUBRICANT	FORCE (kPa)	HEAD ROTATION	Z AXIS (μm)	REFLECTIVE FACTOR	MICROGRAPH YES/NO	IMAGE ANALYSIS YES/NO	REMARKS
	PAPER	P240 SiC WATER	35	COMP	PLANE				

SAMPLE INTEGRITY STAGE	SURFACE and TYPE	ABRASIVE SIZE/TYPE LUBRICANT	FORCE (kPa)	HEAD ROTATION	Z AXIS (μm)	REFLECTIVE FACTOR	MICROGRAPH YES/NO	IMAGE ANALYSIS YES/NO	REMARKS
	S.8*	9 μm WATER	35	CONTRA					
	S.2P	6 μm WATER	35	COMP					

POLISHING STAGE	SURFACE and TYPE	ABRASIVE SIZE/TYPE LUBRICANT	FORCE (kPa)	HEAD ROTATION	TIME	REFLECTIVE FACTOR	MICROGRAPH YES/NO	IMAGE ANALYSIS YES/NO	REMARKS
	PS 7	ALUMINA 0.05 μm	16	COMP	ONE MIN		YES		5.21 (Plate 1)

The micrograph Figure 5.21 (see Plate 1)

52 Surface Preparation and Microscopy of Materials

Example 5.4

Family classification BB5

The material: nimonic 105—super alloy

Super alloy is the name given to those materials developed around 1940 to satisfy the demands for higher operating temperatures in aeroengines and gas turbines. These nickel-based super alloys contain 10 to 20% chromium and smaller amounts of aluminium with even smaller amounts of other alloying elements. Turbine blades are a typical application for this material. Since 1940 there has been an even greater demand for higher operating temperatures in combustion chambers, leading to the thermal barrier coatings described in Chapter 12. Although in the 1980s thermal barrier coatings were heralded as a major breakthrough they are but another step in the march for even higher temperature materials (glass/ceramic matrix composites).

The technique

Again no major problem exists if prepared with sharp abrasives. Silicon carbide can be an excellent choice provided it is not used beyond the critical life of the abrasive. Major structural changes will take place if the surface is rubbed; therefore, blunt abrasives and/or high pressure are to be avoided.

The method

AUDITABLE PREPARATION PROCEDURE   No. 4

CLASSIFICATIONS: BB5									MATERIAL: NIMONIC 105		
SECTIONING TECHNIQUE:	EQUIPMENT TYPE		BLADE						CONCENTRATION		
MOUNTING TECHNIQUE:	METHOD								SAMPLE/MOUNT RATIO		

*Alternatively two different grades of silicon carbide paper could be used.

	SURFACE and TYPE	ABRASIVE SIZE/TYPE LUBRICANT	FORCE (kPa)	HEAD ROTATION	Z AXIS (μm)	REFLECTIVE FACTOR	MICROGRAPH YES/NO	IMAGE ANALYSIS YES/NO	REMARKS
PLANAR GRINDING STAGE	PAPER	P240 SiC WATER	35	COMP	PLANE				

	SURFACE and TYPE	ABRASIVE SIZE/TYPE LUBRICANT	FORCE (kPa)	HEAD ROTATION	Z AXIS (μm)	REFLECTIVE FACTOR	MICROGRAPH YES/NO	IMAGE ANALYSIS YES/NO	REMARKS
SAMPLE INTEGRITY STAGE	S.6*	9 μm OIL	35	CONTRA					
	S.2P	6 μm WATER	35	COMP					

	SURFACE and TYPE	ABRASIVE SIZE/TYPE LUBRICANT	FORCE (kPa)	HEAD ROTATION	TIME	REFLECTIVE FACTOR	MICROGRAPH YES/NO	IMAGE ANALYSIS YES/NO	REMARKS
POLISHING STAGE	PS 7	COLLOIDAL SILICA	16	COMP	TWO MIN		YES		5.22 (Plate 2)

The micrograph Figure 5.22 (see Plate 2). The visual impact of this micrograph can be improved if the reader rotates it through 180°

Example 5.5

Family classification BB1

The material: cast alloy steel

When changing the condition of a steel by, for example, quenching, the rate of cooling from the quick thermal shock at the outside to the slower rate in the middle dictates the structural transformation and hence the hardness of the material. This unequal hardening or thermal shock creates an unequal contraction, inducing residual material stress. By adding alloying elements to the steel it is possible to achieve the hardening transformation at a much reduced temperature, thereby overcoming the different degrees of hardness and the accompanying residual stress. Alloys can also impart qualities such as increased strength and hardness (tool steel), resistance to wear and corrosion (stainless steel)—not forgetting springiness (spring steel). High speed steel (HSS) used in cutting tools is one example where 18% tungsten, 4% chromium, 1% vanadium offered a tremendous increase in tool life and efficient high speed cutting.

The technique

To offer one recipe to cover this range of materials is not possible and, therefore, the enclosed method will require fine tuning. Much will depend on not just the alloying elements but the material hardness. The example given is a relatively soft material. As the hardness increases so the need to use 3 or 1 μm diamond on platen surface 1 will be beckoned. Materials in the very hard condition will require a 10/6/1 application.

The method

AUDITABLE PREPARATION PROCEDURE   No. 5

CLASSIFICATIONS: BB1					MATERIAL: CAST ALLOY STEEL			
SECTIONING TECHNIQUE:	EQUIPMENT TYPE		BLADE		CONCENTRATION			
MOUNTING TECHNIQUE:	METHOD				SAMPLE/MOUNT RATIO			

*Alternatively two different grades of silicon carbide paper could be used.

	SURFACE and TYPE	ABRASIVE SIZE/TYPE LUBRICANT	FORCE (kPa)	HEAD ROTATION	Z AXIS (μm)	REFLECTIVE FACTOR	MICROGRAPH YES/NO	IMAGE ANALYSIS YES/NO	REMARKS
PLANAR GRINDING STAGE	PAPER	P240 SiC WATER	35	COMP	PLANE				

	SURFACE and TYPE	ABRASIVE SIZE/TYPE LUBRICANT	FORCE (kPa)	HEAD ROTATION	Z AXIS (μm)	REFLECTIVE FACTOR	MICROGRAPH YES/NO	IMAGE ANALYSIS YES/NO	REMARKS
SAMPLE INTEGRITY STAGE	S.8*	9 μm OIL	35	CONTRA					
	S.2P	6 μm WATER	35	COMP					

	SURFACE and TYPE	ABRASIVE SIZE/TYPE LUBRICANT	FORCE (kPa)	HEAD ROTATION	TIME	REFLECTIVE FACTOR	MICROGRAPH YES/NO	IMAGE ANALYSIS YES/NO	REMARKS
POLISHING STAGE	PS 7	ALUMINA 0.05 μm	16	COMP	ONE MIN		YES		5.23 (Plate 2)

The micrograph Figure 5.23 (see Plate 2)

Example 5.6

Family classification BB1

The material: sintered iron

Powder metallurgy originated when it was found that metals difficult to melt, due in part to their high melting points, could be made into suitable shapes by compressing metal powders. Compacted components are then sintered, i.e. heated under a controlled atmosphere to temperatures below the melting point. Powder metallurgy is an economical way of shaping usually small components on a mass production scale. It is also a convenient way of mixing other alloys without having them as a solid solution of the matrix. Graphite could be incorporated into bearings and porosity could be evenly distributed as in this illustrated example. Many domestic automotive and DIY products have components made from powder metallurgy.

The technique

The sintered iron in this example also includes particles of copper and has a required porosity level. This material could be prepared using the familiar silicon carbide route. There is a word of warning however—porosity values can vary relative to the pressure and condition of the grinding papers. The pore size can be further changed, viz.:

1. Lidding could occur if high pressure, small abrasives and an oil-based lubricant were to be used on platen surface 1.
2. Pores would be enlarged if prolonged polishing times on a nap cloth were to take place.

The method

AUDITABLE PREPARATION PROCEDURE   No. 6

CLASSIFICATIONS: BB1						MATERIAL: SINTERED IRON			
SECTIONING TECHNIQUE:	EQUIPMENT TYPE		BLADE			CONCENTRATION			
MOUNTING TECHNIQUE:	METHOD					SAMPLE/MOUNT RATIO			

*Alternatively two different grades of silicon carbide paper could be used.

	SURFACE and TYPE	ABRASIVE SIZE/TYPE LUBRICANT	FORCE (kPa)	HEAD ROTATION	Z AXIS (μm)	REFLECTIVE FACTOR	MICROGRAPH YES/NO	IMAGE ANALYSIS YES/NO	REMARKS
PLANAR GRINDING STAGE	PAPER	P240 SiC WATER	35	COMP	PLANE				

	SURFACE and TYPE	ABRASIVE SIZE/TYPE LUBRICANT	FORCE (kPa)	HEAD ROTATION	Z AXIS (μm)	REFLECTIVE FACTOR	MICROGRAPH YES/NO	IMAGE ANALYSIS YES/NO	REMARKS
SAMPLE INTEGRITY STAGE	S.6*	6 μm WATER	35	CONTRA					
	S.1	3 μm WATER	35	CONTRA					

	SURFACE and TYPE	ABRASIVE SIZE/TYPE LUBRICANT	FORCE (kPa)	HEAD ROTATION	TIME	REFLECTIVE FACTOR	MICROGRAPH YES/NO	IMAGE ANALYSIS YES/NO	REMARKS
POLISHING STAGE	PS 7	ALUMINA 0.05 μm	16	COMP	ONE MIN		YES		5.24 (Plate 2)

The micrograph Figure 5.24 (see Plate 2)

Example 5.7

Family classification BB2

The material: SG iron

The name spheroidal graphite cast iron or nodular iron comes from the circular shape of the free graphite. The result is a stronger and more shock resistant material than grey cast iron which exhibits a flake-type structure. SG iron has an improved ductility and is sometimes called ductile iron. Many underground water and gas pipes are made of SG iron; they are also used in preference to steel castings for crankshafts. Cast iron is one of the most complex metal structures in use as the carbon can occur in so many different combinations. The microscopist can always be rewarded with beautiful images by observing the Maltese cross effect between crossed polars or the infinite variations in colours when rotating the polarizer, having first introduced the sensitive tint plate.

The technique

There are many different recipes available for the preparation of SG iron—some complex, some simple. Much depends upon your threshold of expectation. There is little damage that can be done to the metal. The preparation technique will greatly influence the retention of the free graphite. This is very much a low priority to the foundry metallurgist requiring a quick and accurate assessment of the size, distribution and general morphology of the *hole* that once retained the graphite. The method quoted is intended to illustrate the integrity of the sample without destroying the free graphite. The maximum magnification as shown in Figure 5.25 adequately displays this point.

The method

AUDITABLE PREPARATION PROCEDURE   No. 7

CLASSIFICATIONS: BB2									
\multicolumn{5}{l	}{CLASSIFICATIONS: BB2}	\multicolumn{5}{l	}{MATERIAL: SG IRON (GRAPHITE RETENTION)}
SECTIONING TECHNIQUE:	EQUIPMENT TYPE		BLADE		CONCENTRATION				
MOUNTING TECHNIQUE:	METHOD				SAMPLE/MOUNT RATIO				

*Alternatively two different grades of silicon carbide paper could be used.

	SURFACE and TYPE	ABRASIVE SIZE/TYPE LUBRICANT	FORCE (kPa)	HEAD ROTATION	Z AXIS (μm)	REFLECTIVE FACTOR	MICROGRAPH YES/NO	IMAGE ANALYSIS YES/NO	REMARKS
PLANAR GRINDING STAGE	PAPER	P240 SiC WATER	35	COMP	PLANE				

	SURFACE and TYPE	ABRASIVE SIZE/TYPE LUBRICANT	FORCE (kPa)	HEAD ROTATION	Z AXIS (μm)	REFLECTIVE FACTOR	MICROGRAPH YES/NO	IMAGE ANALYSIS YES/NO	REMARKS
SAMPLE INTEGRITY STAGE	S.4*	9 μm OIL	35	CONTRA					
	S.1	COLLOIDAL SILICA 0.06 μm	35	COMP					

	SURFACE and TYPE	ABRASIVE SIZE/TYPE LUBRICANT	FORCE (kPa)	HEAD ROTATION	TIME	REFLECTIVE FACTOR	MICROGRAPH YES/NO	IMAGE ANALYSIS YES/NO	REMARKS
POLISHING STAGE	PS 7	COLLOIDAL SILICA 0.06 μm	16	COMP	ONE MIN		YES		5.25 (Plate 2)

The micrograph Figure 5.25 (see Plate 2)

Example 5.8

Family classification BB12

The material: carbon fibre

Polymer matrices with reinforcing fibres of carbon, boron, kevlar and glass are having significant effects on material strength and stiffness. The fibres can be random oriented, unidirectional, cross-plied and also built up in layers (laminated) to satisfy directional property requirements. The organic matrix can be thermoplastic or thermosetting, depending again on requirements and environmental considerations. Examples are the carbon fibre reinforced thermosetting epoxy materials used in airframe construction or the CFR thermoplastic polyetheretherketone (PEEK) materials designed to be water and kerosene resistant. As with all composite materials the interface, in this case the fibre–matrix bond, is vitally important. If the bond is weak the shear and compressive strengths will be low.

The technique

The information required from the preparation is the fibre orientation, distribution, matrix saturation and interface relationship. An insensitive preparation technique will affect the interface analysis. A good indicator of a well-prepared sample is when the sharp end fibres are retained within the matrix. The planar grinding can best be undertaken using silicon carbide paper; initial sample integrity stages can also be undertaken using another two grades of silicon carbide paper or, alternatively, platen surface 4. Note how the optical technique has identified by colour the orientation of different fibres (caused by bireflectance).

The method

AUDITABLE PREPARATION PROCEDURE No. 8

CLASSIFICATIONS: BB12		MATERIAL: CARBON FIBRE	
SECTIONING TECHNIQUE:	EQUIPMENT TYPE BLADE	CONCENTRATION	
MOUNTING TECHNIQUE:	METHOD	SAMPLE/MOUNT RATIO	

*Alternatively platen surface A could be used.

	SURFACE and TYPE	ABRASIVE SIZE/TYPE LUBRICANT	FORCE (kPa)	HEAD ROTATION	Z AXIS (µm)	REFLECTIVE FACTOR	MICROGRAPH YES/NO	IMAGE ANALYSIS YES/NO	REMARKS
PLANAR GRINDING STAGE	PAPER	P320 SiC WATER	35	COMP	PLANE				

	SURFACE and TYPE	ABRASIVE SIZE/TYPE LUBRICANT	FORCE (kPa)	HEAD ROTATION	Z AXIS (µm)	REFLECTIVE FACTOR	MICROGRAPH YES/NO	IMAGE ANALYSIS YES/NO	REMARKS
SAMPLE INTEGRITY STAGE	PAPER*	P600 SiC WATER	35	CONTRA					
	PAPER*	P1200 SiC WATER	35	COMP					
	S.1	3 µm WATER	35	COMP					
	S.1	GAMMA ALUMINA	70	COMP					

	SURFACE and TYPE	ABRASIVE SIZE/TYPE LUBRICANT	FORCE (kPa)	HEAD ROTATION	TIME (s)	REFLECTIVE FACTOR	MICROGRAPH YES/NO	IMAGE ANALYSIS YES/NO	REMARKS
POLISHING STAGE	PS 7	GAMMA ALUMINA	17	COMP	100		YES		5.26 (Plate 2)

The micrograph Figure 5.26 (see Plate 2)

Example 5.9

Family classification BB1

The material: tool steel

The introduction of alloying elements such as tungsten, chromium, vanadium and many others has produced tools capable of working at very high speeds. Their life and performance far exceed that of simple hypereutectoid steels. The life of cutting tools has been further extended by the addition of a hard outer layer of titanium carbide or nitride. This outer layer is not confined to titanium; other different materials are used, including ceramics. These layers are often submicrometre in size and can be overlayed many times.

The technique

The method is for the softer condition of tool steel and is also designated for routine grain investigation. As the material becomes harder (or is in the hardened condition) it will be necessary to utilize platen surface 1 with $1\,\mu m$ water-based diamond slurry. The polishing stage would in many cases not be necessary. The use of silicon carbide paper would also come into question with an increase in sample hardness. The first sign of incorrect preparation is the breaking and plucking out of hard particles.

The method

AUDITABLE PREPARATION PROCEDURE   No. 9

CLASSIFICATIONS: BB1				MATERIAL: HARDEN TOOL STEEL	
SECTIONING TECHNIQUE:	EQUIPMENT TYPE		BLADE	CONCENTRATION	
MOUNTING TECHNIQUE:	METHOD			SAMPLE/MOUNT RATIO	

*Alternatively two different grades of silicon carbide paper could be used.

	SURFACE and TYPE	ABRASIVE SIZE/TYPE LUBRICANT	FORCE (kPa)	HEAD ROTATION	Z AXIS (μm)	REFLECTIVE FACTOR	MICROGRAPH YES/NO	IMAGE ANALYSIS YES/NO	REMARKS
PLANAR GRINDING STAGE	PAPER	180 SiC WATER	35	COMP	PLANE				

	SURFACE and TYPE	ABRASIVE SIZE/TYPE LUBRICANT	FORCE (kPa)	HEAD ROTATION	Z AXIS (μm)	REFLECTIVE FACTOR	MICROGRAPH YES/NO	IMAGE ANALYSIS YES/NO	REMARKS
SAMPLE INTEGRITY STAGE	S.8*	9 μm WATER	35	COMP					
	S.2P	6 μm WATER	35	COMP					

	SURFACE and TYPE	ABRASIVE SIZE/TYPE LUBRICANT	FORCE (kPa)	HEAD ROTATION	TIME (s)	REFLECTIVE FACTOR	MICROGRAPH YES/NO	IMAGE ANALYSIS YES/NO	REMARKS
POLISHING STAGE	PS 7	0.05 μm ALUMINA	17	COMP	140	/	YES	/	5.27 (Plate 2)

The micrograph Figure 5.27 (see Plate 2)

6

Grinding

6.1 Traditional Metallography with the 'New Concept'

Traditional techniques evolved from the original work of Sorby and have been widely used since that time, primarily because they are simple and easy to apply. They produce acceptable results on a broad range of common metallic alloys and are used here to illustrate the principles of abrasive sample preparation where the operator can exercise maximum control over the outcome.

As illustrated in Figure 6.1, the samples are ground through a series of progressively finer abrasive papers to systematically reduce the depth of deformation so that one or two steps of polishing on cloths will produce the final polish. Note the terms used, such as rough grinding, fine grinding and rough and final polish; all these terms relate to the surface *finish* and not to the surface *condition*. One can therefore see how easy it is to fall into the trap of judging a well-prepared sample by the surface lustre. Table 6.1 gives a comparison of grit sizes shown as CAMI (USA) sizes and their FEPA (Europe) equivalents. From this table it can be seen that the grits are identical at 180. There is little difference below this size

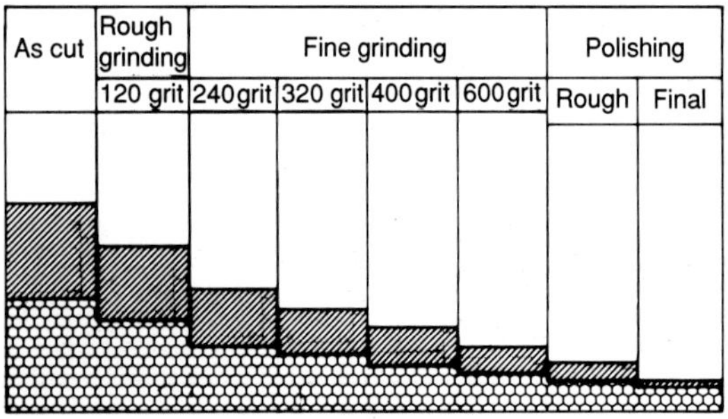

Figure 6.1 Traditional abrasive sample preparation

Table 6.1 Grit size comparison chart

USA		European	
Grit	Average size (μm)	Average size (μm)	Grit
40	428	412	P40
80	192	197	P80
120*	116	127	P120*
180	78	78	P180
240*	53	58	P240*
320*	36	46	P320
400*	23	35	P400*
600*	16	25	P600*
800	12	22	P800
1000	9.2	18	P1000
1200	6.5	15	P1200*
			P2300
		6.5	P4000

but above they progressively differ until grit 600 becomes similar to P1200. To repeat the examples shown in Figure 6.1 without changing the abrasive grit size would require a grit number change from CAMI to FEPA and vice versa, i.e. 120/240/320/400/600 becomes P120/P240/P400/P600/P1200.

Silicon carbide papers liberally flushed with water using the above grades would be common practice for low carbon steel. The general rule is to reduce the number of steps when grinding medium to high carbon steel and increase the number of steps with an increase in ductility of the sample, viz.:

(i) Low carbon steel: P240–P400–P600–P1200
(ii) High carbon steel: P240–P400–P600
(iii) Pure aluminium: P240–P400–P600–P1200–P4000

The explanation for this practice is that it is essential when preparing soft ductile materials to progress the grinding stages as fine as possible, prior to using a polishing cloth. The exception to this rule is when the fine small sized abrasive traps swarf between the sample and abrasive surface or when the abrasive chip becomes embedded in the sample. The reason for not requiring so many steps with the harder materials (high carbon) is due to the reduction in ductility manifesting lower residual deformation. The exception to this rule is when the hardness is associated with brittle qualities, such as minerals or ceramics, when it would be necessary to increase the number of abrasive size stages to reduce induced crack propagation.

Grinding is the stage where the major differences occur between the traditional approach to sample preparation and the 'new concept'. The traditional approach is to use fixed abrasives such as silicon carbide, starting with a coarse grit, followed by progressively smaller grit sizes until reaching, say, P1200 or occasionally P4000. It is then required to use a series of polishing cloths to achieve the sample integrity and accept the induced artefacts that occur at the polishing stage. The 'new concept' overcomes these problems by grinding to the actual sample integrity stage before polishing.

Much has been written about new techniques in specimen preparation and different proprietary preparation surfaces, such surfaces using a method of so-called lapping (lapping occurs as the abrasive is rolling). Lapping does not belong to this discussion and nor does it belong to the special surfaces created for the 'new concept'. These surfaces are based on their ability to cut, which occurs when the abrasive is stationary, i.e. grinding.

The object of making this distinction so early is because with a charged abrasive (one that is not bonded) it is possible to have it either rolling (lapping), which is not necessarily what is wanted, or temporarily stationary (grinding), which *is* probably what is required.

In Chapter 4 (the new concept) it was shown how silicon carbide has a short effective life and how beyond this point structural damage and deformation will occur. It must therefore be remembered when using silicon carbide to ensure that new papers are regularly used.

Traditional metallographic techniques will continue to be used for many years and could in particular conditions and environments always be preferred. It is therefore important not to overlook this aspect when introducing new technology. If it is intended to use, for example, silicon carbide papers exclusively for grinding, it may be that only one grade is necessary after monitoring and control of the sectioning operation. By using sharp abrasives only, the number of preparation steps could be reduced by half. Above all, damage must not be induced because of traditional demands! Figure 6.2 shows how traditional preparation techniques would be displayed within the new concept.

When comparing Figure 6.1 with Figure 6.2 two major differences occur, i.e. instead of

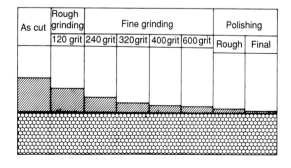

Figure 6.2 Traditional preparation within the new concept

removing all the previous deformation at each stage, the idea is to target a completion position, at or just beyond the original depth of structural Z axis damage. This is a realistic target when dealing with most metals; it is, however, an extended Z axis dimension when dealing with fracture dislocations (ceramics, minerals, etc.). No polishing stage should be used prior to achieving a faithful reproduction of the material structure, i.e. until sample integrity has been achieved.

6.2 The Two Grinding Stages

The grinding section is to be broken into two parts:

(i) Planar grinding
(ii) Sample integrity stage

6.2.1 Planar grinding

The object is to achieve a planar surface in the shortest possible time with minimum damage. When the subject has been carefully sectioned this stage can be omitted for hand preparation. With automatic equipment where more than one sample is clamped into a specimen holder irrespective of the sectioning technique, the planar grinding step is necessary to achieve uniform flatness. When multiple individually loaded samples are prepared, planar grinding is not necessary.

Much care should be taken at this stage. It must not be considered the roughing-out step. The smallest abrasive compatible with stock removal should be employed.

Question. What type of abrasive should be used and should it be bonded or charged?

Taking the last point first; when planar grinding the materials analyst, with all the diverse materials, will always choose a bonded abrasive first because of convenience, speed, cost, etc.

Bonded abrasives are of two types: the fixed diamond or cubic boron nitride or the degrading fixed abrasives such as silicon carbide or aluminium oxide. When grinding materials that exhibit ductility the fixed diamond or CBN wheels should be avoided in preference to the degrading abrasive. Diamond and CBN have obvious advantages with harder materials such as ceramics. When dealing with composites of both hard and soft constituents, degrading abrasives could be used for planar grinding *but* not at the sample integrity stage.

One very quick method of planar grinding with ferrous materials is to use a high speed stone (10 times the surface speed required for, say, silicon carbide papers). The stones, of grit size around 100, are usually vitrified alumina bonded and appear to give a good surface finish. They can, however, induce deep deformation.

When using high-speed stone grinding care must be taken with heat-sensitive materials to ensure an ample water supply, well-dressed stone and reduced pressure. Since deformation-related problems are usually associated with blunt abrasives, it is essential that the grinding stones are not too hard.

Bonded abrasives will cause much greater deformation and structural damage than charged abrasives. It is the degree to which this damage occurs that will dictate whether a change to charged abrasives is necessary. The next question could be 'how do we know if this damage is too severe?' This will reveal itself at the final sample integrity stage when observing optically before polishing. When this occurs the only choice facing the materials analyst (to date) is to use platen surface 10; this is the hardest of the classified surfaces where the abrasive operates predominantly in a grinding mode.

Surface 10 should be used with the smallest abrasive size that can be tolerated, often 30 to 45 μm. Abrasives such as diamond or CBN are employed.

Type

The type of abrasive is very much influenced by the material to be prepared, very hard materials using diamond, the softer more ductile using aluminium oxide, zirconia or silicon carbide. Aluminium oxide is a good stock removal type of abrasive, as is zirconia, but they both produce greater deformation than silicon carbide (assuming all abrasives are used within their efficient cutting lives).

When planar grinding on an automatic machine the question of contra or complementary direction arises. This is the relative direction the specimen takes in relation to the wheel direction. With contra motion the action is more severe, resulting in greater stock removal, but it does inflict a more directional cutting action which could, with very ductile or even brittle and friable subjects, create avoidable structural damage.

Contra motion with its unidirectional attack can be an advantage when grinding plated surfaces and thick coating if the cutting force is directed along the resin-coating substrate (and not vice versa). This reduces the coating damage, in particular with plasma coatings, and is to be recommended. The samples need to be correctly orientated when positioned in the sample holder.

Platen surface choice

The choices of available abrasives and surfaces for planar grinding in order of severity are:

Fixed abrasives

(i) Metal-bonded diamond or CBN, 30 to 45 μm
(ii) Resin-bonded diamond or CBN, 30 to 45 μm
(iii) Plastic-bonded diamond or CBN, 30 to 45 μm

Degrading abrasives

(i) Zirconia papers
(ii) Alumina papers
(iii) Silicon carbide papers

Integrity surfaces

(i) Platen surface 10, 30 to 45 μm
(ii) Any number lower, 30 to 45 μm

Lapping* surface

(i) Metal
(ii) Glass

6.2.2 Sample integrity stage

To coin a prase, 'this is where it all happens'. This is the stage that is the target of the preparation technique to complete sample integrity, observe optically to confirm and either continue with polishing if required or go back to the beginning.

When dealing with a particularly difficult subject the materials analyst is advised to optically observe between each preparation step and not to continue with the next step if improvements from the previous step have not occurred. Too often the temptation to go to the next step in the hope of recovering the lack of integrity is taken, resulting in incorrect preparation and, in consequence, incorrect analysis. It could be argued that without optical observations at the end of the sample integrity stage the investigator cannot confirm that a faithful reproduction of the microstructure has been achieved, i.e. observe prior to polishing.

Choice of abrasives

The choice is once again between bonded or charged abrasives. If the material is hard then there is no reason why fixed bonded abrasives should not be used, but remember that they induce greater deformation (structural damage) than an equivalent charged abrasive which itself can be affected by the shock absorbing characteristic of the platen surface (i.e. its number in relation to the 0 to 10 scale).

Unfortunately fixed bonded abrasives will not in general be 'gentle' enough to complete the integrity stage and will have to be finally supplemented with a charged abrasive on a surface such a platen surface 1 or 2.

To introduce a new system that totally ignores the good points relating to traditional methods would be foolhardy and counter-productive. Silicon carbide papers have been used for many decades and will continue to be used. Silicon carbide papers, if used only within their efficient life, will satisfactorily prepare ductile materials within the, say, 54 to 700 HV figure. Below 54 VHN they will continue to give satisfactory results if extreme care is taken to avoid smearing the top layer and to avoid incurring impressed abrasives. Unfortunately since this technique depends upon the polishing steps to achieve sample integrity it would be unwise to overlook the 'new concept' in the event of a dispute or where material artefacts are thought to be present.

---

*Lapping is the slowest yet the least aggressive and must not be overlooked with very brittle subjects.

Charged platen surfaces

These surfaces are usually metal composites and respond most satisfactorily with a non-degrading abrasive. Diamond has almost exclusively been used but other abrasives such as CBN can have many advantages. As the surfaces approach zero, i.e. surfaces 2 and 1, resin is more prominent in the surface composition with a total absence of metal. This allows the use of alumina/colloidal silica as well as diamond and sometimes a combination of two abrasives together, viz.:

(i) Titanium with its smeary top surface requires colloidal silica with its high pH to activate the surface, allowing efficient cutting.
(ii) Tungsten carbides require diamond to cut efficiently.
(iii) Special silica/alumina is preferred for aluminium, giving an efficient scratch-free cut.
(iv) Diamond and colloidal silica are used together to prepare both soft matrix and hard fibre in an MMC.
(v) Ceramics respond to pH activated silica.

Referring again to Figure 5.11 (making the 'right choice') gives a guide to the appropriate platen surface selection. The surfaces are based on varying shock absorbing characteristics from 10, the hardest, to 1, the softest, where the lowest is targeted to achieve sample integrity. The *primary selection* is based on the specimen hardness, matching the hardness of the platen surface. The surface should be slightly softer than the specimen, giving maximum stock removal with minimum shock absorbing features. As the sample is slightly harder than the appropriate surface, rubbing contact between the two will have no adverse effect, thus allowing a water-based abrasive slurry to be used for maximum stock removal.

This, however, is not always the case with the *secondary mode selection* where the difference in hardness between the specimen and the platen surface can be so great that care has to be taken to avoid the specimen acting as a cutting tool. A membrane is required between the specimen and platen surface; this is achieved by using an oil-based abrasive slurry. From the chart it will be seen how platen surface 4 is used with an oil-based slurry and that it is used for poorly bonded or brittle friable materials.

Surface 4 is therefore being used in general after 10 but could equally be required after surface 8 or 6. It could be questioned why surface 6 is used as a primary choice if it will require surface 4 in the secondary mode, since they are next to each other (in the absence of surface 5). It is quite rare to achieve sample integrity from surface 4 (using conventional 6 μm diamond) and as such a further selection is necessary based on specimen hardness, i.e. either surface 2 or 1. When surface 8 or 6 is used in the primary mode, surface 4 is rarely required (with the quoted exception). The secondary mode is based on specimen hardness as well as structural characteristics.

Abrasive size

In general 6 to 9 μm is adequate for stock removal without undue deformation when using surfaces 10, 8, 6, 4 and 2. If, however, we wish to reduce the structural damage or deformation at any of these stages a smaller sized abrasive would suffice. Equally if the damage from the planar grinding stage is high then 15 μm or higher would be required and if necessary an additional lower micrometre step.

Surface 1 is generally used with finer abrasives. Since surfaces 2 and 1 look like polishing cloths its action must not be confused. They function with high pressure in a grinding, shock absorbing mode.

Platen surface choice

The choice of surfaces in order of severity is:

Fixed abrasives

(i) Resin-bonded diamond or CBN, 15 to 30 μm
(ii) Plastic-bonded, 3 to 15 μm

Degrading abrasives

Silicon carbide papers (within critical life)

Integrity surfaces

Surfaces 10 to 1 (1 to 15 μm)

Lapping* surfaces

(i) Metal
(ii) Glass

Lubricant

To use a non-compatible lubricant would be to totally destroy the abrasive cutting action causing:

(a) skidding—no cutting;
(b) lapping—undesirable hammer finish;
(c) gouging—the swarf sticks to the abrasive.

A very simple test is to listen to the preparation process for the familiar grinding sound and to look at the sample on completion. A well-ground surface is bright, an undesirable lapped surface is dull (matt).

The right amount of lubricant is also important. When using composite surfaces, if the platen surface is too dry it will bind and if too wet it will skid.

As a guide to stock removal, water-based lubricants are more efficient, oil being the least efficient.

Having made 'the right choice' and completed this sample integrity stage the specimens are ready for polishing as necessary. However, before undertaking such steps, *always optically observe the result.*

6.3 Scratch Pattern Monitoring

Microscopic observations of scratch patterns can very often be the clue to indifferent surface preparation. The scratch pattern will indicate whether material is being removed under optimum or critical conditions. It can also be used to give an immediate indication of material removal by grinding, lapping or a combination of both. The lack of scratches has always been an indication of the quality of polishing. To understand scratch patterns is therefore an essential *requirement* in understanding efficient material removal, without which we would never achieve sample integrity. One glance down the microscope can indicate from the quality of scratch, any of the following:

- Incorrect lubricant
- Degradable grinding surface used too long
- Platen speed too high
- Incorrect abrasive type, shape, size or concentration
- Incorrect platen surface

When any of the above conditions occur to an audited preparation procedure, it will usually be associated with an extended time relative to a given Z axis stock removal value. It is very often the extended time relative to a given Z axis that prompts the optical observation. The following examples (Figures 6.3 and 6.4; see Plate 3) are used to draw attention to scratch patterns when using grinding papers (SiC). Both micrographs represent the ground surface of 0.37%C steel, Figure 6.3 after 1 minute and Figure 6.4 after 3 minutes. The clue to good scratch patterns is to look for the strong lines achieved with optimum angle sharp abrasives. Notice how the strength of line is lost in Figure 6.4 when the critical angle abrasives are most prominent. A poorly prepared ground surface can appear brighter than a correct preparation and hence fool the observer. With a loose or charged abrasive the reflectivity will indicate the correct level of grinding activity, i.e. if the abrasive should roll as in lapping, rather than grinding, the brightness will be reduced.

To summarize, for fixed abrasives the brightest surface is not necessarily the best and for loose abrasives the brightest surface is (with ductile materials) generally the best.

---

*A lapped surface will always require a final polish in order to optically observe unless the microscope objective being used is an oil immersion (see explanation in the microscopy section).

7

Polishing

7.1 Effects on Surface and Microstructure

Many definitions for polishing have been propounded, not least of which is a recent publication (1987) which says 'the difference between polishing and grinding is simply of degree; the abrasives are finer for polishing and *are used in a different manner*'. This statement is not entirely true since there is no point in a grinding procedure where it could be said scientifically 'now one is polishing'. It also is not related to the abrasive size but it is related to the action of the abrasive and the surface carrying the abrasive.

From the earlier definitions given in Chapter 5 it is clear that the difference between grinding and polishing is a function of the abrasive action. For example, an abrasive particle size of 6 μm can be used in either a grinding or a polishing mode.

The surface of a well-polished specimen is smooth and reflective and is a function of:

(i) the abrasive characteristics of the cloth,
(ii) abrasive type and shape,
(iii) particle size,
(iv) cloth nap,
(v) lubricant,
(vi) flow characteristics of specimen surface.

Top surface flow would be more pronounced with the softer materials but all materials exhibit a top surface 'shine' when rubbed over a cloth surface. The degree to which this shine (polish) is evident is related to the abrasive function of the cloth (all cloths abrade the specimen in varying amounts) and the force exerted during this 'shining' procedure. One can go further and investigate the polish achieved by burnishing which is a result of top surface flow.

This theory holds good with soft materials but does it apply to hard brittle subjects where surface flow is not likely? The polish in this case is a result of one or more of the following:

(i) Smaller abrasive particle size produces brighter surfaces.
(ii) Cloths that allow the abrasive to rise and fall take off the sharp corners of previous scratches, thus appearing scratch free.
(iii) Some abrasives, such as silica and alumina, because of their shape produce a lesser scratched surface.
(iv) Polishing abrasive slurries often have a chemical attack on the submicrometre surface affecting the finished surface appearance.
(v) Submicrometre particles are said to cause slip plane plastic deformation in ceramic materials producing the surface shine.

Stock removal of ceramics is generally

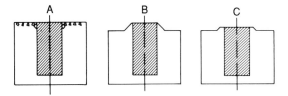

Figure 7.1 The induced relief technique

thought to be confined to brittle fracture but with the smaller sized abrasive (0.25 to 1.0 μm) plastic deformation has a high likelihood of occurring. Chemical mechanical polishing is also successful with ceramics, in particular with smaller sized abrasives.

When polishing composites with very hard and soft constituents polishing relief can be a problem. One example of this would be the polishing of borsic fibres (tungsten core, boron surround with a 1 μm or less silicon carbide outer coating) in a matrix of aluminium. The induced relief technique is used; i.e. the pH of the polishing medium is increased accompanied by the use of colloidal silica on a nap cloth (PS 8). This has the effect of polishing the matrix, leaving the fibres standing proud. The pH is then reduced, using diamond on a non-nap cloth (PS 3) to abrade the fibres. Care must be exercised in order to stop the procedure before abrasion of the matrix occurs. This technique is illustrated in Figure 7.1.

New polishing cloths do not always perform as efficiently as used cloths; i.e. do not expect the best results from new cloths. They have to wear in, allowing the surface to 'even-out' and the abrasive and lubricant to 'bed-in'. Wash contaminated cloths in preference to throwing them away; in some cases use a dummy to condition the cloth before use.

7.1.1 Surface rubbing

This occurs at all polishing stages and is in part one of the necessary attributes to surface reflectivity. This condition is surface type and pressure related in that an increased pressure will increase this condition and changing surfaces (cloths) will also change the degree to which it takes place. The disadvantage is that although it makes the prepared sample look good (no scratches) it destroys the integrity. There are times when this could be desirable in helping to resolve different phases within the material but in general it is to be avoided, in particular with soft smeary materials. Having recently observed a preparation technique for a difficult to prepare MMC it was obvious that poor preparation was disguised at the last polishing step by using high pressure on a nap (PS 9) type cloth. Although the results looked good (no scratches and good interface contact) they could be totally erroneous. For many years the metallographer has insisted upon a good shiny, well-polished sample; this can no longer be tolerated if it occurs at the expense of integrity (as it often does).

Polishing surfaces have been produced for many years (20 at least) where the main, if not total, constituent is resin. These surfaces were originally constructed for use in the preparation of minerals where the high surface rubbing was of little consequence. The benefits were the shock absorbing characteristics of the plastic, allowing the charged abrasives to cut in the grinding mode without the structural damage normally associated with fixed abrasives. These and other similar surfaces are now being used with metals and should be classified as platen surfaces (not polishing surfaces).

7.1.2 Cosmetic finishing

The 'new concept' uses the polishing stage simply for cosmetic or aesthetic purposes. It is not intended to be used in the process of achieving sample integrity.

To quote once again from a recent publication (1987), 'To obtain a specimen surface absolutely scratch free, three polishing stages are recommended for metallic materials. However, for many routine examinations, when the odd scratch can be tolerated only two stages need to be used.' This publication then goes on to name the stages as coarse, routine and final without any reference whatsoever to the function beyond that of removing scratches. This rather sums up the difference between the 'new concept' which is concerned throughout with sample integrity before any polishing takes place and traditional teaching whose objectives are to create a highly polished surface which by definition could obscure the true microstructure.

7.2 Cloths, Abrasives and Carriers

7.2.1 Polishing cloths

Two broad groupings of polishing cloths are available: those having a plain smooth surface and those having a plain short pile (nap). The main advantage of a plain smooth surface is that it provides high cutting rates with low relief. These types of cloths are generally all that are required when final polishing hard materials. They are also advantageous when the subject exhibits soft and hard phases. For all but the hardest materials the final polishing must be carried out on cloths exhibiting varying degrees of nap. Loss of subject integrity can sometimes be tolerated in the interest of resolution; one such example would be tungsten carbide/cobalt where a nap-type cloth will destroy the totally flat surface at the expense of highlighting the hard carbides, i.e. improving the resolution.

7.2.2 Polishing abrasives and carrier

Abrasives used for polishing are generally supplied within a carrier. The object of the carrier is twofold: firstly to keep the small abrasive particles separate and secondly to lubricate during use. Diamond is supplied in one of the following forms in sizes down to $0.25\,\mu$m:

(i) Aerosol spray. Requires lubrication during use.
(ii) Paste. Requires the correct type of lubricant during use. This lubricant must be compatible with the carrier otherwise a dramatic fall-off in stock removal will result.
(iii) Slurry. Often not requiring any additional lubricant but should it be necessary once again the two must be compatible. Slurries are supplied in three types:

    (a) Alcohol based
    (b) Oil based
    (c) Water soluble

Abrasives other than diamond can be used on polishing cloths which have specific advantages in producing the final polish. This is in part due to the smaller sizes available ($0.05\,\mu$m) and also to the abrasive shape and hardness. For example, alumina produces less severe scratches than diamond on soft materials and colloidal silica is exceptionally good for use with titanium and ceramics.

The question arises: what is the lowest size of diamond to use? Since diamond is supplied in sizes down to $0.25\,\mu$m when would it be used? Because of the shape and size of diamond it is not best used as a polishing abrasive; it is the aluminas and silicas that give a better (less scratches) finish.

Diamond of size $0.25\,\mu$m therefore would be of little use with a soft material (impressed abrasives). Although small sized diamond abrasives would not impress with a ferrous material, aluminas or silicas would give a better result. Hard metals generally do not require abrasives less than $1\,\mu$m. It has been established, however, that plastic flow occurs with small sized abrasives when preparing ceramics. This could be one possible application for $0.25\,\mu$m diamond.

7.2.3 Lubrication

The type of lubrication affects the cutting efficiency and also the severity of the cut. With soft materials it is the change from a water-based lubricant to an oil base that improves the surface finish, allowing a thin film of oil to act as a membrane between the polishing cloth and sample. The dry type of lubricant (water, alcohol, etc.) gives maximum material removal; the oil type gives maximum surface lustre.

Lubricants could also be necessary for heat dissipation, in particular when operating at high speeds and pressure.

7.2.4 Polishing speed

There is an optimum speed relative to the materials and abrasive type; in general 240 rev/min for 12 in diameter machines satisfies most demands with diamond. Higher speeds can be tolerated for hard metals but it is unwise to assume that because higher speeds remove stock quicker they can be adopted. The danger with increasing speeds at the polishing stage is the tendency to burnish over the surface with soft materials due to the increased rubbing.

Because of this burnishing effect it is wise to reduce the speed as low as 150 rev/min (12 in diameter wheel) with, say, pure aluminium. The surface speed must also be taken into account when the subject being polished is friable or loosely bonded. Before carrying out the optimization programme it had always been traditionally considered wise to reduce the speed at polishing to roughly half that of grinding. With ferrous materials the reverse was found to be the case: i.e. SiC *grind* at 150 rev/min, diamond polish at 240 rev/min.

7.2.5 Polishing pressure

From our definition of polishing, 'the rise and fall of the abrasive within, say, the nap of a cloth', it follows that the required pressure will be less than the pressure used to grind on that same surface. The general rule is to use half the pressure necessary for grinding, i.e. 17 kPa for polishing, 35 kPa for grinding.

7.2.6 Vibratory polishing

Vibratory polishing employs the use of polishing cloths and abrasives identical to those used on conventional rotational polishing wheels. The difference is in the polishing action: the wheel or platen is caused to vibrate in a horizontal mode at many cycles per second. When the specimen is placed on this vibrating surface, it undergoes a continual series of tiny abrasive cutting movements, which very gently remove surface material. The obvious advantage of such gentle cutting action is that reduced forces are employed, resulting in lower residual stress. With the introduction of the new engineering materials which can react adversely to high stress material removal techniques, vibratory polishing (as it is incorrectly termed) could offer many attractions. This gentle action has two more advantages: i.e. scratch-free surfaces can be produced on soft materials and the sample integrity of many minerals can be improved beyond that of conventional techniques. Purely in terms of polishing, the vibratory action is more faithful than rotary methods in that the 'shine' is not achieved by smearing. The disadvantage of vibratory polishing is time.

It takes anything from half an hour to 24 hours to achieve the desired result. The author is aware of many machines that operate each evening, the specimen being ready for analysis next morning. Some samples are not suitable for vibratory polishing; the best guide is 'if the material is not suitable for ultrasonic cleaning, due to granular disintegration, then it is most likely not suitable for vibratory material removal'.

7.3 Classification

One of the major advantages of the 'new concept' was the introduction of a systematic approach to the selection of platen surfaces for the sample integrity stage. This same philosophy has been extended to polishing to aid in the selection of appropriate polishing cloths.

To date manufacturers list cloths by type (even trade names) but not by classification, viz.:

Selvyt, Nylon, Billiard cloth, Red Felt, silk, Rayvel, canvas, Metcloth®, Microcloth® Texmet®, Mastertex®, Chemomet®, etc.

When attempting to classify cloths, two important functions become relevant: (a) the 'abrasive nature' of the actual cloth surface and (b) the 'cloth nap' or degree to which the applied abrasive is allowed to move (up and down) within the cloth.

Without applying any abrasive to a polishing cloth the actual cloth surface will have a polishing effect. Selvyt, for example, will polish (or shine) the sample to a different degree than would, say, canvas. The texture of the cloth also dictates the way the applied abrasive will perform.

The rate of stock removal is conversely related to the cloth nap.

The 'new concept' relates the shock absorbing characteristics as the basis for classification. With polishing cloths this can be applied equally by relating the degree to which the abrasive is allowed to rise and fall within the polishing cloth. Take, for example, a tightly packed or woven material of short fibre length with that same material of long fibre length; the abrasive will rise and fall more with the latter, giving less stock removal and more relief with an improved surface lustre.

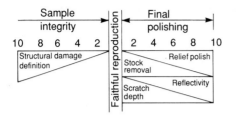

Figure 7.2 Platen surface/polishing surface classification

It can be seen from the above how cloths of similar material exhibiting differing density and nap can be classified as well as cloths differing in type.

Many of the better quality polishing cloths have a plastic membrane between the cloth and the back surface glue. This has the effect of increasing the shock absorbing characteristics of the polishing surface. Another major advantage of this technique is to confine all the polishing abrasives to the cloth and not allow contamination with the platen/polishing cloth adhesive. Since some polishing lubricants will attack the polishing cloth adhesive, it has been necessary to ensure that both adhesive and lubricant are compatible. With the introduction of the 'membrane' this is still necessary for the cloth/membrane interface but not for the membrane/platen interface. This has allowed the introduction of a new breed of adhesives to be used which are not so 'sticky' and therefore have a reduced tensile strength. The advantage of these adhesives is that their shear strength is relatively high. When used on a polishing cloth this means that they will not 'fly off' during use and can be readily removed from the platen when required. This type of adhesive is also used on silicon carbide papers.

From Figure 7.2 it can be seen how a classification of 1 to 10 has been created for the polishing cloths based on two factors, the major factor being the rise and fall of the applied abrasive and the other the abrasive function of the cloth. This classification is not to be confused with the shock absorbing requirements of platen surfaces where the lower the number, the higher the shock absorbing characteristic. Polishing surfaces, if classified similarly, would give a low number to polishing surfaces with a high nap. The assumption made at the polishing stage is that integrity has already been achieved and

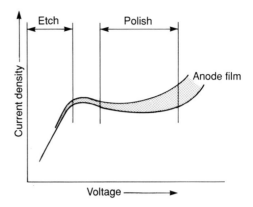

Figure 7.3 Current density curve

therefore reflectivity and preferential polishing increases with an increase in number. The classification of cloths will be covered later, but for the moment we can say that (a) integrity is achieved by progressing platen surfaces towards zero and (b) this achieved integrity is retained by using polishing surfaces close in number to this zero.

7.4 Electrolytic Polishing

Although this book has concentrated on grinding and polishing by mechanical means, it would be remiss not to mention electrolytic polishing. Electrolytic polishing is very suitable for homogeneous materials such as aluminium and copper. In general, the soft single-phase alloys are ideally suitable along with stainless steel and titanium, viz.:

Cu/bronze/brass/Fe/Ni/Al/stainless steel/W/Mo/Ti/V/Zr/U/Pb/Co/Sn/Zn

It is important to use electrolytic polishing when it is essential to eliminate all forms of residual damage or stress from the observed top surface. Electrolytic polishing, or anodic polishing as it is often called, occurs through anodic dissolution of the material surface when used in an electrolytic cell (not unlike electroplating using a cathode and anode within an electrolyte). When a current is passed from one metallic electrode to another, through the electrolyte, metallic ions will move from one electrode to the other, i.e. from the anode to the cathode. During this material removal process (polishing) an anode film forms on the surface of the metal specimen.

When the voltage is reduced a point is reached where the anode film reduces, allowing specimen etching to take place.

Figure 7.3 shows a typical current density curve where the conditions of etching and polishing occur. This curve will change with different materials and electrolytes. The factors affecting results are given below.

7.4.1 Prepreparation

The surface should be flat and preferably ground using P1200 silicon carbide paper. Before polishing, make sure the surface is free from grease or contaminants. The surface can be masked with tape if restricted polishing is required. If it is necessary to encapsulate the sample, this must be carried out using a conductive resin. Alternatively, a contact point can be made after mounting by drilling a hole, later to be plugged with a conductor.

7.4.2 Voltage–current density

High voltages usually result in preferential material removal relative to the variations in the anode film thickness. Anode film thickness variations are a direct result of specimen surface topography. Lowering the voltage can result in undesirable pitting. The ideal current density is therefore influenced by the electrolyte, the surface area and the specimen material.

7.4.3 Temperature

The temperature of the electrolyte must be kept well below any flash point. It is advantageous in the preparation to keep the temperature around 15°C; pitting can be related to temperatures above this figure.

7.4.4 Time

Polishing and etching times should be kept as short as possible (seconds). Many artefacts occur once the polish has been achieved. As with mechanical polishing, time from the optimum will degrade the sample integrity.

7.4.5 Flow

To a small degree flow of the electrolyte can affect the thickness of the anodic layer, though not as much as a voltage change. The flow must be constant and uniform for a balanced metallic ion distribution.

7.4.6 Electrolyte

The balance between pitting and irregular polishing can often be solved by the selection of an electrolyte. General purpose electrolytes are available, covering a wide variety of materials. Specific electrolytes are nevertheless more suitable and reference books are available.

7.4.7 Parameter optimization

From the five variables given above, electrolytic polishing could be wrongly considered as a hit-and-miss technique, but this is not so. Optimize the parameters and one has a quick, reliable inexpensive system for microstructural analysis. The following is intended as a guide to achieve good results:

 (i) Make sure the sample is receptive to electrolytic polishing.
 (ii) If problems could be anticipated, select the most suitable electrolyte, i.e. do not expect general purpose electrolytes to be best for all materials.
 (iii) Select a polishing area, by masking, which is no greater than is required.
 (iv) Ensure that the temperature of the electrolyte is kept low.
 (v) Use the voltage and flow as given in the equipment instructions.
 (vi) The balance between even polishing and pitting often results in a combination of both occurring together. Figure 7.4 shows the conditions prevailing when carrying out centre or peripheral polishing. The object of the exercise is to be somewhere in between. To correct for centre polishing it will be necessary to lower the voltage; this could also require an increase in the flow rate, to eliminate pitting. To correct for peripheral polishing the reverse is required: increase the voltage and, if necessary, decrease the electrolyte flow rate.
 (vii) Having optimized the flow rate and voltage, the time should be reduced to a minimum. The system is now optimized and should work trouble free for considerable periods (until the electrolyte requires changing).

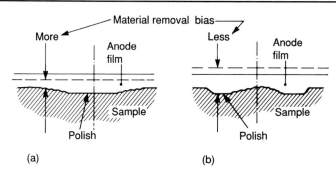

Figure 7.4 Flow rate and voltage adjustments. (a) Centre polishing: (i) lower voltage, (ii) reduced layer, (iii) increase flow (to avoid pitting). (b) Peripheral polishing: (i) increase voltage, (ii) increased layer, (iii) decrease flow (to increase efficiency)

8

Grinding and Polishing Lubricants

8.1 Selection Considerations— Grinding

There are specific considerations that have to be taken into account before selecting the type of diamond suspension or slurry. The following can be used as a general aid in the primary selection to be followed by a verification of selection:

(i) Materials exhibiting high ductility are best suited to oil-based lubricants in order to effect efficient chip removal.
(ii) The most efficient stock removal takes place with the 'thinnest' lubricant, i.e. alcohol-based lubricant will exhibit the highest stock removal characteristics.
(iii) When the hardness of the material is very much greater than that of the platen surface being employed, physical contact must be avoided. This is best achieved using an oil-based lubricant, viz. tungsten carbide sample on platen surface 4. Extreme care must be exercised with, say, a ceramic component exhibiting porosity, if high pressure or insufficient membrane is allowed to exist between the sample and platen surface. Then 'pick-up' occurs, resulting in closed pores from the platen surface material.
(iv) Surface affinity occurs between soft materials of similar type. This puts a restriction on the type of platen surface that can be employed when preparing pure aluminium/zinc/lead and other such materials. If these materials are to be prepared on metal composite platen surfaces then the oil-based lubricant must be used and could require additional oil lubricant to avoid surface rubbing.
(v) Platen surfaces require a charged abrasive in order to operate in the required cutting mode. These abrasives are therefore caused to operate in one or more of the cutting modes by the action of the sample passing over them. It is this action that creates the cutting but it is also this action that moves the abrasive from its cutting mode when effective cutting operation is impaired. It is the latter condition that causes small abrasives to become lodged or impressed into the sample surface. There is therefore a limiting factor on the size of the chosen abrasive relative to the ductility of the chosen material. The degree to which this abrasive attraction

takes place can be reduced by changing from a water- to an oil-based lubricant.

8.2 Selection Verification—Grinding

(i) The scratch pattern is an important guide in detecting the efficiency of the chosen abrasive/lubricant and platen surface combination. There are three types of scratch patterns that can be optically identified:

   (a) Clean fine lined scratches caused by abrasives operating with a sharp point and within the abrasive critical angle.

   (b) Thicker/darker lined scratches where cutting is taking place, the abrasive rake angle being negative or beyond the critical angle.

   (c) Fragmented lines caused by blunt abrasives or abrasives cutting with a swarf buildup. (The scratch pattern may on close inspection be not a line but a series of dots. This is indicative of an undesirable rolling abrasive.)

(ii) Visual inspection of the prepared surface beyond just the scratch pattern and towards a reflectivity measurement can also be a guide to best selection.

(iii) Reduced deformation created at each stage is important and a balance has to be created between this and stock removal. A situation is often found where water-based lubricants are used at the planar grinding stage yet at the first integrity stage oil-based lubricants are used. An oil base is used for ceramics/high ductile metals and a water base for low ductile metals. The final integrity step is water based for ceramics and oil based for high ductile metals.

(iv) To optimize any procedure the stock removal–deformation ratio must go beyond just the type of lubricant and must also take into consideration speeds/directions and pressures. For example, oil-based lubricants could be the best compromise if surface speeds are too high. It is therefore wise when optimizing a cutting procedure to relate the cutting efficiency relative to structural damage; this is done by comparing Z axis dimensions with optical observations.

8.3 Selection Considerations—Polishing

The lubricant used on a polishing cloth has very similar effects in terms of aiding efficient stock removal to the case of grinding. The basic difference lies in designating the polishing stage as cosmetic to an already achieved faithful sample reproduction. Lubricants that aid the 'shining' or dampen the scratching effect from the polishing abrasive will be advantageous here. Since the sample will require to be perfectly clean after completion of this stage, alcohol-based lubricants could be portrayed as the best. Unfortunately, alcohol, although generally more efficient in terms of stock removal, leaves deep scratches on all but the hardest of materials. It is necessary, therefore, to use a slightly more oily based lubricant. As the softness of the specimen increases, so the need for oil increases. It is better to have an oil to remove scratches rather than increase the polishing cloth surface number which will cause relief polishing. When the abrasive is administered via a syringe, as is the case with diamond paste, it is important that the lubricant is compatible with the abrasive carrier. Polishing lubricants can stain the specimen; therefore it is wise to clean the sample as soon as the polishing stage is complete. Cleaning can be via diluted detergents (washing-up liquid) followed by an alcohol spray which is then hot air blown (domestic or industrial hair-dryer). Many ceramic composite materials respond to a chemomechanical polish, the detergent cleaning being necessary *in situ* as the polishing stage concludes.

8.4 Selection Verification—Polishing

Observe the clean polished unetched sample under the microscope. Scratches, if present, will be visible in dark ground and first-order grey DIC. If the sample is only to be observed in bright field the removal of scratches is not so critical. If the polishing stage is to be prolonged then all scratches should disappear. Defects such as relief, pitting or smearing could, however, destroy a previously well-prepared sample. With soft materials the scratches not visible in the unetched condition could reappear when etched. With such materials an etch polish technique is necessary.

Although the polishing lubricant will have an effect on the finished surface, its influence is far less than that of the abrasive type and the quality of the polishing cloth. The author can quote numerous occasions where hours of work to present a totally scratch-free surface for photomicrography have been solved in minutes by a change of polishing cloth rather than any effect attributed to a lubricant.

9

Towards a Metallographic Standard

9.1 Introduction to Grinding Surface Classification

Since the introduction of the 'new concept', a systematic approach to specimen preparation (1987), there has been much progress towards the creation of standards in order to carry out meaningful quality control procedures.

The need for standards and traceability is the concern of all working, in particular, in the field of advanced materials. Due to the nature of new materials this task is not an easy one—hence the reason for such bodies as the Versailles Project on Advanced Materials and Standards and other such organizations. Stumbling blocks to standards in sample preparation have been:

(i) lack of any universal language,
(ii) no system for constructing procedures,
(iii) no means of auditing the results to traceable standards,
(iv) no means of technically challenging empirically derived methods,
(v) no means of creating the best system for individual needs without sacrificing integrity.

The object of this chapter is to find or reference the key to all these five points.

9.1.1 Background

Traditional metallography relied very much on empirically derived methods with no sound technical rules or even a systematic approach beyond grinding with abrasive papers and polishing using polishing cloths. The requirement would be to achieve a so-called polished surface with the assumption that what would be achieved would be a faithful reproduction of the material structure. As has been illustrated in earlier chapters, polishing and sample integrity are mutually incompatible. It followed therefore with the introduction of a systematically derived procedure to target thoughts towards such questions as (a) is there a preferred technical procedure, (b) could alternative procedures also be acceptable and (c) would it be possible to designate terms of reference in order to achieve the ultimate standard procedure to which others would be traceable?

9.1.2 Review

An important aspect that has reinforced the original concept is the relationship between

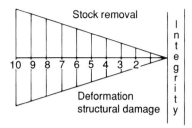

Figure 9.1 Deformation classification

deformation and stock removal. Figure 9.1 illustrates this relationship as being an arithmetical curve. This in practice is true from surfaces 10 to 2, from which point it proves difficult with existing equipment to measure the small micrometre and submicrometre stock removals. This, however, does not restrict the convention relating surface numbers to deformation.

9.1.3 Surface classification

If a system is to be universally adopted the surface classification must be available for all commercially produced platen surfaces where it is advantageous to reference by numbers. For example, silicon carbide paper as well as the new charged platen surfaces should have some relationship. Since a relationship exists between deformation/structural damage and stock removal we are able to designate surface numbers from material removal.

The original concept was to create a series of numbers from 10 to 0, where 10 was the most aggressive—removing the most stock yet yielding the greatest damage. The reducing numbers were related to an arithmetical reduction in damage to the theoretical zero. This classification was achieved by relating stock removal per unit time using the same sized abrasive.

To classify surfaces by stock removal, which we have related to structural damage, is the technically sound approach. For purely commercial reasons there is a case for designating surfaces by their hardness value, which is in part dependent upon the shock absorbing qualities of the surface, thereby satisfying part of the classification requirement. It does, however, ignore the abrasive action that takes place during use which makes the classification slightly subjective.

Some grinding surfaces requiring classification are what the reader could consider to be polishing cloths, i.e. chemotextile material, woven steel mesh cloth, etc. These surfaces are obviously unsuitable for classification via hardness testing. Classification of these materials must therefore be via stock removal. Suppliers of such materials could be reluctant to give a surface classification; this is acceptable providing their designation when used in a procedure does not disguise its origins through the use of trade names. Information such as surface thickness, perforations (if in evidence) and mesh size would be essential.

9.1.4 Datum for classification

Some platen surfaces already exist in the market-place where an attempt has been made to designate resultant specimen integrity via numbers. It was to be expected, therefore, that other suppliers of platen surfaces (of which there are quite a few) would wish to classify their products in order to relate to the 10 to 0 classification. This raised an interesting question of what surface should be used as the datum from which all others are related. The alumina composite platen surface 10 was proposed and is currently being used within some UK organizations until an independent standard can be agreed. An independent standard is being pursued through the European Society for Microstructural Traceability and is hoped to be shortly agreed.

Having established a datum surface such as platen surface 10, the unknown surface can soon be designated as indicated in Figure 9.2 by using the same sized abrasive on both the known and unknown surface for a given time and calibrating the stock removal difference.

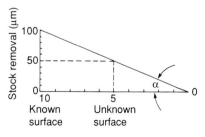

Figure 9.2 Unknown surface classification (α = deformation structural angle)

It was never originally intended to increase the surface number higher than 10. However, in the event of a new surface being created having a greater stock removal with the same abrasive and identical time, then it must be designated accordingly.

In establishing a standard it would also help to carry out the tests using a stable, reproducible material such as, say, 99% dense alumina. When classifying unknown platen surfaces lower than 5 it will be necessary to establish a second known datum reference surface nearer to the unknown. This will also have to be accompanied by a change in specimen material from the ceramic to a softer metal. This is because platen surfaces are more sensitive to classification by specimens of similar hardness; i.e. platen surfaces 3 and 4 would react similarly with a ceramic specimen but very differently with an aluminium alloy as the testpiece. Although the results achieved are independent of the base length it would be advisable to fix this distance at, say, 20 cm with 10 μm stock removal representing 1.0 cm. This then allows the deformation/structural angle to be quoted, soft materials having a larger angle than harder materials. It is envisaged, with more information relating to abrasive types and lubricants, that this figure will be related to actual choice of consumables relative to the material being prepared, i.e.:

(a) blocky diamonds used with curves > angle X,
(b) silicon carbide to be used with curves between X and Y,
(c) alumina to be used with curves < Z, etc.

The 'deformation structural angle' can also be extremely useful in compiling data-based preparation methods. These methods could be technically constructed by multiplying a factor constant K by the deformation structural angle. Take, for example, a three-step method as below (angle $X_1 = 25°$ and constants of 2.4, 1.0, 0.4):

Step 1 $25 \times 2.4 = 60$ μm
Step 2 $25 \times 1.0 = 25$ μm
Step 3 $25 \times 0.4 = 10$ μm

The construction of another method with $X_2 = 30°$ would be

Step 1 $30 \times 2.4 = 72$ μm
Step 2 $30 \times 1.0 = 30$ μm
Step 3 $30 \times 0.4 = 12$ μm

9.2 Deformation Classification

Discussion is ultimately leading up to whether a surface classification covers all materials whereby the roughness average or surface finish takes second place to the resultant damage remaining in the material after each operation. Figure 9.3 illustrates the deformed layer of a carbon steel (see also Figure 1.1 on Plate 1). The object of the exercise is to define a depth of damage and to progress the preparation stages until ultimately a point is reached where the material is faithfully revealed with integrity.

Traditionally, this damaged layer would be removed with the first preparation stage, leaving another smaller damaged layer eventually with successively smaller abrasives achieving integrity. If, however, one considers removing just the amount of material that is the limit of integrity possible with that particular surface and leave remaining existing damage beyond the scope of that particular abrasive, then there is a need to define two factors:

(a) the depth of existing deformation and
(b) the limit from which no improvement takes place with each successive surface and abrasive type.

Consider the first preparation stage with a 0.37 steel as being 320 grit SiC and the estimated total damage to be 120 μm. When estimating the damaged layer in Figure 9.3 there is evidence of the top fragmented surface followed by a definite deformed surface, depicted in the pearlite (as a result of the shear stress). Beyond this point an allowance has to be made for residual stress.

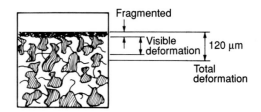

Figure 9.3 Structural damage of a carbon steel

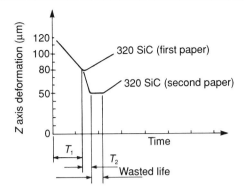

Figure 9.4 Integrity curve for silicon carbide papers

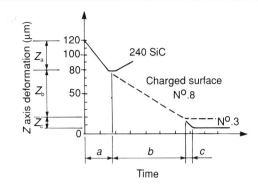

Figure 9.5 Integrity curve for combined surfaces

Residual stress manifests itself more obviously with softer materials, sometimes resulting in twinning; with hard ductile materials add 25% to the deformed layer dimension, with soft materials 50% and with crack propagating materials 100%. If the material had been cut, for example, on an abrasive cut-off machine then the residual damage can be determined in advance for each material, thus avoiding the need to estimate.

From Figure 9.4 it can be seen that more than one sheet of silicon carbide paper would be necessary to achieve the optimum condition for that selected grade. It also shows that continued use of the first paper would result in a deterioration of the damaged layer. Finally, the use of two 320 grit silicon carbide papers would utilize only part of the second paper's useful life.

Clearly from the above analogy it can be seen how inefficient the initial selection had been with an initial damage of 120 μm. Had the initial damage been 100 μm then 320 SiC grade would have been ideal. If the surface chosen was to be a charged platen surface as opposed to a degrading surface then its efficiency is not necessarily time dependent.

Figure 9.5 shows a combination of SiC and charged platen surfaces; notice how each of the two chosen 'charged' surfaces reaches a point from which prolonged time does not improve and nor does it degrade. (Although this is true for this material it must not be assumed to be the case with all materials.) Also notice how the final surface does not give total integrity; it does, however, approach integrity much beyond the scope of degrading abrasives and is often just polished from this condition.

9.3 Dangers Associated with Preparation Time

Preparation procedures quote a given time at each step. This was considered to be a variable depending on type and size of machine, number and area of samples, efficiency and type of abrasive, lubricant, etc. Due to all of these variables it was possible to designate a perfectly sound technically derived procedure yet achieve inconsistent results when emulated.

This unsatisfactory state of affairs existed until time was replaced with the Z axis dimension; this then accommodated the inevitable variables without affecting the result. If, for example, the efficiency of a particular operation was reduced due to, say, incorrect lubricant or a change in abrasive shape or concentration then the required time to overcome this effect would be encompassed within the Z axis dimension.

It can be seen from Figure 9.5 how this weakness in the concept has been overcome. The first step operating for time a is the important dimension of the Z axis, i.e. Za. The preparation procedure would then be correct if Za, Zb and Zc were quoted instead of time.

9.3.1 Interpreting the Z axis stock removal

There is a move within Europe to create data and standards for modern engineering materials (VAMAS, British Standards, NPL, etc.). It follows that a standard must designate that which can be audited and will be traceable. No material preparation technique

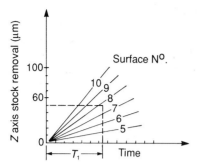

Figure 9.6 The stock removal chart (charged abrasive) showing conversion of the Z axis to operating time

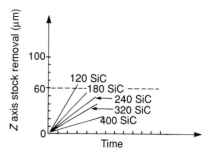

Figure 9.7 The stock removal chart (degrading abrasive) showing conversion of the Z axis for silicon carbide papers

can therefore be data based without the Z axis stock removal at each step.

There is obviously a time lag between the introduction of new machines or modification of existing machines to cope with new technology. This, however, can be overcome. Figure 9.6 shows how a stock removal chart can be created for specific machines with a given set of conditions. If the Z axis was quoted as 60 μm for surface 8 then a time of T_1 would be used. The Z axis can be interpreted from a sample test run simply by noting, from a hardness indent, the reduction in the indent diagonals.

This very useful and accurate way of determining the Z axis stock removal at each stage is to make a hardness impression in the sample after the planar grinding stage. This allows the reduction in the diagonals to be noted after each step. To convert this figure into a depth measurement simply multiply by 0.202; i.e.

$$Z = \frac{\tan 22 \times \text{diagonals}}{2} = 0.202$$

The question relating to the life of degrading surfaces can also be overcome by the use of the Z axis, as shown in Figure 9.7 where the use of two papers would be necessary should 320 grit SiC be chosen.

9.4 Material Cutting Related to Microstructure

In optimizing preparation procedures a set of conventions related to machine functions was written. Those conventions were based on empirical and historical records, which have been superseded by the optimization figures shown earlier. There is still a need, however, for new conventions to suit different machine functions.

It is known that residual deformation is dependent upon the platen surface being used and the shear stress necessary to create the chip and is relatively independent of pressure and speed. There is a case, therefore, for relating these two factors to actual materials being prepared relative to efficient cutting. The cutting of metal by means of a wedge-shaped tool such as an abrasive particle is the result of stress in the metal exceeding its elastic limit, resulting in plastic deformation and cracking in the material being cut. Part of the stressed material separates from the parent metal to form what is called a 'chip'.

Stresses are a combination of compressive, tensile and shear; it is the shear stress and the relative angle of the stress that influences the efficiency of the cut. The distribution of these stresses is such that the parent material can be left with residual stresses after a cutting operation. These stresses can be the cause of structural damage to the microstructure.

When cutting ductile materials such as mild steel, copper or aluminium the presence of slip plain dislocations extending from the tool point is dependent upon the magnitude of the shear stress in the chip. The load, speed and cutting tool geometry have a major effect on this condition, as does to a lesser extent the coefficient of friction between the chip and tool interface or abrasive surface.

The sharpness of the stress applicator (wedge-shaped tool or abrasive) causes a narrow band of high intensity stress in the shear plane, creating the chip by plastic deformation. The blunting of the stress applicator causes redistribution of this stress over a much wider area and in consequence an increase in consumed energy is necessary to achieve the desired magnitude of stress; the area and magnitude of the residual stress in the parent material also increase. The materialographer must therefore always use sharp pointed tools when chip removal is dependent upon plastic deformation (all ductile materials).

Plastic deformation with the face-centred cubic lattice materials requires less energy and can tolerate increased cutting speeds along with the appropriate cutting conditions (rake angle, lubricant). Any alloying, precipitation, heat treatment, etc., will offer obstacles to dislocation movement. As the obstacles to dislocation become greater so the energy consumption is greater—hence the requirement for slower speeds.

Since approximately 30% of the energy consumed is absorbed in friction between the tool and chip interface it follows that there is a need for suitable lubricants to achieve the lowest coefficient of frictional resistance. When using a charged abrasive the chip function is still very important; an incorrect lubricant with ductile materials can also result in impressed abrasives.

9.5 Polishing Cloth Classification

The original surface classification of polishing cloths as described in the December 1988 issue of *Metals and Materials* (numbered 0 to 10 in increased nap and decreased surface abrasion) has raised another factor, that of polishing cloth backing. Currently there is no agreed polishing surface classification; therefore all that can be offered are guide-lines to illustrate the principle. The end user in industry has responded to such ideas and in some cases users are actually designating polishing surface numbers with their specifications.

The polishing surface chart of Table 9.1 compiled at the York College of Arts and Technology is offered as a guide only and is intended to show how constructive such classifications can be—in particular, when the list is more comprehensive. Classification of the polishing chart was made from the results of:

(i) reflectivity measurements from cloths used without any abrasive,
(ii) height measurement of cloth nap,
(iii) depth measurement from a ball test,
(iv) fibre concentration.

Table 9.1 Polishing surface classification chart. (Reproduced by permission of Trevor Bousfield)

PS No.	Leco	Buehler	Struers
1	Silk	Silk	DP Plan PA.W
2	Pan K/Nylon	Nylon	Pan-W
3	Cotton/MM 414	Metcloth®	
	MM 420	Texmet®	DP-DUR (silk)
4	Canvas	Canvas	DP-MOL
5		Chemomet®	
6	Billiard cloth	Billiard cloth	
	Red felt/MM 431	Red felt	
7	Selvyt Suede cloth Imperial	Selvyt Mastertex®	
8	LE cloth B		DP NAP
9		Microcloth®	DP Plus
10	Velvet	Rayvel®	

9.6 Databank by Z Axis Curves

If we consider the preparation possibilities of 0.37 steel and the permutations, there are many, some giving integrity, others not. Those that could faithfully reveal the true structure could be incorrectly used and degrade the surface. Although 0.37 steel is quite accommodating to less than optimum conditions, modern engineering materials are quite sensitive to less than optimum conditions and must therefore be carefully prepared to a correct procedure. The question which must be addressed is, 'What is the correct procedure and how can we resource the information?'

One essential requirement of the correct procedure is that it must result in sample integrity, i.e. it must reveal the true structure. From this point the correct procedure can be defined as that which was carried out in the

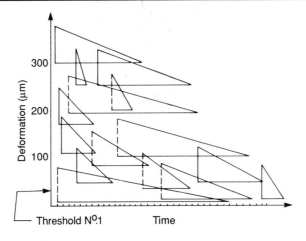

Figure 9.8 Databank for specific material

shortest possible time or that which employed the least number of steps.

One can also see a reason why the best method could be the one that utilized those consumables readily available or those which are the least expensive. There could be occasions where only one method can satisfy the first requirement, that of integrity, in which case there would only be one method.

There can also be a case for having a less than maximum sample integrity threshold for occasions such as grain size counts or hardness penetration. This and other different parameters could be predetermined with values rating from, say, 1 to 9 having been selected from a menu structure prior to computer selection. The results could be referenced as data only or as illustrated with a graphic output. We must be able to resource all the Z axis curves in such a way that we can call up any combination in order to assess its viability.

Figure 9.8 illustrates how the various Z axis material removal curves could be stored relative to their deformation and time ratio. There are two types of curves: the infinite continuous (charged surfaces) and the finite fixed distance (degrading surfaces). The object is to call up the method to the conditions required. In order to be a satisfactory method, data-based preparation methods must, when displayed, represent a continuous line.

Figure 9.9 illustrates (a) an incorrect procedure, (b) an acceptable five-step procedure and (c) an acceptable two-step procedure. The next question is, 'If only one method is to be

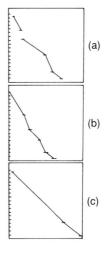

Figure 9.9 Graphic display of preparation procedures. (a) Incorrect, (b) correct five-step short time, (c) correct two-step long time

quoted as a national standard for each material (as should be the case) then which one would be quoted?' If 'integrity in the shortest time' is considered to be the criterion, a totally different curve to, say, the three given in Figure 9.9 may be created.

9.6.1 Comparing techniques

Comparing techniques becomes easy when comparing by integrity curves, which ensures correct interpretation of difficult methods. In order to do this the time/deformation scale must be identical. Figure 9.10 compares two

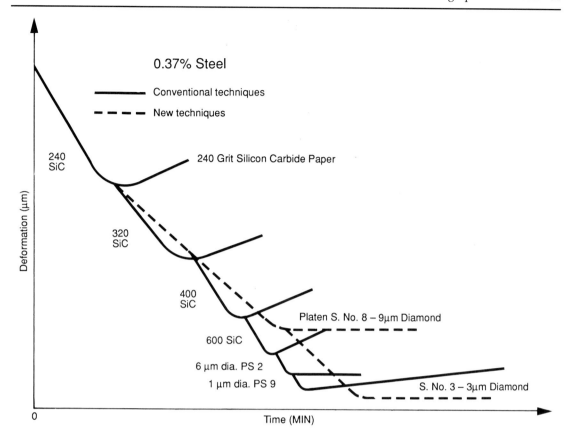

Figure 9.10 Comparing different recipes

alternative recipes for the preparation of 0.37 steel. Note the following (indisputable, non-subjective) points:

(i) New techniques took longer. This is interesting since one could have thought the opposite to be the case. However, remember that the Z axis relates to actual performance time and does not include 'down time' such as changing silicon carbide papers.
(ii) Conventional techniques are based on optimum conditions. The time would increase considerably if the degrading abrasive was to be used too long at any of the four silicon carbide stages.
(iii) The final preparation stage using the conventional technique is time dependent and degrades beyond the optimum.
(iv) The new technique gives improved integrity which could be further improved by using another step such as platen surface 1 with 1 μm diamond.

9.7 Platen Surface Threshold

Platen surface is the name given to the grinding, lapping or polishing surface that is to be the vehicle carrying the fixed or charged abrasive. The 'platen surface threshold' is the limit from which the specimen residual structural damage will not improve. With every platen surface there will be a 'platen surface threshold number'. This number will be a measure of the Z axis structural damage (in micrometres) residual in the specimen after using that specific platen surface. This number will change with different sized abrasives, different types of abrasives, fixed or charged abrasives and different materials. This method

Table 9.2 Material threshold chart for material XYZ where the deformation structural angle K=0.010 62

Surfaces	Threshold No.	Selection	300 mm diameter platen	Time (min)
SIC P240	82	*		
SIC P400	41	*	0.00813	1
Aluminium oxide P400	52			
Charged platen surface 10, 30 μm	46			
Charged platen surface 8, 9 μm	18	*	0.058	4
Charged platen surface 3, 3 μm	3	*	0.045	2
Charged platen surface 1, 1 μm	0.25	*	0.300	2.5

of designation is useful in the absence of a Z axis visual display and allows the systematic selection of platen surfaces to suit particular user requirements. Unlike the Z axis curve it does not relate time but this could be achieved via the deformation/structural angle constant (K).

9.7.1 How would it work?

(i) Define the specimen and appropriate threshold list relating to that specimen.
(ii) Establish the residual specimen damage. This should be available from documented information relating to the specimen condition, i.e. sectioned on a resin-bonded silicon carbide abrasive wheel at 2000 rev/min—92 μm.
(iii) From the threshold list select a suitable series of platen surfaces, ensuring that the first surface to be used has a lower value than the residual value (in the example of part (ii) <92 μm).
(iv) To relate the cycle time for each step multiply the deformation/structural angle K by the Z axis damage (in micrometres) to be removed and then by the area of the sample.

The threshold list given in Table 9.2 for material XYZ is obviously a grossly reduced list but it illustrates how a systematic selection can take place. The time between the first and second step would be $0.00813 \times (82-41) \times$ area of the total sample. Thus, using, say, 3×25 square millimetre samples,

T_1 $0.00813 \times 41 \times 3 = 1$ min
T_2 $0.058 \times (41-18) \times 3 = 4$ min
T_3 $0.045 \times (18 \times 3) \times 3 = 2$ min
T_3 $0.3 \times (3-0.25) \times 3 = 2.5$ min

The method shown above has the added advantage of being able to cope with samples of varying areas; it also makes it easy to translate to a new constant K for different materials. The latter is done by reference to the structural/deformation angle.

In effect what has been created is a series of Z axis curves. This then brings us full circle to the view that all Z axis curves should be stored, allowing an instant visual or statistical selection from a given parameter, such as 'the shortest time' or 'the least number of steps' or whatever is paramount.

9.8 Stumbling Blocks to Standards

(i) *Lack of any universal language.* This has been overcome by the use of time–deformation scaled Z axis curves. By this method any technique can be visually displayed and non-subjective comparisons can be made.
(ii) *No system for constructing procedures.* This is not true; the deformation structural angle allows transfer of techniques between different materials. The designation of platen and polishing surface numbers also introduces a systematic approach and the introduction of deformation/structural damage Z axis information at the initial sectioning stage indicates the remedial action required. A databank of Z axis information will indicate the starting point from which the procedure is constructed.
(iii) *No means of auditing the results to traceable standards.* This can be done via independent professional bodies or internal

company standards, as will be described in the next chapter.

(iv) *No means of technically challenging empirically derived methods*. This is very important if we are to remove the 'black art' or commercially biased information so readily available. By converting all methods into displayed Z axis recipes they can be compared, as has been shown, without recourse to dispute.

(v) *No means of creating the best system to suit individual needs*. Easy, just select or instruct the databank with the specific requirements.

Four of these five points towards a metallographic standard have been resolved in this chapter. Experience has shown how widely differing points of view can soon prove to be extremely similar when methods or recipes are visually Z axis displayed. This is a major step forward towards a metallographic standard.

10

Characterization: Auditing and Traceable Standards

10.1 Introduction

As long as the microstructure is required in the analysis of materials then the *true* microstructure, devoid of induced artefacts, is vitally important. Having established a technically sound procedure it seems a logical step to relate the results to some standard reference. Taking into account the declaration of intent within Europe for common working practices the impetus within industry for traceability is paramount. On a much broader scale there is a need for common working standards with all trading nations; you have multinational organizations working in, say, America, Europe and Japan. Once there is an established database for Z axis curves this question of world standards will be easy to define.

10.1.1 Empirical analysis

The current practice for establishing a standard preparation procedure is to assemble a working group to carry out what is called a 'round-robin' where individual members prepare identical materials then collectively agree which one method produces the best result. The writer considers this to be a total admission of empirical analysis and as long as these practices exist there will remain controversy. A major area of concern is that relating to the integrity of coated materials (plasma, etc.). There have been many working groups and round-robin exercises in this area, the latest (1990) being an important collection of European companies and institutions who concluded that one of two different techniques would satisfy nearly all different types of spray coatings. (If this were the case then the special chapter dedicated to this complex problem would be unnecessary.) The selection of this particular method was based on one (a) for very sensitive coatings or (b) for firmly adhering coatings. It would appear to ignore all other factors such as differing materials, thickness, coating technique, porosity, machine optimization, Z axis, etc. I only wish it were that simple.

It is current practice to observe the microstructure only on completion of the preparation. The trained eye can detect preparation-induced structural damage; this expertise is to be encouraged and developed. True structures, however, are not obvious and the observer must learn, by shape, form of

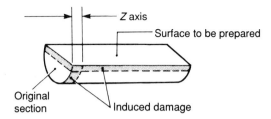

Figure 10.1 Structural damage

phase and grain, whether the preparation has faithfully revealed the sample integrity. Another obvious step is to observe the structure at intermediate preparation stages to identify the inadequate preparation area. Identifying bad preparation techniques is illustrated in Chapter 13. So far we have only discussed the subjective methods of analysis; as important as they are, they can only be of value when supported by good technical and statistical data.

10.2 Creating a Standard Preparation Procedure

(i) *Sectioning*. There are many different ways to take a sample from the parent material. It is the chosen sectioning method that *dictates* the selected specimen preparation technique along with the type or specification of the material. Therefore careful selection of a technique compatible with economics gives the least Z axis induced structural damage.

Assuming there is a databank then simply assign the appropriate Z axis. In the absence of such information then the sectioned specimen is further sectioned across the initial cut as shown in Figure 10.1. The shaded area shows the structural damage caused by the section taken initially transverse followed by the longitudinal cut. The requirement then is to prepare the longitudinal face to reveal the Z axis from the original cut. The optically observed deformation depth must be increased by the appropriate constant for that material. From this dimension (Z) it must be ensured that the total preparation procedure removes at least this amount of material.

(ii) *Mounting*. Encapsulate with minimum shrinkage and compatible abrasion characteristics and define the ratio of sample to resin.

(iii) *Planar grinding*. Specimen orientation when relevant must be noted along with the number of samples. Assuming a databank had been established it is possible, with the initial Z axis, to select the appropriate step for this and all other stages. To create individual procedures, select the smallest abrasive possible which will 'plane' the samples without increasing the Z axis dimension. Ideally it would be expected to reduce this figure.

The planar grinding stage is only necessary if the specimen, for example:

(a) has an excess of mounting resin in front of the sample,
(b) has been mounted on an angle,
(c) is not planar between samples, such as can occur in automatic preparation when the samples are *clamped* in the sample holder.

As with this and all other stages the residual damage after each grinding (or lapping) is related to (a) the type and shape of the abrasive, (b) whether it is bonded or charged, (c) the shock absorbing characteristics of the surface and (d) the lubricant used. The efficiency of the operation is dependent upon using the optimum machine parameters.

(iv) *Sample integrity stage*. Dependent upon the planar grinding stage, if used, all other stages must use a progressively *higher shock absorbing surface* accompanied by a progressively *smaller sized abrasive*. The questions are: (a) which surface, (b) which size of abrasive, (c) for how long?

The surface must have a lower sample integrity number and the choice must be made between fixed or charged abrasives. The object at this stage is to progressively reduce the induced structural damage. The selected surface must be inspected after a short period of operation, say 1 minute. The sample is then optically investigated; if no improvement is evident another lower surface must be selected or the abrasive size reduced. If the induced damage has increased then the sample must be returned to the planar grinding stage. The choice of surface is not confined to integrity numbered surfaces; silicon carbide papers, for example, could be used. To satisfy the shock absorbing requirement it will be necessary to increase the 'backing' when using silicon carbide papers. Unfortunately there is a tendency for the backing to reduce as the abrasive becomes smaller, which is the reverse of our requirements (it can be advantageous with finer grades to always leave a used paper underneath the silicon carbide paper being used). When preparing brittle subjects, the damage from using a large sized abrasive on a high numbered integrity surface can be so severe as to warrant the use of a rolling abrasive, i.e. lapping surfaces.

It is not necessary to progress the sample using lapping at subsequent stages since the higher shock absorbing, smaller sized abrasive will be more 'time efficient'.

Having selected a suitable surface for the initial sample integrity stage the sample must be progressed until the structure will not improve. This can be done by incremental steps, noting the Z axis at each step. When a condition occurs (as it will) when the subject either remains as the previous step or actually degrades then the cycle must be stopped. The cumulative Z axis figure to the best position is the required dimension. At this best position the image must be photographed, a reflectivity measurement made, image analysis taken where necessary and notes taken relating to the structural condition. These notes could relate to the depth of the damage (achieved by using a high NA objective and taking a depth reading using the microscope fine focusing, which is usually graduated in micrometres) and notes could also relate to phase features or just simply to the scratch pattern.

This technique must now be continued until the integrity of the sample has been achieved.

The question to be addressed is, 'How do we know when sample integrity is achieved?' With brittle subjects it can be related to the fracture pattern of the surface; friable elements to phase retention; pores to size, distribution, orientation and morphology; longitudinal fibres to retention at the sharp point; transverse fibres to reproducible shape; particulate to homogeneity, etc. When dealing with ductile materials, from steels to lead, the top surface visual information is often insufficient. Although structural damage at the initial stages of surface preparation extends to, say, 100 μm, which is easy to detect through the cross-sectioning technique, at the final stage this deformation is in the 1 to 10 μm region. There is also another factor when defining integrity. So far the discussion has been about the visual information and defining the achievement of integrity by comparison, but what has not been covered is the presence of residual stresses. The magnitude of residual stresses present in the sample is usually, but not necessarily, indicative of microstructural transformation. One of the most accurate methods of measuring residual stresses in surface preparation is by the use of X-ray diffractometry. A crystalline solid will defract X-rays of given wavelength at specific angles (in accordance with Bragg's law). If a stressed crystal is present there is a corresponding change in angle; the strain within the crystal can be calculated and then related to stress. This system, although limited to a depth penetration of approximately 10 μm, is ideal for this application.

(v) *Polishing*. Remember at this stage that Utopia is to improve the sample aesthetics without loss of integrity; these two features are mutually incompatible with some, in particular, soft materials. Hard materials often require little if any polishing. The polishing surface number (PS number) must be as near zero as possible; as the necessity for higher PS numbers increase so the time must be decreased. It is not necessary at this stage to designate a Z axis figure; time will suffice. Statistical and visual information is as previously, the sample being etched or suitably stained prior to photography.

(vi) *Documentation*. A standard format should be adopted (similar to Figure 10.2) where the statistical information and machine parameters are noted.

10.2.1 Database information

All statistical information should be databased. The previous information derived when creating a standard procedure can be input in its present form; it just requires graphic positioning. The height position is the depth of structural damage remaining after completing that preparation step. This dimension is the summation of subsequent Z axis dimensions required to achieve sample integrity (see Figure 10.3).

The deformation/structural angle alpha is indicative of the rate of progress and is time

AUDITABLE PREPARATION PROCEDURE

CLASSIFICATIONS:		MATERIAL:	
SECTIONING TECHNIQUE:	EQUIPMENT TYPE BLADE	CONCENTRATION	
MOUNTING TECHNIQUE:	METHOD	SAMPLE/MOUNT RATIO	

	SURFACE & TYPE	ABRASIVE SIZE/TYPE LUBRICANT	FORCE kPa	HEAD ROTATION	Z AXIS μm	REFLECTIVE FACTOR	MICROGRAPH YES/NO	IMAGE ANALYSIS YES/NO	REMARKS
PLANAR GRINDING STAGE									

	SURFACE & TYPE	ABRASIVE SIZE/TYPE LUBRICANT	FORCE kPa	HEAD ROTATION	Z AXIS μm	REFLECTIVE FACTOR	MICROGRAPH YES/NO	IMAGE ANALYSIS YES/NO	REMARKS
SAMPLE INTEGRITY STAGE									

	SURFACE & TYPE	ABRASIVE SIZE/TYPE LUBRICANT	FORCE kPa	HEAD ROTATION	TIME (s)	REFLECTIVE FACTOR	MICROGRAPH YES/NO	IMAGE ANALYSIS YES/NO	REMARKS
POLISHING STAGE									

Figure 10.2 Universal language format

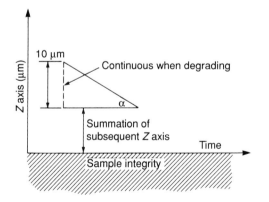

Figure 10.3 Preparation step curve

based. This angle is also specific to particular materials and is therefore important in transferring similar abrasive/surface conditions to different materials. In order to achieve a comparable time-based curve the sample area must be 25 mm^2. Note how the triangle would be closed if the abrasive being used was degrading.

This database will be supplemented as and when different or new methods are created for a particular material. This allows the retrieval of this information and eliminates the need for continuous creation of methods for identical materials by many different people (to invent the wheel once is enough). If in fact it proves possible to prepare a sample in one of many ways and still achieve integrity, the best method for you is the one that best meets your needs.

10.3 The Standard Procedure

From what has been said above, is a standard method needed? Of course there is a need for a standard reference material and documentary evidence; it therefore follows that the procedure should also conform to some such standard. There is a need, when retrieving information, to input the requirement for a 'standard procedure' and it is necessary to input 'integrity' since a standard must give the most faithful result (for production and quality control, absolute integrity is not always required and a lower threshold can be tolerated). The least number of steps could be considered as another input but this could result in extremely long preparation times. It therefore follows that the 'shortest time' must be the required input (hence the reason for inputting the deformation/structural angle).

10.3.1 Auditing

Having achieved a suitable and faithful procedure, one which will reveal the true structure, it must not be assumed that by simply copying the procedure, continuous sample integrity will be guaranteed. With all quality control systems there has to be some auditing from time to time. The preparation procedure must, at intervals, be checked to see that it still conforms to the visual and statistical information relating to that procedure. It is not necessary for industry to use 'standard procedures' unless they are auditing to a 'traceable standard'.

10.3.2 Auditing personnel

To carry out any auditing it is necessary to have trained personnel. Currently there is no recognized body laying down procedures relating to 'elements of competence'. This is shortly to be rectified as the Engineering Training Authority have proposed a two-tier technician level for Microstructural Specimen Preparation to Traceable Standards. The first grade (level II) will be targeted towards the operator; the second grade (level IV) will be for the supervisor. When these courses carry a National Vocational Qualification they will go a long way towards raising the standard of awareness and it is hoped encourage a wide international conformance.

10.3.3 Three steps to traceability

(i) *The procedure.* It is now possible to reference all procedural steps, many without recourse to subjective empirical results. These procedures can now be logged and retrieved as reference becomes necessary. Having a traceable procedure does not necessarily imply that other procedures will not give a similar result, but it does mean that only one procedure is traceable to the reference: *integrity in the shortest time.*

(ii) *The reference material.* This is the specific material from which a traceable procedure is derived. Stocks of these identical standards are retained for subsequent subscribed reference. As with all traceable standards they are only used as verification when auditing the quality control procedures. One important aspect of a traceable reference is in relating reproducible results not conforming to the standard. This could occur when, for commercial reasons, the adopted procedure was not a traceable procedure. One may not, for example, wish to change existing quality controlled reproducible procedures but there is a need to relate the results to that which would be achieved via the traceable procedure using the appropriate reference material.

Established company recipe for example could always give a 5% increase in porosity level from the standard.

(iii) *The custodian.* This is the retainer of specific portfolios relating to the traceable procedures and references. This vehicle is for two-way communications in order to offer the service and ensure its constant update and could be a central body within an organization. It is anticipated that there will be many different industrial standards; these standards, however, can only be traceable to that specific organization and fall short of the major objective, which is total traceability. The European Society for Microstructural Traceability was set up in 1988, its 'logo' being TRAC 92 and its targets being traceability within Europe by 1992. As with many good intentions, time eventually beats you and I have to report, with much regret, its time aspirations were overoptimistic. This does not, however, mean that its objectives are not realistic; *without traceability there is a lack of credibility.*

11

Traditional Methods Only

11.1 Consideration

There is always a tendency for outdating old and traditional ideas in preference for the so-called new technology. Metallography is no exception to this and therefore this section has been introduced to reinforce the benefits of traditional methods and to update these methods with our current findings. Traditional practice is based on the deformation produced by each preceding step which must be completely removed by the succeeding one. Although this is good advice it is hardly an optimized condition and has resulted in the blinkered approach of coarse grind/fine grind/rough polish and final polish.

Take coarse grinding, for example; the size of the abrasive varies from, say, 80 to 120 grit but the important information is the abrasive size which is 200 to 120 μm. Abrasives of this large size will inflict severe structural damage to the specimen and should only be used when high stock removal is required. The level of induced damage could be more severe than facing the specimen on a standard engineering lathe. If specimens have been carefully sectioned then the induced damage will be quite low, much less than that resulting from coarse grinding. It is the depth of this damage that influences the chosen procedure along with the material type. The first requirement of any procedure is (a) to establish a plane surface compatible with the stock removal rate and progress with the least number of steps or time to achieve sample integrity or (b) if the sample is plane, as many are, to select an abrasive size such that the residual damage on completion will be less than when commencing and that the least number of steps or time be employed in progressing the sample to integrity.

11.1.1 Fine grinding practices

Fine grinding utilizes a series of abrasive papers from 220 to P4000 grit, the selection depending on the sample condition and type and not, as is often quoted, on the available equipment. Nor does it mean using every grade in the range. A recent published recommendation was to use grit sizes up to P4000 for composites, copper and its alloys, aluminium and its alloys, graphite, plastics and soft coatings. This sort of advice must be confusing if for no better reason than it conflicts entirely (with the exception of plastics) with the recommendations developed within this book.

Since every material responds differently to mechanical working it would seem logical to expect the fine grinding operations to reflect this fact. In order to express a broad opinion on fine grinding techniques another authority was investigated. The published information for ferrous materials, copper-based alloys, aluminium/magnesium and their alloys, titanium, zirconium, hafnium, niobium, molybdenum, tungsten, vanadium and tantalum was P240/400/600/1200 silicon carbide paper. (The same sized abrasive, i.e. 6 μm diamond, on the same type of cloth was also recommended for the very rough polishing step.)

If one was to reinterpret this well-intended advice, it would be 'forget the different mechanical responses with this vastly differing range of materials; just treat them all the same'. This, of course, would be totally unacceptable and would contribute nothing to furthering our understanding of material removal for microstructural analysis.

11.1.2 Fine grinding recommendations

In view of the published material the following guide-lines are recommended in practising traditional metallography:

(i) Establish the depth of induced damage within the plane faced sample as per (a) and (b) above.
(ii) Select the appropriate grit abrasive based on the micrometre size as shown in Chapter 6.
(iii) With general ferrous materials stop at P600 or P800 grit abrasive.
(iv) With non-ferrous materials stop at P1200.
(v) With subjects such as printed circuit boards finish at P600 or P800.
(vi) With very soft materials stop much earlier.
(vii) With lamellar boards or the non-ductile plastic go to P4000.
(viii) Having established the starting and finishing abrasive use the least number of steps: viz. starting at 220 or 320 and finishing at P1000 include P600 only; starting at 240 and finishing at P600 include P400 if necessary. Remember that some materials could be satisfactorily fine ground using the last grade only if the residual damage was, say 30 μm or less. P1200 grit size is 15 μm; P4000 grit size is 6 μm.

11.1.3 Rough polishing

This is a description the author wishes to *abolish* since it is not a polishing operation. It is a grinding function and as such the surface or polishing cloth should be of zero nap; the abrasive particle size is usually 3 to 9 μm (diamond). The following is a guide to abrasive size selection:

(i) General ferrous   6 to 9 μm followed by 1 to 3 μm
(ii) Brass   1 to 3 μm
(iii) Al alloy   6 then 1 μm
(iv) PCBs   3 or 6 and 1μm
(v) Hard materials   9 then 3 μm
(vi) Very hard materials 15 then 6 then 1 μm
(vii) Very soft materials 6 μm

It is possible to *section* material leaving a residual damage considerably lower than that which would result from using abrasives up to P1200. On such occasions P4000 silicon carbide paper or the appropriate diamond size on a zero nap surface would be used.

11.1.4 Polishing

Although this chapter relates to traditional methods the polishing stage should, as the word implies, be reserved for just polishing the prepared surface. The integrity of the preparation will, in the majority of cases, be achieved before commencing the polishing step.

There are exceptions when it is necessary to etch polish at the final stage, in which case the integrity will not have been achieved prior to polishing. With very hard materials (ceramics, tungsten carbide, etc.), it is not necessary to incorporate a polishing step. With materials where polishing is required, only one step is necessary using a medium to high nap cloth, depending upon the material and the surface scratch expectations. Softer materials require softer cloths to reduce scratches. High relief can be desirable to give improved subject resolution, in which case use a high nap cloth. The choice of abrasive will generally be sized 1 μm or less and could be diamond for the harder materials with slurries of alumina, colloidal silica, magnesium oxide, cerium oxide, iron oxide and even mixed oxides.

11.2 Preparation Examples

Example 11.1

Family classification BB1

The material: 0.37% steel

Steel is a combination of iron and carbon, the amount of carbon influencing the hardness, strength and ductility of the material. Iron, being quite ductile, loses these qualities as the carbon increases. Although an increase in carbon results in an increase in hardness and strength, it is the condition of this combined carbon that influences characteristics along with the ability to further harden by quenching. Carbon up to the eutectoid (0.89) manifests itself as pearlite and beyond the eutectoid as pearlite and cementite. This is often referred to as hypoeutectoid and hypereutectoid. When heated, the carbon combines to form austenite. If quenched from high temperatures the partial transformation creates the very hard bainite, or martensite when quenched at a slightly lower temperature.

As a general guide the following are typical uses.

Carbon (%)	Qualities	Application	Carbon (%)	Qualities	Application
0.05	Ductility	Chains	0.60	Wear and strength	Tracks and rollers
0.10	Pressing	Cans	0.80	Hardness without ductility	Screwdrivers
0.25	Stiffness	Sheets and plates	0.90	Hardness and ductility	Saws
0.40	Strength	Shafts	1.1	Hardness	Chisels

The technique

This is a good example of traditional metallography brought up-to-date. Notice how the silicon carbide stages are restricted to two, from which the sample is progressed to integrity on platen surface 1. When the threshold level can be reduced, one surface 1 stage, using 3 μm diamond, will suffice.

The method

AUDITABLE PREPARATION PROCEDURE No. 1

CLASSIFICATIONS: BB1							MATERIAL: 0.37% STEEL			
SECTIONING TECHNIQUE:		EQUIPMENT TYPE		BLADE			CONCENTRATION			
MOUNTING TECHNIQUE:		METHOD					SAMPLE/MOUNT RATIO			

	SURFACE and TYPE	ABRASIVE SIZE/TYPE LUBRICANT	FORCE (kPa)	HEAD ROTATION	Z AXIS (μm)	REFLECTIVE FACTOR	MICROGRAPH YES/NO	IMAGE ANALYSIS YES/NO	REMARKS
PLANAR GRINDING STAGE	PAPER	240 SiC WATER	35	COMP	PLANE				

	SURFACE and TYPE	ABRASIVE SIZE/TYPE LUBRICANT	FORCE (kPa)	HEAD ROTATION	Z AXIS (μm)	REFLECTIVE FACTOR	MICROGRAPH YES/NO	IMAGE ANALYSIS YES/NO	REMARKS
SAMPLE INTEGRITY STAGE	PAPER	P600 SiC WATER	35	COMP					
	S.1	6 μm WATER	35	CONTRA					
	S.1	1 μm WATER	35	CONTRA					

	SURFACE and TYPE	ABRASIVE SIZE/TYPE LUBRICANT	FORCE (kPa)	HEAD ROTATION	TIME (s)	REFLECTIVE FACTOR	MICROGRAPH YES/NO	IMAGE ANALYSIS YES/NO	REMARKS
POLISHING STAGE	PS 7	0.05 μm ALUMINA	16	COMP	120		YES		11.1 (Plate 3)

The micrograph Figure 11.1 (see Plate 3)

Example 11.2

Family classification BB4

The material: Brass

These copper/zinc alloys have been a familiar part of everyday life for some 300 years. Zinc up to 36% can be in solid solution with the copper. In this condition the brass is quite ductile. Students often use the low zinc brasses to study twin bands caused by dislocation when cold working. Above 36%, the beta phase is present, i.e. alpha–beta brass. This increase in zinc is associated with an increase in hardness and a reduction in ductility. To *work* the alpha–beta brass it is wise to elevate the temperature, which increases the plasticity. This type of brass can also be strengthened by the addition of other alloying elements, such as aluminium.

The technique

This material is quite easy to prepare using the classical approach of silicon carbide papers, followed by what has traditionally been considered as two polishing cloths. It has been known for only one polishing cloth to be used, with 'Brasso' as the abrasive. This is very good from a surface finish point of view, but must raise doubts about total integrity.

It is wise to use a vibratory polisher when a scratch-free surface is desired; this avoids any smearing. For automatic sample preparation, the last two silicon carbide stages could be replaced with 15 μm diamond on a platen surface 2 or 3.

The method

AUDITABLE PREPARATION PROCEDURE   No. 2

CLASSIFICATIONS: BB4			MATERIAL: BRASS	
SECTIONING TECHNIQUE:	EQUIPMENT TYPE	BLADE	CONCENTRATION	
MOUNTING TECHNIQUE:	METHOD		SAMPLE/MOUNT RATIO	

	SURFACE and TYPE	ABRASIVE SIZE/TYPE LUBRICANT	FORCE (kPa)	HEAD ROTATION	Z AXIS (μm)	REFLECTIVE FACTOR	MICROGRAPH YES/NO	IMAGE ANALYSIS YES/NO	REMARKS
PLANAR GRINDING STAGE	PAPER	P320 SiC WATER	35	COMP	PLANE				

	SURFACE and TYPE	ABRASIVE SIZE/TYPE LUBRICANT	FORCE (kPa)	HEAD ROTATION	Z AXIS (μm)	REFLECTIVE FACTOR	MICROGRAPH YES/NO	IMAGE ANALYSIS YES/NO	REMARKS
SAMPLE INTEGRITY STAGE	PAPER	P600 SiC WATER	35	COMP					
	PAPER	P1200 SiC WATER	35	COMP					
	S.1	3 μm OIL	35	CONTRA					

	SURFACE and TYPE	ABRASIVE SIZE/TYPE LUBRICANT	FORCE (kPa)	HEAD ROTATION	TIME (s)	REFLECTIVE FACTOR	MICROGRAPH YES/NO	IMAGE ANALYSIS YES/NO	REMARKS
POLISHING STAGE	PS 7	0.05 μm SILICA OR ALUMINA	16	COMP	120		YES		11.2 (Plate 3)

The micrograph Figure 11.2 (see Plate 3). The visual impact of this micrograph can be improved if the reader rotates it through 180°

Example 11.3

Family classification BB17

The material: printed circuit board

The PCB and the microchip have revolutionized the electronics industry beyond recognition, within two decades. More and more information is being stored and transmitted in less and less space.

The technique

Because the ratio of soft plated metal to the substrate fibre is small, the use of sharp silicon carbide papers is most suitable. These papers can be as fine as P1200 and even finer if the plated material is minimal. Avoid using a high contraction resin and ensure the viscosity allows through-hole saturation. A well-prepared dimensionally correct layer can be destroyed by overpolishing, using a high nap cloth or exerting high polishing pressures. If many samples are to be prepared there are fixtures available that allow multiple alignment and built-in stops to aid centre line grinding.

The method

AUDITABLE PREPARATION PROCEDURE No. 3

CLASSIFICATIONS: BB17		MATERIAL: PCB	
SECTIONING TECHNIQUE:	EQUIPMENT TYPE BLADE	CONCENTRATION	
MOUNTING TECHNIQUE:	METHOD	SAMPLE/MOUNT RATIO	

	SURFACE and TYPE	ABRASIVE SIZE/TYPE LUBRICANT	FORCE (kPa)	HEAD ROTATION	Z AXIS (µm)	REFLECTIVE FACTOR	MICROGRAPH YES/NO	IMAGE ANALYSIS YES/NO	REMARKS
PLANAR GRINDING STAGE	PAPER	P320 SiC WATER	35	COMP	PLANE				

	SURFACE and TYPE	ABRASIVE SIZE/TYPE LUBRICANT	FORCE (kPa)	HEAD ROTATION	Z AXIS (µm)	REFLECTIVE FACTOR	MICROGRAPH YES/NO	IMAGE ANALYSIS YES/NO	REMARKS
SAMPLE INTEGRITY STAGE	PAPER	P600 SiC WATER	35	COMP					
	S.1	6 µm OIL	35	CONTRA					
	S.1	1 µm OIL	35	CONTRA					

	SURFACE and TYPE	ABRASIVE SIZE/TYPE LUBRICANT	FORCE (kPa)	HEAD ROTATION	TIME (s)	REFLECTIVE FACTOR	MICROGRAPH YES/NO	IMAGE ANALYSIS YES/NO	REMARKS
POLISHING STAGE	PS 7	0.06 µm SILICA	16	COMP	100		YES		11.3 (Plate 3)

The micrograph Figure 11.3 (see Plate 3)

Example 11.4

Family classification BB3

The material: aluminium alloy

Aluminium is next to steel in the world usage of material. Like steel it is also commonly recycled from scrap. Unlike steel it does not corrode (beyond the surface aluminium oxide) and it is very light. Pure aluminium is used for reflecting mirrors (aluminized), replacing the traditional silvered mirror. Thin foil is another use of the ductile soft pure aluminium. Silicon is a major alloying element and this improves the flow when casting, in particular, intricate shapes and the new squeeze casting technology. The addition of copper to the silicon proves extremely advantageous in improving machining characteristics. The phenomenon of age hardening can be manifest by alloying aluminium with copper, magnesium, nickel and/or iron. Precipitation that takes place at room temperature after first quenching can be induced by preheating to a much lower temperature for a few hours. The technically interesting point is the relationship between slip plane dislocations and the strength and hardness of materials. When intermetallic compounds are present, resistance to slip plane dislocations are high, the material being hard and strong. When all elements are in solid solution, there is less resistance to slip plane dislocation, the material being softer and more ductile.

The technique

The quoted method is suitable for both pure and alloyed conditions. The harder aluminium alloys could well be prepared using silicon carbide paper prior to the last sample integrity stage.

The method

AUDITABLE PREPARATION PROCEDURE   No. 4

CLASSIFICATIONS: BB3						MATERIAL: ALUMINIUM ALLOY			
SECTIONING TECHNIQUE:	EQUIPMENT TYPE		BLADE			CONCENTRATION			
MOUNTING TECHNIQUE:	METHOD					SAMPLE/MOUNT RATIO			

*Use two grades of silicon carbide paper with the harder aluminium alloys.

	SURFACE and TYPE	ABRASIVE SIZE/TYPE LUBRICANT	FORCE (kPa)	HEAD ROTATION	Z AXIS (µm)	REFLECTIVE FACTOR	MICROGRAPH YES/NO	IMAGE ANALYSIS YES/NO	REMARKS
PLANAR GRINDING STAGE	PAPER	P320 SiC WATER	35	COMP	PLANE				

	SURFACE and TYPE	ABRASIVE SIZE/TYPE LUBRICANT	FORCE (kPa)	HEAD ROTATION	Z AXIS (µm)	REFLECTIVE FACTOR	MICROGRAPH YES/NO	IMAGE ANALYSIS YES/NO	REMARKS
SAMPLE INTEGRITY STAGE	S.2P*	30 µm OIL	35	CONTRA					
	S.1	3 µm OIL 35	35	COMP					

	SURFACE and TYPE	ABRASIVE SIZE/TYPE LUBRICANT	FORCE (kPa)	HEAD ROTATION	TIME (s)	REFLECTIVE FACTOR	MICROGRAPH YES/NO	IMAGE ANALYSIS YES/NO	REMARKS
POLISHING STAGE	PS 7	0.05 µm ALUMINA OR SILICA	16	COMP	100		YES		11.4 (Plate 3)

The micrograph Figure 11.4 (see Plate 3)

Example 11.5

Family classification BB8

The material: electronic component

These components are so diverse in the different materials one can expect to find. The automotive spark plug can be taken as an example, where ferrous, non-ferrous and ceramic materials are involved, or the component illustrated with the ceramic and silicon joint surrounded by a lead fillet with gold connections. The object of microsectioning electronic components is often to reveal the integrity of the component hidden within the assembly. This being a destructive test, it is often used in failure analysis.

The technique

With all delicate components or when investigating hidden detail, the sectioning technique is often the most critical step. The use of incorrect wheel, load, speed and lubricant can destroy the information beyond recovery. It is particularly wise to grind with the smallest sized abrasive possible after sectioning. When the component includes a ceramic or silicon, as in the example quoted, care must be taken at the planar grinding stage not to induce severe cracking.

Note: planar grinding is only necessary if the specimen is to be secured into a multisample holder.

The method

AUDITABLE PREPARATION PROCEDURE — No. 5

CLASSIFICATIONS: BB8 ELECTRONIC COMPONENT			MATERIAL: SILICON/GOLD/LEAD/TIN	
SECTIONING TECHNIQUE:	EQUIPMENT TYPE	BLADE	CONCENTRATION	
MOUNTING TECHNIQUE:	METHOD		SAMPLE/MOUNT RATIO	

PLANAR GRINDING STAGE	SURFACE and TYPE	ABRASIVE SIZE/TYPE LUBRICANT	FORCE (kPa)	HEAD ROTATION	Z AXIS (µm)	REFLECTIVE FACTOR	MICROGRAPH YES/NO	IMAGE ANALYSIS YES/NO	REMARKS
	PAPER	P400 SiC WATER	35	COMP	PLANE				

SAMPLE INTEGRITY STAGE	SURFACE and TYPE	ABRASIVE SIZE/TYPE LUBRICANT	FORCE (kPa)	HEAD ROTATION	Z AXIS (µm)	REFLECTIVE FACTOR	MICROGRAPH YES/NO	IMAGE ANALYSIS YES/NO	REMARKS
	S.2P	15 µm OIL	35	CONTRA					
	S.1	3 µm OIL + 0.06 µm SILICA	35	COMP					

POLISHING STAGE	SURFACE and TYPE	ABRASIVE SIZE/TYPE LUBRICANT	FORCE (kPa)	HEAD ROTATION	TIME (s)	REFLECTIVE FACTOR	MICROGRAPH YES/NO	IMAGE ANALYSIS YES/NO	REMARKS
	PS 7	0.06 µm SILICA	16	COMP	120		YES		11.5 (Plate 4)

The micrograph Figure 11.5 (see Plate 4)

Example 11.6

Family classification BB2

The material: nodular cast iron

This iron, often called ductile iron, is used where shear stress or shock loading is likely. Grey cast iron (flakes) is excellent in compression but not in tension. Under these circumstances SG iron (nodular) would be better.

The technique

There are more published methods for the preparation of cast iron than, perhaps, any other material. The method given is based on the use of traditional materials. This five-step procedure to integrity will be reduced to three by the foundry metallurgist when total retention of free graphite is unnecessary. The point to note about this procedure is the relatively large sized abrasive (3 μm) used at the final sample integrity step. The reason for this is twofold:

(i) The ferrite is relatively soft and is therefore susceptible to top surface smearing and impressed abrasives.
(ii) When preparing a loose friable material (graphite) surface rubbing must be avoided if loose particles are to be retained. The tendency for surface rubbing increases as the abrasive size reduces.

The method

AUDITABLE PREPARATION PROCEDURE   No. 6

CLASSIFICATIONS: BB2		MATERIAL: CAST IRON (NODULAR)	
SECTIONING TECHNIQUE:	EQUIPMENT TYPE BLADE	CONCENTRATION	
MOUNTING TECHNIQUE:	METHOD	SAMPLE/MOUNT RATIO	

*Three-step method for lower threshold.

	SURFACE and TYPE	ABRASIVE SIZE/TYPE LUBRICANT	FORCE (kPa)	HEAD ROTATION	Z AXIS (μm)	REFLECTIVE FACTOR	MICROGRAPH YES/NO	IMAGE ANALYSIS YES/NO	REMARKS
PLANAR GRINDING STAGE	PAPER*	240 SiC WATER	35	COMP	PLANE				

	SURFACE and TYPE	ABRASIVE SIZE/TYPE LUBRICANT	FORCE (kPa)	HEAD ROTATION	Z AXIS (μm)	REFLECTIVE FACTOR	MICROGRAPH YES/NO	IMAGE ANALYSIS YES/NO	REMARKS
SAMPLE INTEGRITY STAGE	PAPER*	P400 SiC WATER	35	COMP					
	PAPER	P800 SiC WATER	35	COMP					
	S.1	6 μm OIL	35	CONTRA					
	S.1*	3 μm OIL	35	COMP					

	SURFACE and TYPE	ABRASIVE SIZE/TYPE LUBRICANT	FORCE (kPa)	HEAD ROTATION	TIME (s)	REFLECTIVE FACTOR	MICROGRAPH YES/NO	IMAGE ANALYSIS YES/NO	REMARKS
POLISHING STAGE	PS 7	ALUMINA 0.05 μm	16	COMP	100		YES		11.6 (Plate 4)

The micrograph Figure 11.6 (see Plate 4)

12

Preparation of Spray Coatings

12.1 Background

Surface coatings is an area where microsectioning is an essential part of the analysis (in particular plasma sprayed coatings), yet if five people were to prepare identical coatings in five different locations you could expect five different results (at best four incorrect results). The only answer previously submitted as a solution to this problem was to aim for the most sophisticated degree of machine automation to achieve the much desired reproducibility. This, however, has proved to be unsatisfactory since consistently reproducing incorrect structures is not at all satisfactory. Another disadvantage of the blinkered automation approach has led to operators reproducing the results they want to see. Subcontractors are often given 'go/no go' photomicrographs of what will be tolerated; to this end the operator often prepares the sample in a variety of ways until it satisfies the 'go' requirements.

This is a classic case where technology is required to replace the so-called round-robin approach to satisfactory preparation. A recent seventeen company working group investigating the preparation of sprayed alumina coatings published (1990) their subjective findings:

(i) When dealing with 'very sensitive coatings' use a five-step method taking in total 11.5 minutes (excluding planar grinding). It then goes on to say, with 'firmly adhering coatings' use an eight-step method taking 28 minutes. Surely, the very sensitive coatings require the greater care and the requirements of a firmly adhering coating would be less than that of the former.

(ii) Both of the published methods made use of silicon carbide paper in the sample integrity stage. This is not wise when cutting aluminium oxide. No reference was made to the fact that silicon carbide is an extremely degrading abrasive, particularly when cutting alumina.

(iii) The Z axis was not referred to at any stage; however, time was. Time has proved to be the major downfall of any preparation route.

(iv) If one accepts the point that smaller sized abrasives would induce less damage than larger sized abrasives, then the very sensitive coatings would utilize the smaller abrasives. This was not the finding of this working group.

These four points are but a few of many issues that could be taken with the findings of this working group. The reason for this illustration is to establish the need for publishing recipes that are technically sound and have credibility. It is possible to prepare a satisfactory micro in one of many ways. It is this fact alone that has obscured the progress towards technically sound procedures.

One could have a situation where one manufacturer could withhold the issue of specifications because of the lack of traceable and auditable standards. Others could take the view that their specification is acceptable. This clearly is a case for having a preparation system that is non-commercial and traceable to sound technology as opposed to the empirically derived methods currently used. It also necessitates the use of references not least of these being:

(i) Sample material
(ii) Photomicrograph
(iii) Image analysis
(iv) Reflectivity measurement
(v) Z axis

The European Society for Microstructural Traceability was formed to investigate and implement this need, one requirement being that subcontractors would have to conform to such standards and tests. It is understood that a major coating manufacturer has voiced his objections to such systems on the grounds that they have standardized world wide an internally empirically derived procedure.

Although internal procedures within individual companies are to be encouraged, they are of little value without the supporting statistical information and above all a *standard* material to make the procedure traceable. It is important to stress at this stage that the procedure is *not* what is being standardized; it is the achievement of a faithful reproduction, i.e. sample integrity that is the standard. The procedure is simply a vehicle in achieving integrity, the sample and visual and statistical information being the standard.

Since many committees throughout the world are currently investigating the techniques related to the preparation of coatings, examples of how the 'new concept' can overcome many of the problems and a review of different techniques is included in Section 12.3 on 'Aids to metallography'.

12.1.1 Properties

Coatings are often complex composites and as such the chosen preparation procedure should be based on maximum retention of these composites even, if necessary, at the expense of the substrate or surface aesthetics. Since the coating and substrate will be dissimilar care must be taken with the mounting medium to ensure even abrasion between the coating and mount, and not the substrate and mount. To ensure maximum adhesion between the coating and substrate a grit blast and/or bond coat is used. The grit blast (often alumina) effectively 'roughs up' the surface to improve adhesion, leaving on the surface small particles of the blast material. Care must be exercised in preparation otherwise the grit will break away, giving the impression of poor adhesion (delamination). The bond coat, on the other hand (often nickel based), is usually easier to prepare with little if any porosity and with well-bonded melted particles. There could, however, be secondary phases at the interface.

Sprayed coatings exhibit varying degrees of porosity—some intentional, others unavoidable; some interconnected, others not. Since porosity is one of the controlling factors affecting the performance of the coating then it is important that all porosity visible in the finished microsection be true and not preparation induced. As an aid to reducing induced porosity the coatings are often vacuum impregnated with ultra thin (low viscosity) resin prior to preparation. Unfortunately the resin does not always penetrate throughout the coating, resulting in pores that are impregnated and those that are not. Many results based on ingression of resin have resulted in erroneous analysis. This does not mean that the prepared sample is without integrity just because some pores are without resin; conversely it does not imply that the prepared sample exhibits integrity just because resin is evident in every pore.

To summarize, coatings are often brittle, friable, porous, poorly bonded composites.

12.1.2 Coating specifications

One would hope that future specifications for the control of coated layers would involve the use of traceable procedures and traceable

Preparation of Spray Coatings

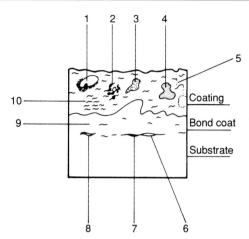

Figure 12.1 Metallurgical requirements for spray coatings

referenced standards in order to ensure integrity of the preparation. The microstructural analyst is generally concerned with the structural interpretation of the visible grains but on this occasion the following integrity checks are usually more critical:

(i) Coatings shall be uniform and free from delamination and cracks.
(ii) Porosity of the bond coat should be controlled to a given limit.
(iii) Porosity of the top coat could have a limit in minimum as well as maximum acceptance.
(iv) Voids in the bond coat substrate interface and the bond layer top-coat interface should satisfy given constraints.
(v) There should be a limit on tolerated unmelted spheroidal particles.
(vi) There should be a limit on the oxides in the bond coat.
(vii) Penetration of the aluminium oxide grit blast should be evident

12.2 How to Identify a Faithful Reproduction

The materialographer's skill is tested to the limit when identifying sample integrity from preparation-induced structural damage. Before this judgement can take place it is necessary to know how integrity and induced damage manifest themselves in the prepared sample. Since they can both look very similar clues are needed to give a guide towards the correct judgement. Figure 12.1 shows how a sample of integrity could look to the untrained eye as specimen-induced artefacts.

12.2.1 Metallurgical requirements

(1) *Globular particles* associated with oxides and porosity are often rejectable if on the outside of the coating. To retain particles surrounded by oxides or porosity is extremely difficult and could result in pull-out. When pull-out occurs it is difficult to differentiate from massive porosity. Observations in first-order grey DIC should indicate which of the two conditions is most likely.
(2) *Porosity* is part of the controlled conditions when spraying; therefore, limits are necessary. Porosity as in the example illustrated is of greater concern when associated with globular structures. Higher levels of porosity can be tolerated in areas such as corners. The materialographer has to be an expert on porosity and porosity morphology if making an incorrect analysis is to be avoided. This skill can best be developed by constant use of the first-order grey DIC technique.
(3) *Unreacted particles* are always difficult to retain and sometimes not easily distinguishable from holes. Crossed polars for alumina or high magnification for metals can often be the clue to complex identification.
(4) *Massive porosity* is sometimes acceptable within given constraints. Massive porosity could be preparation induced. The clue is to look very closely around the perimeter of the hole and into the hole. Clean non-jagged edges could be indicative of integrity. If the hole is associated with oxide clusters or unreacted particles, clues could be found in the vicinity of the hole or by the hole shape.
(5) *Shrink cracks*, when small, could be acceptable. Small cracks are usually connected to oxides; before rejecting those samples above the limit, it is wise to ensure they have not been preparation induced. The main cause of preparation-induced cracks occurs when hot mounting, but could also result from incorrect clamping or sectioning.

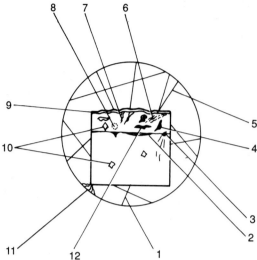

Figure 12.2 Preparation-induced structural damage

require careful grinding to avoid pull-out. They can be retained with progressive shock absorbing grinding. If they appear in the proximity of a hole, after careful grinding, it is usually indicative of globular particle pull-out.

12.2.2 Preparation-induced structural damage

There is often little difference between a metallurgically rejectable feature and a preparation-induced feature. Figure 12.2 is indicative of induced damage; this has to be identified in order to make a correct analysis.

The following list is of different types of preparation-induced damage:

(1) *Scratched resin surface.* This is often unavoidable when using thermoplastic or cold setting resins. They should be ignored rather than attempted to be polished out since this could destroy the coating integrity.
(2) *Delamination* must always be considered as preparation induced until the sample has been clamped, sectioned and ground under carefully controlled conditions.
(3) *Grit blast* can be pulled out, enlarging the occupied interface length.
(4) *Axial porosity* as opposed to the more frequent longitudinal porosity is a candidate for induced porosity.
(5) *Interface related scratches.* It is often wise to traverse rogue scratches since they can be a vital clue to the specimen surface condition. It could be caused by particles breaking away, in particular at the coating/resin interface. When this takes place it is usually due to cold mounting contraction or an uncured phenolic hot mounting resin. If the rogue scratch is seen to be both sides of what looks like a pore it is probably caused by a blunt abrasive. If the scratch starts in, say, the middle of the coating without an associated aberration, it is again connected with the abrasive becoming blunt or starting to lap. When the rogue scratch is caused by the abrasive it is wise to investigate the type of lubrication and perhaps change to a more oily version.

(6) *Delamination* is normally rejectable.
(7) *Interface bond length*, a percentage length of the interface, shall be occupied by grit blast, e.g. 15 to 20%. The size of particles can be described as fine or coarse. The retention of grit blast is often an indication of good preparation. When this grit is cracked and still retained it is further evidence of careful preparation.
(viii) *Oxides and separations at the interface.* Delamination or a combination of oxides at the interface combine to reduce the bond strength of the coating. Delamination or separation is normally not tolerated and oxides will have a prescribed limit. Relief polishing or staining often exaggerates this condition and is therefore to be avoided. Although delamination is a rejectable fault it could have been caused by incorrect clamping and/or sectioning.
(ix) *Bond coat and coating* are to be well bonded. This is sometimes described as undercoat and coating. Usually this presents few problems with the specimen preparation since the two materials are selected to be compatible, one with the other.
(x) *Oxide clusters* result in poor coating integrity and as a result are subject to limits. Higher levels are often acceptable on corners. Oxide clusters being friable

(6) *Smearing* is to be avoided. Its presence can readily be observed with the aperture diaphragm stopped down or alternatively the surface investigated in DIC. It is usually associated with unidirectional preparation or high pressure. Any preparation technique employing a polishing step that is not *less* than a normal grinding pressure must be looked upon with great caution.

(7) *Radial cracks* caused by pressure mounting often appear like shrink cracks and could result in an incorrect coating rejection.

(8) *Lidding* takes place when blunt abrasives or abrasives of similar hardness to, say, the particulate are being employed. A close investigation of the pore surface wall will soon reveal such a defect.

(9) *Contraction* by the mounting resin can be overcome by using an epoxy resin either in total or simply by using the drip-dry technique.

(10) *Impressed abrasives* caused by the grinding abrasive lodging into the softer phase of the specimen are frequently noticed when the charged abrasive technique is used. This is usually indicative of either incorrect platen surface, lubricant or machine parameters. It could be a major problem if the mounting resin has not been correctly cured. A great majority of materialographers have not been trained to look for this fault and in consequence have to accept an inferior preparation.

(11) *Poor mounting technique.* Use a larger diameter mount, a smaller sample or centralize the specimen.

(12) *Longitudinal or round porosity* is the more familiar true porosity. When there is a predominance in the outer layer, the preparation technique including the mounting must be questioned. To have a predominance in the inner layer would be indicative of poor resin ingress and would require an improved ingress or a change in the preparation procedure.

12.3 Aids to Metallography

12.3.1 Sectioning

One classic quotation (Thermal Spray Conference in October 1988, Ohio), 'the cutting did not have any visible influence on the final result', was in relation to a WC/Co plasma coating. In metallography it is most important, in analysing each step, to investigate the effect each step has on the structure and not just to check at the completion of the polishing step. It is important to define the type of wheel and the cutting parameters in order to evaluate the depth of structural damage.

Sectioning aids

(i) Would a rubber/resin bonded abrasive wheel be any improvement and how strong should the bond be?

(ii) A thin wheel will always result in less damage than a thicker wheel.

(iii) There is an optimum speed relative to each type of wheel. The thin CBN low deformation slitting wheels have an efficient cutting speed one-tenth that of the abradable type of wheels when sectioning ductile materials.

(iv) When using the abradable type of wheel should silicon carbide or alumina be chosen? The majority of users would choose alumina yet silicon carbide could result in less coating damage.

(v) When sectioning a coating always go from coating to substrate and not vice versa.

(vi) When clamping the coated material in the sectioning machine *do not* clamp the piece that is to be used for your sample; simply retain it.

(vii) Avoid the use of cutting lubricants where degreasing the pores prior to mounting could prove ineffective.

Before any suitable preparation procedure can be selected it is *essential* to have some idea of the Z axis induced structural damage resulting from the sectioning operation. It was shown in Chapter 2 that the Z axis damage on a carbon steel could vary from 10 to 900 μm. With a plasma coating the situation is more grave since incorrect sectioning can cause irrevocable damage.

12.3.2 Mounting

It is wise with all coating materials to use a low to nil contracting mounting resin. This usually implies the use of cold setting epoxy resins. For those users who prefer to use a hot mounting press, the sample must endure the epoxy drip-dry technique prior to mounting. Since many coatings exhibit porosity, some discrete others are interconnected; vacuum impregnation using a low viscous resin (250 cP) will be necessary.

The degree to which this resin penetrates the coating will depend on:

(a) coating thickness,
(b) vacuum,
(c) resin viscosity.

Total resin ingress is not possible with the thicker coatings, and the chosen preparation technique will have to take this into account. When this proves to be a problem it will be necessary to reimpregnate the sample after the planar grinding stage. When looking down the microscope, it is necessary to train the eye to recognize pores with resin ingress and those without. First-order grey differential contrast will be a big help in this respect. When optically observing porosity the following points need clarification:

(i) Differentiate between resin ingressed and non-ingressed pores.
(ii) An ingressed pore does not imply integrity. It could still exhibit structural damage from the sectioning/mounting stage; i.e. confirm that the combined Z axis is beyond the sectioning Z axis.
(iii) Unmelted ceramic particles can look like non-ingressed pores. Use the cross-polars technique to identify such situations.
(iv) Non-ingressed interconnecting pores and discrete pores must be separated from specimen preparation-induced pores. Preparation-induced pores can be identified (with care) by shape and size. First-order grey allows the observer to *look into* the subject. This again will help in the identification. When using a *standard* traceable technique that has been audited, it is still possible to manifest preparation-induced porosity, the level of which must be recorded separately from true porosity. The alternative is to create another standard.

The following is a quotation from a recent round-robin: 'Fluorescent dye used for vacuum impregnation is one of the few possibilities to give positive indication of porosity, and it was the only help to make sure we were talking about porosity and not pull-outs.' I hope the reader can see all the pitfalls in such a statement, i.e.:

(i) Discrete pores would be called pull-out.
(ii) Non-ingressed would be called pull-out.
(iii) Total ingression could simply represent structural damage, etc.

Mounting aids

(i) Always degrease, being particularly diligent with samples exhibiting porosity.
(ii) If it is necessary to impregnate the sample with resin use the ultra-thin resins currently available with the maximum vacuum possible. The resin will bubble when beyond the pressure limit.
(iii) Strong coloured dyes can be used in the resin to aid optical observations.
(iv) Fluorescent dye can be used as (iii) above for those occasions when identification of resin becomes difficult (see the special section on 'fluorescent microscopy'—Section 12.4.1).
(v) Where feasible, encapsulate prior to cutting.
(vi) Medium hard to soft coatings are usually dense and as such could be hot mounted; others would be vacuum impregnated and cold mounted. Use the preload facility in the mounting press.
(vii) Choose where possible a mounting medium with the abrasion characteristics similar to the coating. (*Note:* this is usually expressed as having a similar hardness but there is little point in having the same hardness if the abrasive characteristics differ.)
(viii) Reimpregnate after planar grinding when necessary.
(ix) If (vii) above necessitates an acrylic or polyester resin, the sample must be precoated with an epoxy resin.

12.3.3 Grinding

Grinding abrasives must be much harder than the material being cut. This in principle rules out the use of silicon carbide papers when preparing ceramic coatings. Unfortunately many users ignore this point and continue to use various grades of silicon carbide papers. I am not aware of any round-robin that has outlawed such practices. Silicon carbide could, if necessary, be used at the planar grinding stage providing it was at least three times as coarse as the sample material particles. Since charged platen surface 10 using 30 to 40 μm diamond will remove stock without a high associated induced damage it is the obvious choice for ceramic coatings.

Since most coatings could include particles that are brittle, poorly bonded or friable it is necessary when grinding to absorb the impact shock without creating relief as would occur from using a polishing cloth. Platen surface 3 or 4 satisfies this requirement and is included as part of the preparation sequence.

The metal sprayed coatings are generally easier to prepare than the ceramic versions and are capable of being adequately planar ground using silicon carbide paper. The choice for subsequent steps would be to progress on, say, three stages of silicon carbide paper or one stage using charged platen surface 4, 5 or 6 (the latter being the favoured choice). Whether the choice is silicon carbide or a charged platen surface, it will be necessary to use surface 1 to faithfully reveal the true structure.

Grinding aids

(i) Avoid the use of high speed grinding stones although they appear to give a good finish. They cause gross structural damage.
(ii) Ensure that the abrasive is twice as hard as the particles being ground. The exception to this is when hard particles are small and bonded in a soft matrix. It is then possible to use silicon carbide paper providing the silicon carbide particles are much greater in size than the specimen particulate and providing it is only used at the planar grinding stage.
(iii) With coarser grinding stages use a unidirectional approach (resin–coating–substrate). This requires correct orientation of the specimens in the sample holder, prepared in contra motion. The example in Figure 12.3 shows where grinding at (a), the coating, has been broken away at the mounting–resin interface. This has been overcome at (b) by reversing the grinding direction.

(a)

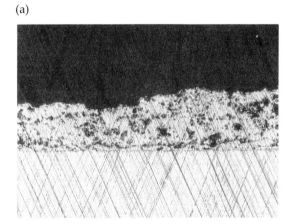

(b)

Figure 12.3 Grinding damage. (a) Substrate–coating–resin (×400). (b) Resin–coating–substrate (×400)

(iv) Grind to integrity before polishing, i.e. SiC/6/1 or SiC/4/1 for medium to soft coatings, 10/4/1 or 10/6/1 for hard coatings.
(v) Avoid grinding beyond the depth of impregnation (preparation procedures are currently based on a time factor at each stage—future methods quoting Z axis factors will overcome some of these problems).
(vi) Avoid ultrasonic cleaning between stages.

12.3.4 Polishing

This stage is generally for aesthetic purposes only and care should be exercised in not inducing damage due to the cloth nap or relief through prolonged times. Colloidal silica/

alumina abrasives are suitable as slurries on polishing surface 7.

12.4 Microscope Techniques

12.4.1 Fluorescence

A recommendation from some commercial suppliers of preparation equipment and many users is to make use of the fluorescence phenomenon of excitation to reveal ingressed resin in coated surfaces. Since the writer is against such practices a further explanation is required.

Excitation occurs in specific materials that have been mixed with a fluorescent dye (secondary fluorescence) causing the short wavelength of light to reemit into the longer more visible wavelength. If, for example, we were to excite the dyed subject with blue 435 nm light we could find a colour change to the green/yellow part of the spectrum (500 to 600 nm) being emitted from the dyed area only. The non-dyed area reflects back the short wavelength blue. Because the short wavelength light is harmful it is necessary to cancel this light before it enters the observer's eyes. Barrier filters are used to cancel out specific wavelengths and can be positioned anywhere in the optical train after specimen excitation (usually below the eyepiece). To summarize, the light from the lamp is isolated to a specific wavelength; when reflected from the sample this light combines with the reorientated visible light and the harmful short wavelength light, the latter being barred from entry (cancelled) into the observed optical image. In theory this means that only the fluorescing subjects will be visible. This is not the case, however, since for the materials application a broad band excitation is used allowing the background also to be observed. Although the *reflected* harmful wavelengths are barred in the observed image the *scattered* harmful rays are not and the microscopist must be protected from these carcinogenic rays, a point often overlooked.

The image observed from the excited area is poor in resolution due to:

(a) the image being constructed from long wavelengths and

(b) the 'glow' always in evidence reducing subject quality.

Fluorescence excitation is good for identifying the existence of resin but there are so many other ways to do this without incurring the penalties of expensive, dangerous and poorly resolved results.

The implication that a black hole must be a pull-out is a *very* dangerous assumption since it could be and often is a discrete or non-ingressed pore. The existence of the dye gives a positive indication of porosity, again a *very* dangerous assumption; it could be resin ingressed structural damage.

Before any combination of filters can be constructed (excitation and barrier) it is important to know the excitation wavelength of the dye. All too often the blue excitation is confused with the violet and ultraviolet.

Commercially available dyes respond to the blue 435 nm excitation. A recently published paper (February 1989) suggests excitation with ultraviolet light (365 nm). Ultraviolet light is very expensive to achieve, i.e. the microscope would require a mercury vapour lamp; it is also dangerous, but above all it is not the required excited wavelength for the dye. As a consequence this could give inferior results to the standard quartz halogen lamp supplied with the microscope. Care must be exercised to ensure that the barrier filter removes 365 nm as well as 435 nm wavelengths.

Simple blue light fluorescence can be adequately excited with a standard quartz halogen microscope bulb operating at maximum voltage, the microscope being fitted with 435 nm excitation filter and the appropriate barrier filter. Improvements would be made to the observed image if the microscope internal reflector was the dichroic type matched to the blue excitation.

12.4.2 First-order red excitation

When it is necessary to investigate resin ingress or resin condition within a pore then the first-order red technique offers many attractions without the disadvantages manifest in fluorescent dye techniques. To investigate with first-order techniques the microscope should first be set to extinction (the polarizer and analyser crossed) and then the appropriate

Preparation of Spray Coatings 105

first-order wedge inserted at 45° to the extinction angle. When the first-order red is used the background information takes on a relaxing red colour, resin-filled pores appear black and unfilled pores appear green. Because this method is a total resolution technique the quality of the observed image far exceeds that of a fluorescing image. Equally the wrong conclusions must not be drawn from this technique; though quality and clarity is greater it still must not be concluded that unfilled pores will be pull-outs. What it does do is give the observer an image quality that allows the observer to make a better subjective judgement.

The example in Figure 12.4 (see Plate 4) shows an alumina coating in bright field and first-order red sensitive tint.

Analysis from micrographs

A careful study of these two examples with identical fields of view reveals only one green pore that has been preparation induced. The other unfilled pores shown in green are within a black boundary and as such are indicative of incomplete resin ingress. Even with this information there is still a need to address the question 'Accepting the single preparation-induced pore, are the others true and faithful?' Providing the initial Z axis structural damage dimension has been exceeded then they are faithful. From the orientation and morphology of the pores it would appear that without the Z axis information we are looking at preparation-induced structural damage.

12.4.3 Crossed polars

When the coating consists of materials that are anisotropic then the mineral content will be made visible against an extinguished background. (This means that the light reflecting from the material is bireflecting; it reflects in *two* discrete directions only. This is not the case when reflecting light from a metal, where the light reflects in *all* directions.) Many minerals are called birefringent materials since the light not only reflects in two directions (at 90° to each other) but in passing through one direction is retarded in relation to the other due to having a different refractive index. If the first-order red plate is introduced then many different subject colours can be introduced by rotating the polarizer. One example

of this technique is when graphite is included to make a nickel/graphite layer; the graphite will then appear coloured.

Plasma-sprayed alumina particles often exhibit a second phase particle. It is difficult to differentiate between porosity, resin-filled pores and this particle since they can all appear the same shade. The same subject observed in polarized light clearly identifies this second phase, said to be unmelted particles.

Crossed polars is also a useful technique in identifying refractive index changes, providing the material being observed is semi-transparent. The different refractive index interface causes some low level light to be reflected back into the optic axis; this is usually indicative of hidden porosity.

The example given in Figure 12.5 is an alumina coating shown in bright field and between crossed polars.

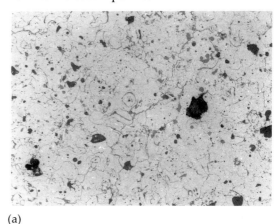

(a)

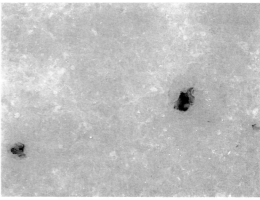

(b)

Figure 12.5 Artefact identification. (a) Bright field (×400) and (b) crossed polars (×400)

Analysis from micrographs

It will be clear when comparing these two micrographs how easy it is to identify the unmelted particle or non-bireflecting secondary phase. With care one can also make out a 'mottled' lighter background in the crossed polar micrograph. Information relating to subsurface grains and porosity can be derived from this.

12.4.4 Differential interference contrast (DIC)

This technique gives a depth to the normally flat observed image. The information available for interpretation is greatly increased from the normal image and allows the observer to look *into* the subject and not *at* it. The writer recommends this technique as a *must* for all final investigations. The information manifests itself in such a manner as to allow true analysis; without such information it would be guess work. This technique also shows all the bad points relating to the surface preparation: it exaggerates the surface topography and those points most investigators do not wish to display. They must, however, if they wish to make a proper analysis, reveal *all* the information.

The following example illustrates the applications of first-order grey DIC.

12.5 Plasma-sprayed Preparation Examples

Example 12.1

Family classification BB16

The material: plasma-sprayed zirconia/yttria

Plasma-sprayed zirconia with the appropriate bond coat is used as a thermal barrier in, for example, engine combustion chambers. The efficiency of a gas turbine depends upon the temperature of the operating gas stream. It is anticipated that the need to operate at higher temperatures will in the future make our current materials redundant (nickel-based and titanium alloys).

Thermal barrier coating on traditional materials allows an increase in the operating temperature, resulting in improved engine efficiency. When the thermal barrier is used at existing temperatures the life of the combustor is improved.

Porosity in the coating has a profound effect on thermal conductivity and strength. These pores must therefore exhibit an optimum uniform distribution.

The technique

Most spray coatings including zirconia/yttria can be destroyed beyond recovery by incorrect sectioning and unwise hot mounting. When clamping the material for sectioning, tighten the non-sample side of the machine vice, the vice on the sample side just making contact. When selecting the correct abradable wheel to use it is wise to section in the direction of coating/bond coat/substrate and not vice versa.

When mounting, avoid pressure cracking by either cold mounting or protecting the sample with an epoxy coat before hot mounting. It will be necessary in any event to vacuum impregnate the sample prior to any mounting procedure. As can be seen from the method, preparation-induced damage is reduced when progressing towards sample integrity by the use of platen surfaces.

Preparation of Spray Coatings 107

The method

AUDITABLE PREPARATION PROCEDURE	No. 1
CLASSIFICATIONS: BB16/PLASMA SPRAY	MATERIAL: ZIRCONIA/YTTRIA–STEEL SUBSTRATE
SECTIONING TECHNIQUE: LOW DEFORMATION SAW	BLADE
MOUNTING TECHNIQUE: EPOXY COLD MOUNT	CONCENTRATION

	SURFACE and TYPE	ABRASIVE SIZE/TYPE LUBRICANT	LOAD (kPa)	HEAD ROTATION	Z AXIS (μm)	REFLECTIVE FACTOR	MICROGRAPH YES/NO	IMAGE ANALYSIS YES/NO	REMARKS
PLANAR GRINDING STAGE	S.10	30 μm OIL	35	COMP	PLANE				

	SURFACE and TYPE	ABRASIVE SIZE/TYPE LUBRICANT	LOAD (kPa)	HEAD ROTATION	Z AXIS (μm)	REFLECTIVE FACTOR	MICROGRAPH YES/NO	IMAGE ANALYSIS YES/NO	REMARKS
SAMPLE INTEGRITY STAGE	S.4	9 μm OIL	35	CONTRA					
	S.1	1 μm HIGH pH	70	CONTRA					

	SURFACE and TYPE	ABRASIVE SIZE/TYPE LUBRICANT	LOAD (kPa)	HEAD ROTATION	TIME (s)	REFLECTIVE FACTOR	MICROGRAPH YES/NO	IMAGE ANALYSIS YES/NO	REMARKS
POLISHING STAGE	PS 7	0.06 μm COLLOIDAL SILICA	17	CONTRA	120		YES		12.6

The micrograph Figure 12.6

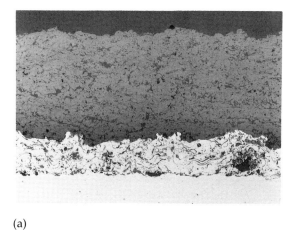 

(a) (b)

Figure 12.6 Looking within. (a) Bright field (×200) and (b) first-order grey DIC (×200)

Analysis from micrographs

Bond coat/substrate interface This looks quite good, the DIC micrograph showing how the grit blast has been retained (half way along).

Bond coat Notice the black area at the bottom of the bright field image. This looks like gross porosity or some artefact. The DIC micrograph clearly reveals it to be *not* porosity or preparation-induced artefact but retained grit.

Bond coat/coating interface This is clearly defined in DIC.

Coating The bright field image indicates a faithful reproduction. The pores have a uniformity about their size and morphology, their direction being longitudinal. This view is confirmed in the DIC micrograph where the three-dimensional information available when looking down the microscope has to be seen to be believed.

This subjective visual information accompanied by Z axis and reflectivity values could form the basis of a traceable standard.

Example 12.2

Family classification BB16

The material: thin plasma-sprayed alumina–nickel alloy substrate—resin ingressed

The technique

The method given is for a relatively thin layer of plasma-sprayed alumina fully ingressed with epoxy resin poured under vacuum and sectioned transversely. With the structure fully ingressed with resin the preparation is relatively short and simple. Diamond is used on platen surface 10 for the planar grind as silicon carbide would simply smash the layer, not cut it. Subsequent stages use progressively smaller particle sizes and progressively higher shock absorbing surfaces.

The method

AUDITABLE PREPARATION PROCEDURE   No. 2

CLASSIFICATIONS: BB16/PLASMA SPRAY—THIN	MATERIAL: ALUMINA–NICKEL ALLOY SUBSTRATE
SECTIONING TECHNIQUE: HIGH SPEED—LOW DEFORMATION	BLADE THIN SILICON CARBIDE ABRADABLE
MOUNTING TECHNIQUE: VACUUM EPOXY—FULLY INGRESSED	CONCENTRATION

	SURFACE and TYPE	ABRASIVE SIZE/TYPE LUBRICANT	LOAD (kPa)	HEAD ROTATION	Z AXIS (μm)	REFLECTIVE FACTOR	MICROGRAPH YES/NO	IMAGE ANALYSIS YES/NO	REMARKS
PLANAR GRINDING STAGE	S.10	30 μm DIAMOND WATER	35	CONTRA	PLANE				

	SURFACE and TYPE	ABRASIVE SIZE/TYPE LUBRICANT	LOAD (kPa)	HEAD ROTATION	Z AXIS (μm)	REFLECTIVE FACTOR	MICROGRAPH YES/NO	IMAGE ANALYSIS YES/NO	REMARKS
SAMPLE INTEGRITY STAGE	S.4	9 μm OIL	35	CONTRA					
	S.1	1 μm WATER	50	CONTRA					
	S.1	0.06 μm SILICA	50	COMP					

	SURFACE and TYPE	ABRASIVE SIZE/TYPE LUBRICANT	LOAD (kPa)	HEAD ROTATION	TIME (min)	REFLECTIVE FACTOR	MICROGRAPH YES/NO	IMAGE ANALYSIS YES/NO	REMARKS
POLISHING STAGE	PS 7	0.06 μm SILICA	16	COMP	1		YES		12.7

The micrograph (Figure 12.7)

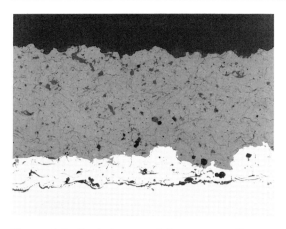

Figure 12.7 Resin-ingressed thin coating. Plasma-sprayed alumina (×400)

Analysis from micrographs

Bond coat/substrate interface From the information this bright field micrograph displays, the coating would be rejected. This, however, would be unwise without a clearer look *within* this interface to eliminate staining, lack of resolution, etc.

Bond coat Although this is quite good, there is an area towards the right-hand side where the black dots require further investigation.

Bond coat/coating interface This looks very good.

Coating This coating without the black dots looks quite good and faithfully revealed. There are two transverse pores that would require investigating in DIC. The black dots when observed between crossed polars proved to be unmelted particles, leaving only one transverse pore unaccounted for (the one near the centre of the micrograph). Since this specimen was not prepared to any Z axis dimension, it could represent damage prior to impregnation. Another 5 μm removal would clarify the situation.

Example 12.3

Family classification BB16

The material: thick plasma-sprayed alumina–nickel alloy substrate—partial resin ingress

The technique

The inner region of thick sprayed alumina often reveals areas of no resin even with interconnecting pores. Since the resin has not totally filled the pores it is more difficult to know for sure whether the preparation route has achieved integrity or not. An indication of integrity can be made by comparing the number of unfilled pores with those nearer the outer surface that have been filled. Another indication is the shape of the pores; they should not exhibit fragmented peripheries.

When optically observing such areas without resin ingress the networks between the particles appear black, as do all larger areas of porosity.

The preparation route for this specimen is more complex and considerably longer than the route in Example 12.2 for the fully ingressed sample. This is because in the fully ingressed sample, the resin acts as a binder holding the particles in place and supporting the edge of the particles. Non-ingression of resin makes the layer much more friable. This means that the depth of damage caused at each stage is higher than in the ingressed sample. This generally leads to more steps and longer times in the preparation procedure.

110 Surface Preparation and Microscopy of Materials

The method

AUDITABLE PREPARATION PROCEDURE   No. 3

CLASSIFICATIONS: BB16/PLASMA SPRAY—THICK	MATERIAL: ALUMINA–NICKEL ALLOY SUBSTRATE
SECTIONING TECHNIQUE: HIGH SPEED—LOW DEFORMATION	BLADE THIN SILICON CARBIDE ABRADABLE
MOUNTING DIRECTIONALLY MOUNTED TECHNIQUE: VACUUM EPOXY—PARTIAL INGRESS	CONCENTRATION

	SURFACE and TYPE	ABRASIVE SIZE/TYPE LUBRICANT	LOAD (kPa)	HEAD ROTATION	Z AXIS (µm)	REFLECTIVE FACTOR	MICROGRAPH YES/NO	IMAGE ANALYSIS YES/NO	REMARKS
PLANAR GRINDING STAGE	S.10	30 µm DIAMOND WATER	35	CONTRA	PLANE				

	SURFACE and TYPE	ABRASIVE SIZE/TYPE LUBRICANT	LOAD (kPa)	HEAD ROTATION	Z AXIS (µm)	REFLECTIVE FACTOR	MICROGRAPH YES/NO	IMAGE ANALYSIS YES/NO	REMARKS
SAMPLE INTEGRITY STAGE	S.4	9 µm OIL	35	CONTRA					
	S.1	6 µm WATER	35	CONTRA					
	S.1	1 µm WATER	50	CONTRA					
	S.1	0.06 µm SILICA	50	COMP					

	SURFACE and TYPE	ABRASIVE SIZE/TYPE LUBRICANT	LOAD (kPa)	HEAD ROTATION	TIME (min)	REFLECTIVE FACTOR	MICROGRAPH YES/NO	IMAGE ANALYSIS YES/NO	REMARKS
POLISHING STAGE	PS 7	0.06 µm SILICA	16	COMP	1		YES		12.8

The micrograph Figure 12.8

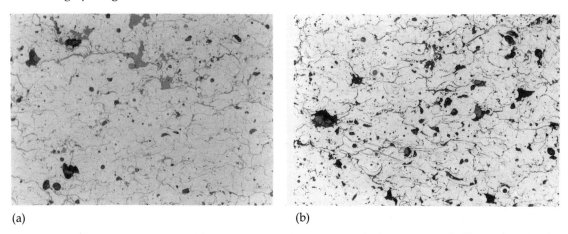

(a)   (b)

Figure 12.8 Partial resin-ingressed thick coating. Plasma-sprayed alumina–nickel alloy substrate. (a) Ingressed outer area (×400) and (b) non-ingressed inner area (×400)

Analysis from micrographs

It is clear when comparing the two micrographs that the non-ingressed version looks darker. When observing the two images orthoscopically those pores filled with resin are easily distinguished from those that are not. Some of this information is lost in the micrographs and it would therefore be of benefit in terms of visual display to use a coloured dye. The reader is asked to look at the largest black hole in the non-ingressed micrograph. This is in fact not a hole but an unmelted particle, which can be identified as such when observed between crossed polars.

Example 12.4

Family classification BB16

The material: plasma-sprayed 88% tungsten carbide, 12% cobalt–nickel alloy substrate

The technique

Vacuum impregnation, as has been mentioned, does not always fill every pore. Those not filled with resin are either *closed* pores or preparation-induced damage or both. If on optical observation there still remains doubt then the first-order grey DIC technique will often indicate what is true or induced porosity.

Diamond is used predominantly in the preparation route; silicon carbide (coarse grade) could be used for planar grinding but subsequent steps would incur prolonged grinding, in particular with platen surface 4. The second step on surface 1 using colloidal silica merely develops crispness of the edge of the resin-filled pores. It can be carried out on the same surface simply as an extension to the 1 μm diamond stage.

The short final polish in a complementary direction develops a small amount of relief between the carbide particles and the cobalt matrix, allowing better resolution of the particles, i.e. achieving resolution at the expense of integrity.

The method

AUDITABLE PREPARATION PROCEDURE — No. 4

CLASSIFICATIONS: BB16/PLASMA SPRAY	MATERIAL: 88% Wc 12Co–N: ALLOY SUBSTRATE
SECTIONING TECHNIQUE: HIGH SPEED—LOW DEFORMATION	BLADE THIN SILICON CARBIDE ABRADABLE
MOUNTING TECHNIQUE: VACUUM EPOXY—FULLY INGRESSED	CONCENTRATION

	SURFACE and TYPE	ABRASIVE SIZE/TYPE LUBRICANT	LOAD (kPa)	HEAD ROTATION	Z AXIS (μm)	REFLECTIVE FACTOR	MICROGRAPH YES/NO	IMAGE ANALYSIS YES/NO	REMARKS
PLANAR GRINDING STAGE	S.10	30 μm WATER	35	CONTRA	PLANE				

	SURFACE and TYPE	ABRASIVE SIZE/TYPE LUBRICANT	LOAD (kPa)	HEAD ROTATION	Z AXIS (μm)	REFLECTIVE FACTOR	MICROGRAPH YES/NO	IMAGE ANALYSIS YES/NO	REMARKS
SAMPLE INTEGRITY STAGE	S.4	9 μm OIL	35	CONTRA					
	S.1	1 μm OIL	35	CONTRA					
	S.1	0.06 μm SILICA	50	COMP					

	SURFACE and TYPE	ABRASIVE SIZE/TYPE LUBRICANT	LOAD (kPa)	HEAD ROTATION	TIME (min)	REFLECTIVE FACTOR	MICROGRAPH YES/NO	IMAGE ANALYSIS YES/NO	REMARKS
POLISHING STAGE	PS 4	0.06 μm SILICA	16	COMP	1		YES		12.9

The micrograph Figure 12.9

Figure 12.9 Questionable integrity (×400)

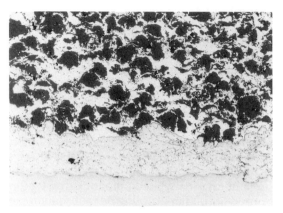

Figure 12.12 Sprayed aluminium, silicon, polyester on steel substrate (×200)

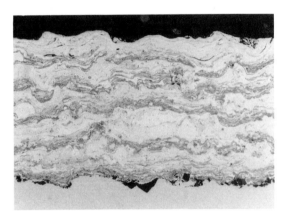

Figure 12.10 Sprayed copper, nickel, indium on titanium alloy substrate (×400)

Figure 12.13 Sprayed nickel, graphite on steel substrate (×200)

Analysis from micrograph (Figure 12.9)

This micrograph is a classic example of the often published plasma-sprayed tungsten carbide/cobalt coating. It represents, for example, the best of a round-robin exercise, but is it a faithful reproduction? If the reader has agreed with the foregoing information in this chapter, then he or she will, without doubt, be saying 'this micrograph appears to exhibit gross preparation-induced damage'. The size, shape and direction of the dark pores are examples of structural damage. The flat bright field image allows little more than this subjective analysis. To make a more positive analysis, differing optical techniques would have to be employed, not least of these being the first-order grey DIC. This sample would

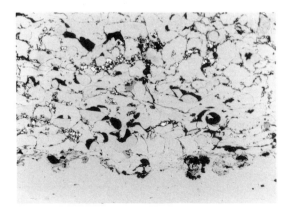

Figure 12.11 Complex nickel-based coating (×400)

require repreparation to some auditable condition, the induced structural damage being investigated after each preparation step. The quoted preparation procedure (10-4-1-1) is not necessarily where the problem lies; it is in the parameters and functions of the abrasive surfaces, the lack of conformity to the Z axis dimensions and, in particular, the reflectivity or scratch pattern. This example is used to illustrate the tremendous shortfall that occurs (a) when the illustrated micrograph is incorrectly quoted as a faithful reproduction of the coating, and (b) how futile it is to just quote a preparation method without any means of auditing.

Soft coatings examples

Figures 12.10 to 12.13 show examples of coatings where it is possible to use silicon carbide papers at the planar grinding stage. Also note the low pressure on polishing surfaces to reduce the risk of smearing the soft phase.

Note how two of the methods use platen surface 6 as the primary surface; the other two use surface 4. The clue to the reason is to be found in the material, graphite being friable and the complex nickel-based coating being brittle. Sectioning of these materials is also less sensitive to induced damage. Care must still be taken when the sample is directionally cut.

AUDITABLE PREPARATION PROCEDURE — No. 5

CLASSIFICATIONS: BB16/PLASMA SPRAY	MATERIAL: Cu/Ni/In–Ti ALLOY SUBSTRATE
SECTIONING TECHNIQUE: HIGH SPEED ABRADABLE	BLADE SILICON CARBIDE—THIN
MOUNTING TECHNIQUE: EPOXY DIP—HOT MOUNTED PHENOLIC	CONCENTRATION

PLANAR GRINDING STAGE	SURFACE and TYPE	ABRASIVE SIZE/TYPE LUBRICANT	LOAD (kPa)	HEAD ROTATION	Z AXIS (μm)	REFLECTIVE FACTOR	MICROGRAPH YES/NO	IMAGE ANALYSIS YES/NO	REMARKS
	PAPER	320 SiC WATER	35	CONTRA	PLANE				

SAMPLE INTEGRITY STAGE	SURFACE and TYPE	ABRASIVE SIZE/TYPE LUBRICANT	LOAD (kPa)	HEAD ROTATION	Z AXIS (μm)	REFLECTIVE FACTOR	MICROGRAPH YES/NO	IMAGE ANALYSIS YES/NO	REMARKS
	S.6	6 μm OIL	35	CONTRA					
	S.1	1 μm OIL	35	CONTRA					

POLISHING STAGE	SURFACE and TYPE	ABRASIVE SIZE/TYPE LUBRICANT	LOAD (kPa)	HEAD ROTATION	TIME (min)	REFLECTIVE FACTOR	MICROGRAPH YES/NO	IMAGE ANALYSIS YES/NO	REMARKS
	PS 7	0.06 μm SILICA	16	COMP	2		YES		12.10

114 Surface Preparation and Microscopy of Materials

AUDITABLE PREPARATION PROCEDURE — No. 6

CLASSIFICATIONS: BB16/PLASMA COAT	MATERIAL: COMPLEX Ni-BASED—STEEL SUBSTRATE
SECTIONING TECHNIQUE: HIGH SPEED ABRASIVE	BLADE SILICON CARBIDE
MOUNTING TECHNIQUE: EPOXY DIP—HOT MOUNTED PHENOLIC	CONCENTRATION

	SURFACE and TYPE	ABRASIVE SIZE/TYPE LUBRICANT	LOAD (kPa)	HEAD ROTATION	Z AXIS (µm)	REFLECTIVE FACTOR	MICROGRAPH YES/NO	IMAGE ANALYSIS YES/NO	REMARKS
PLANAR GRINDING STAGE	PAPER	320 SiC WATER	35	CONTRA	PLANE				

	SURFACE and TYPE	ABRASIVE SIZE/TYPE LUBRICANT	LOAD (kPa)	HEAD ROTATION	Z AXIS (µm)	REFLECTIVE FACTOR	MICROGRAPH YES/NO	IMAGE ANALYSIS YES/NO	REMARKS
SAMPLE INTEGRITY STAGE	S.4	9 µm OIL	35	CONTRA					
	S.1	1 µm WATER	35	CONTRA					
	S.1	0.06 µm SILICA	35	COMP					

	SURFACE and TYPE	ABRASIVE SIZE/TYPE LUBRICANT	LOAD (kPa)	HEAD ROTATION	TIME (min)	REFLECTIVE FACTOR	MICROGRAPH YES/NO	IMAGE ANALYSIS YES/NO	REMARKS
POLISHING STAGE	PS 7	0.06 µm SILICA	16	COMP	2		YES		12.11

AUDITABLE PREPARATION PROCEDURE — No. 7

CLASSIFICATIONS: BB16/PLASMA SPRAY	MATERIAL: Al/Si/POLYESTER—STEEL SUBSTRATE
SECTIONING TECHNIQUE: HIGH SPEED ABRASIVE	BLADE SILICON CARBIDE—THIN
MOUNTING TECHNIQUE: EPOXY RESIN	CONCENTRATION

	SURFACE and TYPE	ABRASIVE SIZE/TYPE LUBRICANT	LOAD (kPa)	HEAD ROTATION	Z AXIS (µm)	REFLECTIVE FACTOR	MICROGRAPH YES/NO	IMAGE ANALYSIS YES/NO	REMARKS
PLANAR GRINDING STAGE	PAPER	320 SiC WATER	35	CONTRA	PLANE				

	SURFACE and TYPE	ABRASIVE SIZE/TYPE LUBRICANT	LOAD (kPa)	HEAD ROTATION	Z AXIS (µm)	REFLECTIVE FACTOR	MICROGRAPH YES/NO	IMAGE ANALYSIS YES/NO	REMARKS
SAMPLE INTEGRITY STAGE	S.6	6 µm OIL	35	COMP					
	S.1	1 µm OIL	35	CONTRA					
	S.1	0.06 µm SILICA	35	COMP					

	SURFACE and TYPE	ABRASIVE SIZE/TYPE LUBRICANT	LOAD (kPa)	HEAD ROTATION	TIME (min)	REFLECTIVE FACTOR	MICROGRAPH YES/NO	IMAGE ANALYSIS YES/NO	REMARKS
POLISHING STAGE	PS 7	0.06 µm SILICA	16	COMP	1		YES		12.12

AUDITABLE PREPARATION PROCEDURE	No. 8
CLASSIFICATIONS: BB16/PLASMA SPRAYED	MATERIAL: Ni/GRAPHITE–STEEL SUBSTRATE
SECTIONING TECHNIQUE: EPOXY VACUUM COAT HIGH SPEED—LOW DEFORMATION	BLADE SILICON CARBIDE—THIN
MOUNTING TECHNIQUE:	CONCENTRATION

PLANAR GRINDING STAGE	SURFACE and TYPE	ABRASIVE SIZE/TYPE LUBRICANT	LOAD (kPa)	HEAD ROTATION	Z AXIS (µm)	REFLECTIVE FACTOR	MICROGRAPH YES/NO	IMAGE ANALYSIS YES/NO	REMARKS
	PAPER	320 SiC WATER	35	CONTRA	PLANE				

SAMPLE INTEGRITY STAGE	SURFACE and TYPE	ABRASIVE SIZE/TYPE LUBRICANT	LOAD (kPa)	HEAD ROTATION	Z AXIS (µm)	REFLECTIVE FACTOR	MICROGRAPH YES/NO	IMAGE ANALYSIS YES/NO	REMARKS
	S.4	9 µm OIL	35	CONTRA					
	S.1	1 µm WATER	35	CONTRA					

POLISHING STAGE	SURFACE and TYPE	ABRASIVE SIZE/TYPE LUBRICANT	LOAD (kPa)	HEAD ROTATION	TIME (min)	REFLECTIVE FACTOR	MICROGRAPH YES/NO	IMAGE ANALYSIS YES/NO	REMARKS
	PS 7	0.06 µm SILICA	16	COMP	2		YES		12.13

13

Preparation of Composites

13.1 Background

Composites by definition, being a combination of different materials, offer a particular challenge when preparing a surface or cross-section for analysis.

Metal matrix composites (MMC) often consist of a soft aluminium or titanium matrix with brittle particles of, say, silicon carbide. These particles can vary in size, distribution and concentration. MMCs could also consist of a soft matrix with reinforced strands or whiskers of, say, silicon carbide or alumina, these strands being aligned longitudinally and/or transversely to the direction of investigation. One of the advantages of composite materials is that they can be *engineered* for specific applications. A necessary requirement therefore is an investigation of not just the microstructure of individual elements but their relationship with one another when combined. The materials scientist has always related the preparations of a material as a direct consequence of the microstructural features of the material. In a composite this by itself, without regard to fibre or particulate volume fraction, distribution and interface condition, would be incomplete.

As composites are inhomogeneous they raise the problem of differential shrinkage with a cast matrix; high volume fraction fibres may also impede the flow of liquid matter—hence the reason for the frequent use of aluminium–silicon alloys. Second-phase distribution can be arranged throughout the matrix or concentrated at the fibre/particulate matrix interface, giving control through the careful choice of alloying elements.

Composites are not restricted to a metal matrix. They can be polymer (PMC) or glass (GMC), refractor or ceramic. They are all very critical to the correct specimen preparation techniques and it is this that is to be pursued in this chapter.

13.1.1 Considerations

Before one selects a preparation procedure it is wise to understand the characteristics of the composite constituent and how they react to material removal. To simplify classification the elements will be considered to be hard, soft or friable.

Hard constituents

(i) Subjects tend to be brittle.
(ii) Long transverse fibres can be cracked beyond recovery, in cutting or grinding.
(iii) Subjects tend to chip.
(iv) High pH silica grinding improves the final integrity.

(v) Before this stage, 1 μm diamond with a high pH slurry improves the cutting effect.
(vi) Smaller charged abrasives can grind when larger abrasives would roll (lapping being undesirable).
(vii) Stock removal by cracking changes to plastic dislocation with submicrometre grinding.
(viii) Small grinding abrasives can go round, rather than cut across, very hard particles (the moat effect).
(ix) When very hard subjects are found in a very soft matrix the charged abrasive will impress at the subject/matrix interface (ultrasonic cleaning necessary).
(x) Blocky grinding abrasives are quite suitable.

Soft constituents

(i) Prone to impressed chips when grinding with silicon carbide papers.
(ii) Grinding abrasives can impress when using charged integrity surfaces.
(iii) Oil slurries improve cutting efficiency and reduce the tendency for impressed chips and abrasives.
(iv) Angular abrasives are more suitable.
(v) High grinding and polishing pressures must not be used.
(vi) Unidirectional preparation is to be avoided.
(vii) Platen surface 4 using 9 μm diamond, although correct for hard constituents, is too hard for pure soft metals; surface 2 would be preferred.
(viii) pH control (high and low) can be effective, in particular at the final preparation stage.

Friable constituents

(i) 'Nap'-type cloths (PS 8 to 10) pluck out phases.
(ii) Lapping is undesirable.
(iii) Platen surface 3 or 4 is advantageous.
(iv) Rubbing contact is to be avoided.

The points raised above will be taken into consideration when formulating techniques, as illustrated later.

13.2 Identifying Bad Techniques

Unfortunately bad techniques do not always reveal themselves in the final microstructure. This has been one of the reasons why empirically derived methods have survived as long as they have. Take, for example:

(i) A tungsten/carbide spray coating, an incorrect preparation method could give a porosity level less than the more correct method and in consequence the former would be favoured.
(ii) A sintered iron prepared to the 'new concept' showed a porosity level as much as 25% less than had been universally accepted by traditional preparation techniques (it took nearly 2 years to convince the supplier).
(iii) Continuous fibre matrix material depends upon a good matrix bond to the fibre whisker. Unfortunately the matrix is often very ductile; hence, a poor bond can appear to be a good bond when incorrectly prepared, i.e. using high pressure and/or napped cloth.

The first three examples below are used to illustrate bad preparation techniques that are visually explicit, but perhaps of greater importance is the need to be aware of possible preparation-induced structural changes when the prepared sample looks good.

13.3 Preparation Examples

Example 13.1

Figure 13.1 shows what happens when the conventional approach using four silicon carbide steps and two polishing steps are employed. The silicon carbide particulate is clearly cracked and plucked out. An important characteristic of the microstructure is the particulate size, bond and distribution.

The second micrograph allows a clear analysis of the matrix, matrix/particulate interface and the level of porosity. Incorrect preparation could portray an image looking even better than this micrograph, yet destroying the vital information. In order to identify a true and faithful prepared surface it is necessary to

118 Surface Preparation and Microscopy of Materials

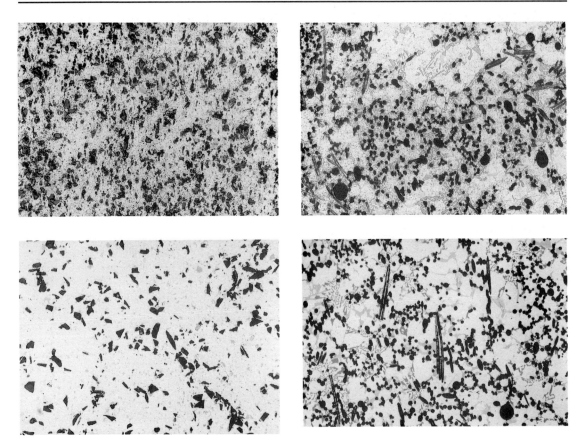

Figure 13.1 Silicon carbide particles in aluminium (×400)

Figure 13.2 Alumina fibres in aluminium (×200)

investigate the preparation technique even when the visual information looks good.

A similar effect to that shown in the first micrograph would also occur if a blunt diamond was to be employed. If, for example, an alcohol-based lubricant was used, the sharp diamond would quickly 'bind' with the soft 'swarf' thus becoming blunt.

Example 13.2

The conventional preparation techniques produced a finish as in the first micrograph in Figure 13.2 where the general morphology of the alumina is poor and the integrity at each interface looks questionable.

If this specimen, in this condition, was to have the polishing stage repeated, only this time with double pressure, the end result would look *similar* to the second micrograph with the following reservation. Very careful optical inspection would show a slight loss of clarity at the fibre/matrix interface; optical observations using a ×100 (1.3 NA) objective would clearly illustrate a flowed, and therefore false, top surface.

Specimens prepared to the 'new concept' (the second micrograph) overcome the preparation problems, showing good clarity of alumina and matrix.

Example 13.3

It is essential to implement the good practice of using sharp wedge-shaped abrasives with friable subjects otherwise the damaged structure as shown in the first micrograph of Figure 13.3 will always be prevalent. Silicon carbide papers are suitable providing they are used in the sharp condition *only*; the gross damage as shown in the first micrograph would suggest that, beyond the planar

Preparation of Composites 119

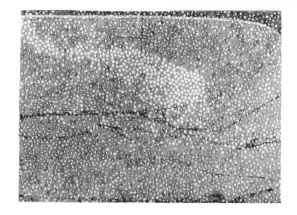

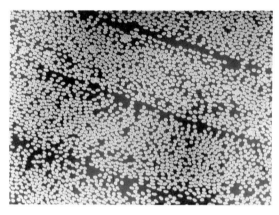

Figure 13.3 Carbon fibres in polymer matrix (×200)

grinding stage, diamond can be used on platen surfaces. The second micrograph shows the excellent results achieved when using these new cutting surfaces.

It is important to emphasize that the results achieved when using silicon carbide paper within the optimum life are equally good.

13.3.1 Preparation technique

Although the 'new concept' gives a systematic approach to the preparation technique it does require an understanding of the way poor selections and combinations are manifest. Ductile materials, for example, can be judged on the scratch pattern relative to the Z axis rate, brittle subjects such as the example given in the chapter on ceramics can be judged by crack propagation and the Z axis rate. The example to be illustrated here is very different and raises some interesting points which should prove helpful to those preparing difficult materials. This material is silicon carbide fibres in a heat-resistant glass matrix.

In developing a technique for the preparation of this material some very important points emerged confirming two basic statements so closely associated with the origins of the 'new concept':

(i) Structural damage is reduced in relation to the ability of the abrasive to absorb impact shock.
(ii) That abrasive has the ability to grind and not lap.

Silicon carbide is very hard and the heat-resistant matrix, although not quite so hard, did prove to be very friable in areas of least fibre concentration. It was suggested that preparations using a diamond abrasive in a lapping mode for a duration of some *30 hours* had given satisfactory results; this clearly was not the way to go. Since longitudinal fibres of silicon carbide had a tendency to break away from the matrix, it was obvious that relatively large stock removal was required (30 μm) to achieve sample integrity. This stock removal if carried out with a rolling abrasive would take too long (30 hours) and if carried out with a fixed abrasive would create continued structural damage. The object then was to find the most efficient charged platen surface that would *cut* the specimen and to continue with the preparation by subsequently choosing lower numbered surfaces operating in a grinding mode.

The test for efficient grinding with these surfaces was to optically observe the scratch pattern at each preparation stage. Since this material was hard it follows that surface 10 would be selected. This proved unsatisfactory since it was not possible with this material to achieve a grinding condition.

Surface 8 proved equally unsatisfactory for the same reasons. Surface 6, when used in the nominally stationary mode (with specimens rotating in contra motion, the platen rotating at 50 rev/min), did show evidence of grinding, but this grinding condition although adequate did not conform to the machine conventions and proved no better than surface 4 (operating to conventions). Surface 4 was therefore adopted using 9 μm diamond as the abrasive.

This surface progressed the preparation up to 4 minutes, after which time no further improvement was achieved. The condition at this stage was a 30% satisfactory scratch pattern, the other 70% being structural damage, predominantly in areas of low fibre density. One could at this stage have used a smaller sized abrasive on the same surface to progress the preparation further, but this would have extended the preparation time without achieving major stock removal which was still necessary.

It was concluded that in order to progress the preparation and still remove adequate stock one would have to use the same sized abrasive (9 μm) but absorb the impact shock to a greater extent than did surface 4. To this end surface 2 was used, progressing the sample to its limit of improvement, which again occurred after about 4 minutes. As the sample still required further grinding, surface 1, with on this occasion 1 μm diamond, again showed improvement to near integrity after, say, 4 minutes. Since there were no more integrity numbers it was not possible to progress to another surface in the grinding mode. Ideally, integrity is to be achieved by grinding (not polishing). One therefore has to make the abrasive on the last surface operate more efficiently, which was achieved by adding a high pH solution to the 1 μm diamond and progressing the preparation another 4 minutes, achieving sample integrity. The final polishing stage proved to be unnecessary for optical analysis (being included only for reasons of photomicrography or image analysis).

This example actually quotes time, which of course would be of little value in formulating a recommended procedure, the Z axis being the required dimension.

13.3.2 Fixed or charged abrasives

Although the 'systematic approach' to surface preparation has gone a long way to helping the materialographer's choice, the basic question of fixed or charged (loose) abrasives has to be addressed at every stage. If one takes the planar grinding stage with an aluminium matrix/silicon carbide particulate as an example, the choice of surface would be in order of structural damage:

(i) Nickel-plated diamond on metal platen
(ii) Metal-bonded diamond
(iii) Resin-bonded diamond
(iv) Platen surface 10 (50 μm)

From the above, if one were to choose the surface manifesting the least structural damage, platen surface 10 would be chosen. The question to be addressed is whether to use fixed or charged abrasive.

Fixed abrasive platen surface 10

This will be diamond or CBN (50 μm) abrasive embodied within some plastic (shock absorbing) medium and preferably in 'circular spots' or some similar shape to allow good lubrication, reduced surface tension and space for swarf to temporarily collect. This system is most efficient when the particulate or fibre Vf is high, since it acts as a built-in dressing stick, cleaning the matrix clogged abrasives. When the Vf is low the abrasives must be liberally oiled to avoid abrasive ploughing.

Charged abrasive platen surface 10

This could be a ceramic/resin type composite platen where the abrasive (diamond/CBN) is charged or sprayed onto the surface. This surface is continually charged with a small amount of lubricant and abrasive, allowing the abrasives to perform in a *grinding* mode. This is slightly more gentle (less residual damage) than the fixed abrasive platen surface 10. The problems occur with high and low Vf ceramics. Low Vf ceramics means that the cutting abrasives will become 'clogged' quickly and will, due to increased shear load, roll off the grinding platen, i.e. into the grooves provided. Although this gives a satisfactory result it can be inefficient, requiring constant abrasive replenishment. High Vf ceramics will create greater mobility in the charged abrasive and therefore increase the tendency for impressed abrasives.

13.3.3 Future predictions

Surface preparation of composites is going to increase the urgency for better, more efficient, reliable and repeatable preparation techniques. This, the author considers, will result from the use of *fixed* abrasives on shock absorbing surfaces, the abrasive being appropriate to the material being cut, utilizing the correct abrasive rake angle. The cutting parameters, such as speed, direction and force, will be optimized and the lubricant will be compatible.

In the meantime the materialographer must balance the two conditions as described above.

This must be done at every stage as the sample progresses towards integrity, viz. the surface preparation method 10/4/1 could be:

Charged abrasive surface 10
Fixed abrasive surface 4
Charged abrasive surface 1

or any other permutation.

13.3.4 Further examples

Example 13.4

Family classification BB12

The material: carbon fibre reinforced polyetheretherketone (PEEK)

Advanced carbon fibre reinforced plastics are light, stiff and strong and are finding increasing applications in aircraft structures, domestic appliances and many other areas, such as sports equipment. Practical laminates are made up from carbon fibres aligned at 0° in direction for the main load with other laminations at 45° and 90°. Carbon fibre composites can be moulded into complex shapes with the 0° plies orientated according to the application. To improve the properties of organic matrix composites high strain carbon fibres have been developed. New resin combinations for improved toughness and increased moisture resistance along with high temperature resins have also been developed. The matrix PEEK used in this example is thermoplastic and can be formed into complex shapes, being resistant to water and kerosene.

The technique

When selecting the carbon fibre, care must be exercised in order to restrict the depth of damage in the transverse fibres. This can be cut using the high speed abradable wheels or with delicate sections using the low speed diamond blade. Planar grinding should equally present few problems and the appropriate grade of silicon carbide would be suitable. Platen surface 4 with $9\,\mu m$ diamond should remove the major damage caused at the planar grinding step.

When the preparation is complete, close optical observation of the longitudinal fibres at the infinity point (fibre end) will confirm or otherwise the integrity of the preparation.

The method

AUDITABLE PREPARATION PROCEDURE   No. 4

CLASSIFICATIONS: BB12/PMC	MATERIAL: CARBON FIBRE IN PEEK
SECTIONING TECHNIQUE: EQUIPMENT TYPE BLADE	CONCENTRATION
MOUNTING TECHNIQUE: METHOD	SAMPLE/MOUNT RATIO

PLANAR GRINDING STAGE	SURFACE and TYPE	ABRASIVE SIZE/TYPE LUBRICANT	FORCE (kPa)	HEAD ROTATION	Z AXIS (μm)	REFLECTIVE FACTOR	MICROGRAPH YES/NO	IMAGE ANALYSIS YES/NO	REMARKS
	PAPER	P320 SiC WATER	35	COMP	PLANE				

SAMPLE INTEGRITY STAGE	SURFACE and TYPE	ABRASIVE SIZE/TYPE LUBRICANT	FORCE (kPa)	HEAD ROTATION	Z AXIS (μm)	REFLECTIVE FACTOR	MICROGRAPH YES/NO	IMAGE ANALYSIS YES/NO	REMARKS
	S.4	9 μm OIL	35	CONTRA					
	S.1	15 μm WATER	35	COMP					
	S.1	GAMMA ALUMINA	70	COMP	1		YES		13.4 (Plate 5)

POLISHING STAGE	SURFACE and TYPE	ABRASIVE SIZE/TYPE LUBRICANT	FORCE (kPa)	HEAD ROTATION	TIME (s)	REFLECTIVE FACTOR	MICROGRAPH YES/NO	IMAGE ANALYSIS YES/NO	REMARKS

The micrograph Figure 13.4 (see Plate 5)

Example 13.5

Family classification BB11

The material: alumina fibres in aluminium

The search for new materials that are lighter will always be paramount to the designer of power driven vehicles. To save weight is to save money; reduce vehicle emission and save the world's diminishing resources. It is to this end that the alumina fibres offer attractions in the automotive industry, the attraction being lighter weight and more fuel efficient engines. Reinforced aluminium alloys with short-staple alumina fibres offer enhanced modulus, improved tensile and fatigue strength, lower coefficient of thermal expansion and superior wear resistance. The concept of selected area reinforcement, using fibre preforming technology and squeeze casting, is proving attractive. Future development could include the lighter magnesium alloys. Also of future interest will be the development of new alloying possibilities, via powder metallurgy.

The technique

The preparation technique is influenced by the fibre volume fraction and the matrix alloy. The quoted preparation example is for a 0.20 Vf in a 10 to 13% silica aluminium matrix. Impressed grinding abrasives can be a problem, in particular as the silicon content reduces; on such occasions it will be necessary to:

(a) avoid the use of charged abrasives or
(b) increase the charged abrasive size or
(c) only use small charged abrasives in a high pH solution.

Some procedures recommend the use of fixed abrasives at every sample integrity stage; due to the structural damage evident in the alumina fibres a more suitable solution was sought. It will be noticed from the enclosed method how this has been overcome by absorbing the grinding shock through the use of charged platen surfaces. Also of importance is the use of low pressure at the polishing stage. To use a normal pressure would be to disguise poor preparation and destroy the fibre interface integrity by smearing.

Sectioning alumina fibres can be carried out using the silicon carbide, resin bonded, abradable high speed wheel. Mounting requires that the abrasive characteristics of the specimen and the mounting material be compatible; to this end, an epoxy thermosetting hot mounting resin was used.

Planar grinding is best achieved (least damage) using diamond as the abrasive on surface 10. Silicon carbide if used would create deep damage, difficult to remove at subsequent stages.

The carbon fibres in Example 13.4 were polished using alumina. With this example, colloidal silica is used. Colloidal silica with its high pH has a gentle polishing action on the aluminium matrix, yet still has an effective grinding action on the alumina fibres.

The matrix often varies from the near pure aluminium to include silicon or copper; this will have an effect on the rate of progress and the material removal parameters. When using the same technique for this group of materials, the Z axis must be adhered to. If this proves difficult it is indicative of the need for parameter changes, as indicated above.

The method

AUDITABLE PREPARATION PROCEDURE — No. 5

CLASSIFICATIONS: BB11/MMC		MATERIAL: ALUMINA FIBRES IN LM13	
SECTIONING TECHNIQUE:	EQUIPMENT TYPE	BLADE	CONCENTRATION
MOUNTING TECHNIQUE:	METHOD		SAMPLE/MOUNT RATIO

	SURFACE and TYPE	ABRASIVE SIZE/TYPE LUBRICANT	FORCE (kPa)	HEAD ROTATION	Z AXIS (μm)	REFLECTIVE FACTOR	MICROGRAPH YES/NO	IMAGE ANALYSIS YES/NO	REMARKS
PLANAR GRINDING STAGE	S.10	30 μm OIL	35	CONTRA	PLANE				

	SURFACE and TYPE	ABRASIVE SIZE/TYPE LUBRICANT	FORCE (kPa)	HEAD ROTATION	Z AXIS (μm)	REFLECTIVE FACTOR	MICROGRAPH YES/NO	IMAGE ANALYSIS YES/NO	REMARKS
SAMPLE INTEGRITY STAGE	S.4	9 μm OIL	35	CONTRA					
	S.1	1 μm OIL	35	COMP					
	S.1	0.06 μm COLLOIDAL SILICA	35	COMP					

	SURFACE and TYPE	ABRASIVE SIZE/TYPE LUBRICANT	FORCE (kPa)	HEAD ROTATION	TIME (s)	REFLECTIVE FACTOR	MICROGRAPH YES/NO	IMAGE ANALYSIS YES/NO	REMARKS
POLISHING STAGE	PS 7	0.06 μm COLLOIDAL SILICA	17	COMP	100		YES		13.5 (Plate 5)

The micrograph Figure 13.5 (see Plate 5). The visual impact of this micrograph can be improved if the reader rotates it through 180°

Example 13.6

Family classification BB11

The material: aluminium matrix/silicon carbide particulate

Modern aluminium alloys can exhibit high tensile strengths, challenging both alloys of steel and titanium. Unfortunately, however, they are not comparable in other properties such as in stiffness, temperature stability, coefficient of expansion and wear resistance. By using a particulate ceramic such as silicon carbide or alumina to create a composite material, the situation is very different.

Ceramic particles in metals are becoming more than just experimental materials; their inclusion in aluminium, magnesium and titanium has resulted in improved component performance. The high performance characteristics of particulate reinforced MMCs are proving attractive for aerospace, automotive and domestic appliances.

Although particulate reinforcement does not offer the same significant improvements in properties as does fibre reinforcement, improvement is nevertheless substantial and it does offer significant savings in production costs. Where price constraints take a high priority, aluminium-based MMC reinforced with 10 to 20% by volume of SiC or alumina particulate seems most favourable. These cost effective materials exhibit a high tensile strength with an associated high modulus of elasticity; they also have good frictional wear properties and are light in weight. One other interesting feature is the ability to decrease the coefficient of thermal expansion with an increase in particulate concentration. This is important for applications where dimensional thermal matching is necessary. This is also a technique the metallographer uses in reducing thermal contraction when hot mounting. These materials can be cast, extruded or forged. The performance of these materials depends upon the particulate size, concentration, distribution, matrix/particulate interface, and additionally on the matrix and alloying elements.

The technique

Sectioning MMCs is best carried out using a diamond wheel, in particular when sectioning large dimensioned materials. Recently a 70 mm diameter component was sectioned within 2 minutes using resin-bonded diamond blade; to use a silicon carbide or alumina bonded wheel with this large diameter component would prove nearly impossible. Mounting presents little problem and where possible hot mounting is to be recommended.

The sample integrity stage requires a careful study when selecting the appropriate recipe if structural damage to the particulate and impressed particles or impressed abrasives in the matrix are to be avoided. Planar grinding can be carried out using 180 to 240 grit silicon carbide paper, providing the specimen particulate is one third (or less) of the grit size used on the papers, i.e. particles of 25 μm and less.

If structural damage is to be avoided at the sample integrity stage, the cutting abrasive must not only be hard but must also absorb the impact shock. Choice of the first charged integrity surface will very much depend upon the composite matrix. A high alloy matrix could, for example, tolerate surface 6 without exhibiting impressed grinding abrasives. Where the matrix is pure (softer) it may be necessary to use surface 2. The alternative particulate to silicon carbide is alumina; such materials could also necessitate the use of platen surface 2 as the primary surface. The chosen abrasive size is governed by the particulate size, i.e. the larger the particulate the larger the grinding abrasive size.

In the example given, the first stage employs platen surface 4 with a 9 μm diamond oil-based slurry. Sample integrity is completed using platen surface 1, with 1 μm diamond and colloidal silica. To use 1 μm diamond on surface 1 without a high pH solution would encourage unwanted impressed abrasive. To use 1 μm, adopting the 'relief polishing' technique, on high density fibre polishing cloths would be another alternative. Polishing is achieved by reducing the pressure to half the grinding force; failure to do this will smear the matrix surface.

The method

AUDITABLE PREPARATION PROCEDURE   No. 6

CLASSIFICATIONS: BB11/Metal Matrix Particulate		MATERIAL: SiC IN ALUMINIUM MATRIX
SECTIONING TECHNIQUE:	EQUIPMENT TYPE BLADE	CONCENTRATION
MOUNTING TECHNIQUE:	METHOD	SAMPLE/MOUNT RATIO

PLANAR GRINDING STAGE	SURFACE and TYPE	ABRASIVE SIZE/TYPE LUBRICANT	FORCE (kPa)	HEAD ROTATION	Z AXIS (μm)	REFLECTIVE FACTOR	MICROGRAPH YES/NO	IMAGE ANALYSIS YES/NO	REMARKS
	S.10 LARGE PART SiC SMALL PART	30 μm OIL 180 SiC WATER	35	CONTRA	PLANE				

SAMPLE INTEGRITY STAGE	SURFACE and TYPE	ABRASIVE SIZE/TYPE LUBRICANT	FORCE (kPa)	HEAD ROTATION	Z AXIS (μm)	REFLECTIVE FACTOR	MICROGRAPH YES/NO	IMAGE ANALYSIS YES/NO	REMARKS
	S.4	9 μm OIL	35	CONTRA					
	S.1	1 μm DIAMOND OIL 0.06 μm SILICA (HIGH pH)	35	COMP					

POLISHING STAGE	SURFACE and TYPE	ABRASIVE SIZE/TYPE LUBRICANT	FORCE (kPa)	HEAD ROTATION	TIME (MIN)	REFLECTIVE FACTOR	MICROGRAPH YES/NO	IMAGE ANALYSIS YES/NO	REMARKS
	PS 7	0.06 μm SILICA	16	COMP	2	/	YES	/	13.6 (Plate 5)

The micrograph Figure 13.6 (see Plate 5)

Example 13.7

Family classification BB11

The material: Borsic fibres in aluminium silicon matrix

This is another example of high strength composite materials where the accompanying low density characteristics make it a serious contender in the automotive industry and many others.

The technique

This material can be prepared in such a manner as to induce irrevocable damage. If this should occur at the first step (cutting), then subsequent preparation is of little point. The tungsten boron fibre with its often submicrometre silicon carbide coat will fracture along its length if incorrectly ground. It must therefore be sectioned on a low deformation saw, using a thin diamond blade. The planar grinding stage is also critical to longitudinal fibre fracture—hence the reason for a resin-bonded diamond wheel (nickel-plated diamond wheels are too severe).

Platen surface 10, normally used for planar grinding, has to be used in the integrity mode for the first step. The tendency for abrasive particles to impress into a soft matrix increases as the diamond size reduces; in consequence, the use of metal-based platen surfaces with small abrasives must be avoided. The photomicrograph taken with a ×100 objective illustrates the outer silicon carbide layer normally obscured through bad preparation techniques.

Careful observation of the micrograph will reveal a host of small impressed diamonds in the aluminium matrix. This can be avoided by using the 'relief-polishing' technique.

When preparing materials such as these, there are many pitfalls to be avoided:

(i) Small abrasives (1 μm) will cut the fibre but will be attracted to the matrix (impress).
(ii) Medium sized abrasives (3 μm) will cut the matrix but will tend to go around the fibre, creating a moat.
(iii) Large sized grinding abrasives (15 μm) will be dislodged from the charged integrity surface.
(iv) Fixed abrasive platen surfaces of the same size (15 μm) will crack the fibres.
(v) Grinding abrasives can lodge at the fibre interface (ultrasonic cleaning is necessary).
(vi) The 'relief-polishing' technique can help.

The method

AUDITABLE PREPARATION PROCEDURE   No. 7

CLASSIFICATIONS: BB11/Metal Matrix Fibre Composite		MATERIAL: BORSIC FIBRE IN ALUMINIUM
SECTIONING TECHNIQUE:	EQUIPMENT TYPE BLADE	CONCENTRATION
MOUNTING TECHNIQUE:	METHOD	SAMPLE/MOUNT RATIO

	SURFACE and TYPE	ABRASIVE SIZE/TYPE LUBRICANT	FORCE (kPa)	HEAD ROTATION	Z AXIS (μm)	REFLECTIVE FACTOR	MICROGRAPH YES/NO	IMAGE ANALYSIS YES/NO	REMARKS
PLANAR GRINDING STAGE	RESIN BOND OIL	320 GRIT WATER FIXED DIAMOND	35	CONTRA	PLANE				

	SURFACE and TYPE	ABRASIVE SIZE/TYPE LUBRICANT	FORCE (kPa)	HEAD ROTATION	Z AXIS (μm)	REFLECTIVE FACTOR	MICROGRAPH YES/NO	IMAGE ANALYSIS YES/NO	REMARKS
SAMPLE INTEGRITY STAGE	S.10	6 μm OIL	35	CONTRA					
	S.1	3 μm WATER	50	COMP					
	S.1	1 μm high pH	50	CONTRA					

	SURFACE and TYPE	ABRASIVE SIZE/TYPE LUBRICANT	FORCE (kPa)	HEAD ROTATION	TIME (s)	REFLECTIVE FACTOR	MICROGRAPH YES/NO	IMAGE ANALYSIS YES/NO	REMARKS
POLISHING STAGE	PS 7	0.06 μm COLLOIDAL SILICA	16	COMP	300		YES		13.7 (Plate 5)

The micrograph Figure 13.7 (see Plate 5)

Example 13.8

Family classification BB13

The material: silicon carbide fibres in heat-resistant glass

Research into materials capable of withstanding stress at temperatures in excess of 1400°C favours the ceramic/glass matrix composites. One of the restrictions imposed on the use of brittle fracture materials is crack propagation. This can be restricted through careful selection of matrix structures and particulate or fibre interface strength. Tensile strength is improved in fibre matrix composites when the continuous fibre is unidirectionally orientated.

The technique

This is not the easiest of materials to prepare due to the brittle and poor abrasive features of both the silicon carbide fibres and the glass matrix. Transverse fibres split longitudinally and longitudinal fibres break away at the infinity point (fibre end). Since glass is poor in shear stress the matrix can be constructed to reduce crack propagation. With this matrix it is possible to see rosettes previously obscured (plucked out). This is a characteristic of the glass and is not to be interpreted as crack inhibitors. With hard brittle subjects it is possible to adopt a lapping technique which has proved very successful, but unfortunately the procedure is measured in many hours, rather than minutes.

The cutting and mounting are as in Example 13.7 with the additional requirement of aligning the longitudinal fibres parallel to the direction of abrasive force. (*Note:* all procedures use contra motion head direction.) From the planar grinding stage it is necessary to use a charged platen surface 4. This is because surfaces 10, 8 and 6 will only operate in a lapping mode—considered undesirable. Note again the use of a high pH at the last integrity stage, proving advantageous with ceramics.

128 Surface Preparation and Microscopy of Materials

The method

AUDITABLE PREPARATION PROCEDURE No. 8

CLASSIFICATIONS: BB13/Glass Matrix Fibre Composite	MATERIAL: SILICON CARBIDE FIBRES IN HEAT RESISTANT GLASS
SECTIONING TECHNIQUE: EQUIPMENT TYPE BLADE	CONCENTRATION
MOUNTING TECHNIQUE: METHOD	SAMPLE/MOUNT RATIO

PLANAR GRINDING STAGE	SURFACE and TYPE	ABRASIVE SIZE/TYPE LUBRICANT	FORCE (kPa)	HEAD ROTATION	Z AXIS (µm)	REFLECTIVE FACTOR	MICROGRAPH YES/NO	IMAGE ANALYSIS YES/NO	REMARKS
	RESIN BOND OIL	320 GRIT WATER FIXED DIAMOND	35	CONTRA	PLANE				

SAMPLE INTEGRITY STAGE	SURFACE and TYPE	ABRASIVE SIZE/TYPE LUBRICANT	FORCE (kPa)	HEAD ROTATION	Z AXIS (µm)	REFLECTIVE FACTOR	MICROGRAPH YES/NO	IMAGE ANALYSIS YES/NO	REMARKS
	S.4	6 µm OIL	35	CONTRA					
	S.2P	6 µm WATER	35	CONTRA					
	S.1	1 µm OIL pH ACTIVE	35	CONTRA	3		YES		13.8 (Plate 5)

POLISHING STAGE	SURFACE and TYPE	ABRASIVE SIZE/TYPE LUBRICANT	FORCE (kPa)	HEAD ROTATION	TIME (s)	REFLECTIVE FACTOR	MICROGRAPH YES/NO	IMAGE ANALYSIS YES/NO	REMARKS

The micrograph Figure 13.8 (see Plate 5)

Example 13.9

Family classification BB22

The material: carbon–carbon composites

Carbon–carbon is a composite material consisting of carbon fibres in a matrix of carbon. By selecting the carbon fibres, the form of fibre substrate, the amount of fibre and the processing technology of matrix manufacture, a range of properties can be produced. The materials have the following basic characteristics:

(i) Thermal stability in inert atmospheres up to 2500°C
(ii) Retention of strength in the absence of oxygen
(iii) High thermal and shock resistance
(iv) Low coefficient of thermal expansion
(v) Non-toxic
(vi) Good wear resistance
(vii) Low density.

The main uses of carbon–carbon are aircraft brake discs and rocket components with a lower volume use in nuclear fusion reactors, prosthetic implants, furnace furniture and racing car brakes.

Microscopy is a key control technique where the CVD (chemical vapour deposition) process is used. The microstructure must show a high level of optical activity and 'rough laminar' growth cones to produce a matrix that will graphitize on heat treatment at over 2000°C.

The optical birefringence of these features manifests itself when using reflected light polarizing microscopy. To introduce the first-order red sensitive tint plate between the polarizer and analyser will reward the observer with a very colourful subject. An infinite variety of colours can be achieved by rotating the polarizer.

The technique

The preparation technique for carbon–carbon can be very critical, in particular if the porosity level is high. The high optical activity, although very informative, with a well-prepared sample can be destroyed by preparation-induced structural damage. As with all specimen preparation techniques the need for appropriate care starts from the very first stage and must be tailored to suit the material condition. If, for example, the fibres are well bonded then a normal high speed slitting wheel can be used for sectioning. If this should not be the case, then the slow speed low deformation wheels must be used. Encapsulation must also be appropriate to the material condition; those carbon–carbons exhibiting porosity must be vacuum impregnated with an epoxy resin whose viscosity is in the region of 250 cP (centipoise). If the material is extremely friable it could be necessary to impregnate prior to sectioning, but this generally is not necessary. The abrasion rate of carbon–carbon is very high, in comparison with epoxy which is much lower (especially when polishing). By implication, this means that when mounting a small area of carbon–carbon in a much greater area of resin, a multimoulding technique must be applied, i.e. the specimen must be surrounded with epoxy resin and the whole then encapsulated in, say, phenolic resin.

When vacuum impregnating, the depth of resin ingress is related to not just the achieved vacuum but also the individual pore dimensions. It is important at the planar grinding and sample integrity stage not to exceed this ingressed depth. This necessitates a very short step at the silicon carbide stage, using the finest grade compatible with stock removal. The sample integrity stage must remove any residual preparation damage, faithfully revealing the true structure. Platen surface 4, using 9 μm oil-based diamond, progresses the preparation to within 1 μm (Z axis) of integrity. Integrity is achieved using gamma alumina on platen surface 1, followed by a short polishing step, again using gamma alumina at lower pressure on polishing surface 7.

From the micrograph of Figure 13.9 (Plate 5), observe the various 'Maltese-cross' effects; this is synonymous with both good fibre growth and exceptionally good sample preparation. Observe the longitudinal extinguished fibre core with the resolved structure of the carbon growth. The translucent effect normally associated with transversed fibres has been optically 'compensated out', resulting in a good material, well presented and caringly photographed.

The method

AUDITABLE PREPARATION PROCEDURE No. 9

CLASSIFICATIONS: BB22/COMPOSITE		MATERIAL: CARBON–CARBON	
SECTIONING TECHNIQUE:	EQUIPMENT TYPE BLADE	CONCENTRATION	
MOUNTING TECHNIQUE:	METHOD	SAMPLE/MOUNT RATIO	

	SURFACE and TYPE	ABRASIVE SIZE/TYPE LUBRICANT	FORCE (kPa)	HEAD ROTATION	Z AXIS (µm)	REFLECTIVE FACTOR	MICROGRAPH YES/NO	IMAGE ANALYSIS YES/NO	REMARKS
PLANAR GRINDING STAGE	PAPER	P320 SiC WATER	35	COMP	PLANE				

	SURFACE and TYPE	ABRASIVE SIZE/TYPE LUBRICANT	FORCE (kPa)	HEAD ROTATION	Z AXIS (µm)	REFLECTIVE FACTOR	MICROGRAPH YES/NO	IMAGE ANALYSIS YES/NO	REMARKS
SAMPLE INTEGRITY STAGE	S.4	9 µm OIL	35	CONTRA					
	S.1	GAMMA ALUMINA	35	COMP					

	SURFACE and TYPE	ABRASIVE SIZE/TYPE LUBRICANT	FORCE (kPa)	HEAD ROTATION	TIME (s)	REFLECTIVE FACTOR	MICROGRAPH YES/NO	IMAGE ANALYSIS YES/NO	REMARKS
POLISHING STAGE	PS 7	COLLOIDAL SILICA	17	COMP	180		YES		13.9 (Plate 5)

The micrograph Figure 13.9 (see Plate 5)

Example 13.10

Family classification BB19

The material: silicon carbide/graphite/paint layer

The method

AUDITABLE PREPARATION PROCEDURE		No. 10	
CLASSIFICATIONS: BB19/MIXED COMPOSITE		MATERIAL: SiC/GRAPHITE–PAINT LAYER	
SECTIONING TECHNIQUE:	EQUIPMENT TYPE BLADE	CONCENTRATION	
MOUNTING TECHNIQUE:	METHOD	SAMPLE/MOUNT RATIO	

	SURFACE and TYPE	ABRASIVE SIZE/TYPE LUBRICANT	FORCE (kPa)	HEAD ROTATION	Z AXIS (μm)	REFLECTIVE FACTOR	MICROGRAPH YES/NO	IMAGE ANALYSIS YES/NO	REMARKS
PLANAR GRINDING STAGE	RESIN BOND	320 GRIT DIAMOND WHEEL	35	COMP	PLANE				

	SURFACE and TYPE	ABRASIVE SIZE/TYPE LUBRICANT	FORCE (kPa)	HEAD ROTATION	Z AXIS (μm)	REFLECTIVE FACTOR	MICROGRAPH YES/NO	IMAGE ANALYSIS YES/NO	REMARKS
SAMPLE INTEGRITY STAGE	S.2P	9 μm OIL	35	COMP					
	S.1	3 μm WATER	35	COMP					
	S.1	1 μm WATER	35	COMP					

	SURFACE and TYPE	ABRASIVE SIZE/TYPE LUBRICANT	FORCE (kPa)	HEAD ROTATION	TIME (min)	REFLECTIVE FACTOR	MICROGRAPH YES/NO	IMAGE ANALYSIS YES/NO	REMARKS
POLISHING STAGE	PS 7	0.06 μm SILICA	16	COMP	5		YES		13.10 (Plate 6)

The micrograph Figure 13.10 (see Plate 6)

Example 13.11

Family classification BB11

The material: alumina fibres in aluminium/copper matrix

The method

AUDITABLE PREPARATION PROCEDURE   No. 11

CLASSIFICATIONS: BB11/MMC				MATERIAL: ALUMINA FIBRES IN ALUMINIUM COPPER			
SECTIONING TECHNIQUE:	EQUIPMENT TYPE		BLADE	CONCENTRATION			
MOUNTING TECHNIQUE:	METHOD			SAMPLE/MOUNT RATIO			

	SURFACE and TYPE	ABRASIVE SIZE/TYPE LUBRICANT	FORCE (kPa)	HEAD ROTATION	Z AXIS (µm)	REFLECTIVE FACTOR	MICROGRAPH YES/NO	IMAGE ANALYSIS YES/NO	REMARKS
PLANAR GRINDING STAGE	S.10	45 µm OIL	35	CONTRA					

	SURFACE and TYPE	ABRASIVE SIZE/TYPE LUBRICANT	FORCE (kPa)	HEAD ROTATION	Z AXIS (µm)	REFLECTIVE FACTOR	MICROGRAPH YES/NO	IMAGE ANALYSIS YES/NO	REMARKS
SAMPLE INTEGRITY STAGE	S.4	9 µm OIL	35	CONTRA					
	S.1	1 µm OIL pH ACTIVE	35	COMP					

	SURFACE and TYPE	ABRASIVE SIZE/TYPE LUBRICANT	FORCE (kPa)	HEAD ROTATION	TIME (s)	REFLECTIVE FACTOR	MICROGRAPH YES/NO	IMAGE ANALYSIS YES/NO	REMARKS
POLISHING STAGE	PS 7	COLLOIDAL SILICA	17	COMP	120		YES		13.11 (Plate 6)

The micrograph Figure 13.11 (see Plate 6)

Example 13.12

Family classification BB11

The material: silicon carbide fibres/graphite core in aluminium matrix

The method

AUDITABLE PREPARATION PROCEDURE									No. 12	
CLASSIFICATIONS: BB11/MMC						MATERIAL: SiC FIBRES/GRAPHITE CORE/Al MATRIX				
SECTIONING TECHNIQUE:	EQUIPMENT TYPE		BLADE			CONCENTRATION				
MOUNTING TECHNIQUE:	METHOD					SAMPLE/MOUNT RATIO				

	SURFACE and TYPE	ABRASIVE SIZE/TYPE LUBRICANT	FORCE (kPa)	HEAD ROTATION	Z AXIS (μm)	REFLECTIVE FACTOR	MICROGRAPH YES/NO	IMAGE ANALYSIS YES/NO	REMARKS
PLANAR GRINDING STAGE	S.10	30 μm OIL	35	CONTRA	PLANE				

	SURFACE and TYPE	ABRASIVE SIZE/TYPE LUBRICANT	FORCE (kPa)	HEAD ROTATION	Z AXIS (μm)	REFLECTIVE FACTOR	MICROGRAPH YES/NO	IMAGE ANALYSIS YES/NO	REMARKS
SAMPLE INTEGRITY STAGE	S.4	9 μm OIL	35	CONTRA					
	S.1	1 μm OIL	70	COMP					

	SURFACE and TYPE	ABRASIVE SIZE/TYPE LUBRICANT	FORCE (kPa)	HEAD ROTATION	TIME (s)	REFLECTIVE FACTOR	MICROGRAPH YES/NO	IMAGE ANALYSIS YES/NO	REMARKS
POLISHING STAGE	PS 7	COLLOIDAL SILICA	17	COMP	240		YES		13.12 (Plate 6)

The micrograph Figure 13.12 (see Plate 6)

Example 13.13

Family classification BB11

The material: ceramic particles/steel matrix

The method

AUDITABLE PREPARATION PROCEDURE			No. 13	
CLASSIFICATIONS: BB11/MMC			MATERIAL: CERAMIC/STEEL MATRIX	
SECTIONING TECHNIQUE:	EQUIPMENT TYPE	BLADE	CONCENTRATION	
MOUNTING TECHNIQUE:	METHOD		SAMPLE/MOUNT RATIO	

	SURFACE and TYPE	ABRASIVE SIZE/TYPE LUBRICANT	FORCE (kPa)	HEAD ROTATION	Z AXIS (µm)	REFLECTIVE FACTOR	MICROGRAPH YES/NO	IMAGE ANALYSIS YES/NO	REMARKS
PLANAR GRINDING STAGE	S.10	45 µm OIL	35	CONTRA	PLANE				

	SURFACE and TYPE	ABRASIVE SIZE/TYPE LUBRICANT	FORCE (kPa)	HEAD ROTATION	Z AXIS (µm)	REFLECTIVE FACTOR	MICROGRAPH YES/NO	IMAGE ANALYSIS YES/NO	REMARKS
SAMPLE INTEGRITY STAGE	S.6	9 µm OIL	35	CONTRA					
	S.2P	3 µm OIL	35	CONTRA					
	S.1	1µm OIL pH ACTIVE	35	COMP	1	/	YES	/	13.13 (Plate 6)

	SURFACE and TYPE	ABRASIVE SIZE/TYPE LUBRICANT	FORCE (kPa)	HEAD ROTATION	TIME (s)	REFLECTIVE FACTOR	MICROGRAPH YES/NO	IMAGE ANALYSIS YES/NO	REMARKS
POLISHING STAGE									

The micrograph Figure 13.13

Figure 13.13 Ceramic particulates in steel (×200)

Example 13.14

Family classification BB11

The material: titanium carbide in titanium

The method

AUDITABLE PREPARATION PROCEDURE			No. 14	
CLASSIFICATIONS: BB11/METAL MATRIX PARTICULATE COMPOSITE			MATERIAL: TITANIUM CARBIDE IN TITANIUM	
SECTIONING TECHNIQUE:	EQUIPMENT TYPE	BLADE	CONCENTRATION	
MOUNTING TECHNIQUE:	METHOD		SAMPLE/MOUNT RATIO	

PLANAR GRINDING STAGE	SURFACE and TYPE	ABRASIVE SIZE/TYPE LUBRICANT	FORCE (kPa)	HEAD ROTATION	Z AXIS (μm)	REFLECTIVE FACTOR	MICROGRAPH YES/NO	IMAGE ANALYSIS YES/NO	REMARKS
	S.10	30 μm OIL	35	CONTRA	PLANE				

SAMPLE INTEGRITY STAGE	SURFACE and TYPE	ABRASIVE SIZE/TYPE LUBRICANT	FORCE (kPa)	HEAD ROTATION	Z AXIS (μm)	REFLECTIVE FACTOR	MICROGRAPH YES/NO	IMAGE ANALYSIS YES/NO	REMARKS
	S.4	9 μm OIL	35	CONTRA					
	S.1	1 μm WATER	70	COMP					

POLISHING STAGE	SURFACE and TYPE	ABRASIVE SIZE/TYPE LUBRICANT	FORCE (kPa)	HEAD ROTATION	TIME (s)	REFLECTIVE FACTOR	MICROGRAPH YES/NO	IMAGE ANALYSIS YES/NO	REMARKS
	PS 7	0.06 μm COLLOIDAL SILICA	17	COMP	100		YES		13.14 (Plate 6)

The micrograph Figure 13.14 (see Plate 6)

136 Surface Preparation and Microscopy of Materials

Example 13.15

Family classification BB11

The material: silicon carbide particles in aluminium/silicon matrix

The method

AUDITABLE PREPARATION PROCEDURE   No. 15

CLASSIFICATIONS: BB11/MMC						MATERIAL: SiC PARTICULATE IN ALUMINIUM/SILICON MATRIX			
SECTIONING TECHNIQUE:	EQUIPMENT TYPE		BLADE			CONCENTRATION			
MOUNTING TECHNIQUE:	METHOD					SAMPLE/MOUNT RATIO			

	SURFACE and TYPE	ABRASIVE SIZE/TYPE LUBRICANT	FORCE (kPa)	HEAD ROTATION	Z AXIS (µm)	REFLECTIVE FACTOR	MICROGRAPH YES/NO	IMAGE ANALYSIS YES/NO	REMARKS
PLANAR GRINDING STAGE	PAPER OIL	180 SiC WATER	35	COMP	PLANE				

	SURFACE and TYPE	ABRASIVE SIZE/TYPE LUBRICANT	FORCE (kPa)	HEAD ROTATION	Z AXIS (µm)	REFLECTIVE FACTOR	MICROGRAPH YES/NO	IMAGE ANALYSIS YES/NO	REMARKS
SAMPLE INTEGRITY STAGE	S.4	9 µm OIL	35	CONTRA					
	S.1	1 µm OIL	35	COMP					
	S.1	COLLOIDAL SILICA	35	COMP					

	SURFACE and TYPE	ABRASIVE SIZE/TYPE LUBRICANT	FORCE (kPa)	HEAD ROTATION	TIME (s)	REFLECTIVE FACTOR	MICROGRAPH YES/NO	IMAGE ANALYSIS YES/NO	REMARKS
POLISHING STAGE	PS 7	COLLOIDAL SILICA	17	COMP	90		YES		13.15 (Plate 6)

The micrograph Figure 13.15 (see Plate 6)

13.4 Biomaterials

13.4.1 The material

Materials selected for implantation into the human body are extremely important and are subjected to what could be described as a hostile environment. Assuming that the selected material satisfies the requirements of strength and is corrosion resistant, then the interaction of the body chemistry and cell growth must be evaluated. The compatibility of various materials used in implants is very often the subject of microstructural analysis. In biomedical studies the microstructure is important at the preimplant stage; it is also necessary to retrieve implants after clinical trials to examine the microstructure at the interface between the implant and its host. This important area of bone integration (osseointegration) has resulted in much research into implant materials and material porosity. Metal, although not the most desirable material (it does not biodegrade), is used because of its tensile, shear and compressive strength; it can be made rigid to the bone through the use of pins and screws. The tissue exposed metal, once the bone has healed, is redundant. Unfortunately, in some cases, it has to remain to endure the hostile environment and perhaps weaken the natural bone characteristics. Ideally the material would be biodegradable or a more compatible material exhibiting similar characteristics to the bone, such as Young's modulus, would be desirable. In the case of hip replacement, for example, the femur fitting material must be strong, the bearing surface of the prosthesis exhibiting low coefficients of wear and friction. Alumina has been helpful in reducing wear. Another area of concern is the fixation of the material prosthesis to the bone. Bone cement and uncemented fixation are both practised methods. Titanium and its alloys represent extreme inertness and are used in bone replacement. This material with a porous surface is also the subject of investigation. Bioactive materials such as a calcium hydroxyapatite (HA) offer many attractions, in particular when combined with a stronger substrate. The substrate could be an inert material with a double-layer coating, the first to bond the metal interface, the second layer being a porous HA which encourages bone growth within the interconnecting pores and voids (approximately 100 to 150 μm diameter). These two layers are sometimes plasma sprayed to the substrate; the condition of these layers affects the donor response or perhaps rejection. It is for this reason that a close check on the microstructure is essential. HA surface coatings are still very much in the experimental stage in terms of specification and it could be that the future surfaces will be, say, 50 μm thick, totally dense and produced by a non-plasma technique, giving an extremely smooth surface. The desirable pore size for encouraging bone growth is said to be between 100 and 150 μm diameter. If this is the case, then the coating thickness, for minimum strength, would need to be 500 μm, which is probably unacceptable.

13.4.2 Materials for investigation

Dense or porous coatings can be applied to implant materials via the powder sintering or the powder plasma spray technique. Metal sintering has been used on screws and pins, plasma spraying being a favoured route for larger implants requiring a coating of hydroxyapatite (HA). The two examples chosen for investigation are (a) plasma-sprayed HA on a titanium 6A1–4V substrate and (b) two different species of coral.

The examples quoted are preimplant conditions; there is obviously a need to investigate the interfacial bond after varying periods of use. The rate of bonding and the strength and stability of the bond vary with the composition and microstructure of the bioactive material. The surface preparation techniques given will also apply at the clinical trial stage. Techniques using cross-sections in reflected light are recommended when investigating materials exhibiting brittle fracture mechanisms (bone, coral, hydroxyapatite, etc.). When it is necessary to investigate tissue ingress from the implant, as opposed to implant ingress, it is necessary to prepare thin sections for transmitted light observation.

13.5 Material Investigations

13.5.1 Hydroxyapatite (Example 13.16)

This material is non-toxic yet biologically active and is used to create an interfacial bond between the host and donor. The material in this example has been plasma sprayed onto titanium alloy substrate. HA bonds to the substrate by mechanical, not chemical or diffusion, means and therefore requires the substrate to be roughened or porous. In this example, the interface surface has been grit blasted prior to coating; it could alternatively have been titanium powder plasma coated. The sprayed coating should make a good mechanical bond to the interface surface, the spraying parameters controlling the porosity level. The requirements of the coating could be fully dense or open structured. The microsectioning technique will reveal the integrity of the coating manifesting any artefacts undesirable to its performance in use.

Surface preparation techniques

(i) *Sectioning*. Use a thin blade on a low deformation cut-off machine. Suitable blades would be the high speed abradable resin bonded silicon carbide or the low speed cubic boron nitride (CBN) wheel. Always section from the direction of coating into the substrate and do not clamp the portion designated to be the sample.

(ii) *Mounting*. It is necessary to encapsulate the sample to protect the coating; vacuum impregnation using a low viscous (250 cP) epoxy resin is necessary with porous coatings. Compression mounting is not to be recommended for these materials unless a protective cold mounted film is first applied.

(iii) *Grinding*. Directional preparation with a progressively smaller abrasive on a reducing shock absorbing surface will reveal a faithful reproduction of the microstructure. The method chosen for this material is illustrated in Table 13.1.

Table 13.1 Method

AUDITABLE PREPARATION PROCEDURE No. 16

CLASSIFICATIONS:			MATERIAL: HYDROXYAPATITE	
SECTIONING TECHNIQUE:	EQUIPMENT TYPE ISOMET PLUS	BLADE	CONCENTRATION	
MOUNTING TECHNIQUE:	METHOD VACUUM IMPREGNATED EPOXY		SAMPLE/MOUNT RATIO	

*Samples mounted directionally.

	SURFACE and TYPE	ABRASIVE SIZE/TYPE LUBRICANT	FORCE (kPa)	HEAD ROTATION	Z AXIS (μm)	REFLECTIVE FACTOR	MICROGRAPH YES/NO	IMAGE ANALYSIS YES/NO	REMARKS
PLANAR GRINDING STAGE	SiC	P240 WATER	35	CONTRA	PLANE				

	SURFACE and TYPE	ABRASIVE SIZE/TYPE LUBRICANT	FORCE (kPa)	HEAD ROTATION	Z AXIS (μm)	REFLECTIVE FACTOR	MICROGRAPH YES/NO	IMAGE ANALYSIS YES/NO	REMARKS
SAMPLE INTEGRITY STAGE	S.4	9 μm OIL	35	CONTRA	24				
	S.1	1 μm HIGH pH	70	CONTRA	6				

	SURFACE and TYPE	ABRASIVE SIZE/TYPE LUBRICANT	FORCE (kPa)	HEAD ROTATION	TIME (s)	REFLECTIVE FACTOR	MICROGRAPH YES/NO	IMAGE ANALYSIS YES/NO	REMARKS
POLISHING STAGE	PS 7	0.06 μm COLLOIDAL SILICA	17	CONTRA	120		YES		13.16 (Plate 7)

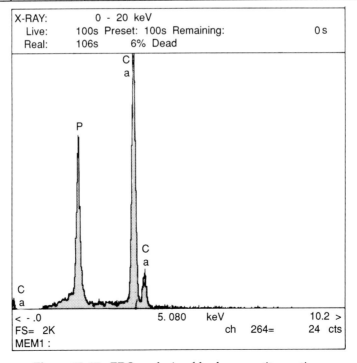

Figure 13.17 EDS analysis of hydroxyapatite coating

 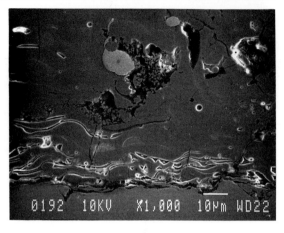

(a) (b)

Figure 13.18 Regions of porosity in hydroxyapatite coating. (a) Unmelted particles of sponge-like appearance. (b) Globular copper contamination

Auditable preparation procedures (see Table 13.1) will require completion of all monitoring parameters shown to the right-hand side of the thick central line. It is also assumed that machine operating conditions such as surface speed will be optimized. Time is not included in the method since this will vary depending upon the size of the sample, ratio of mounting resin, number of samples in an automatic head, efficiency of grinding parameters, etc. Auditing therefore utilizes the Z axis material removal, reflectivity factor, micrograph, image analysis and such factors as scratch pattern.

The analysis of the prepared surface requires that it is a faithful reproduction presented in the most favourable optical condition. From Figure 13.16 (see Plate 7) the coating has been displayed in (a) bright field to show the planar information and (b) first-order grey DIC to give an in-depth picture of the same field of view. Notice the radial cracks, probably caused by differential thermal contraction or thermal shocking. Without a trained eye the porosity level is not easy to detect; this can be overcome by using a fluorescent dye in the mounting resin.

A requirement of the coating shown in Figure 13.16 is to have a uniformly distributed porosity of between 15 and 20%. This clearly is not the case, the level within the field of view being nearly zero. From Figure 13.16(b) there is a strong indication that all is not well at the coating/substrate interface. Figure 13.16(c), taken at ×400 magnification, reveals unmelted particles as well as confirming the unresolved structure at the coating/substrate interface. Using a ×100 objective (×1000 magnification) this interface exhibits a 2 μm undesirable fragmented layer. The resolution of the layer was not clear, which was thought to be a result of preferential grinding between the HA and the substrate or some type of chemical attack, due to using water during the preparation. To confirm the observations made from the optical microscope, it was decided to examine the coating using an electron microscope.

The specimen was sputter coated with a 100 Å layer of gold to ensure conductivity and to prevent charging of the epoxy mounting resin. The back-scattered electron image indicated that the coating was of consistent composition, confirming the optical analysis of a dense layer. Energy dispersion X-ray analysis (EDS) also confirmed the consistent composition as having a basic composition of calcium and phosphorus (see Figure 13.17).

Additional examination was carried out on what appeared to be areas of possible low level porosity. One such example is shown in Figure 13.18(a). This appears to be unmelted particles, as shown in the optical micrograph of Figure 13.16(c).

Another feature found in the vicinity of an unmelted particle was the globular light grey

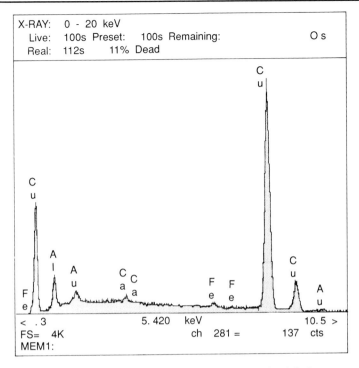

Figure 13.19 EDS analysis of copper-rich globule

subjects shown in Figure 13.18(b). Subsequent analysis indicated these features to be copper rich with some trace of iron (see Figure 13.19). The existence of copper was thought to be from the spray gun.

The use of an electron microscope has added to the information derived from the optical analysis, if for no better reason than to confirm the original diagnosis.

13.5.2 Coral (Example 13.17)

Two different species of coral have been studied (see Figure 13.20 on Plate 7), porites (Figure 13.20(a)) and acropora (Figure 13.20(b)). These samples were prepared as part of an investigation into microporosity. Figure 13.20 (a) and (b) clearly illustrate the vastly different porous structure of the two species. Unlike the pores with interconnecting channels, which occur in plasma-coated materials, coral as it grows creates a honeycomb structure with continuous tunnels slowly varying in size and shape. Therefore black shapes within the micrograph illustrate areas of poor resin ingress. This can be overcome by using a lower viscosity resin or increasing the vacuum pressure.

The procedure for coral looks and is very simple (see Table 13.2), but this does not necessarily imply that any alternative method would yield the same information. To use, as is traditional, a diamond abrasive in the integrity stages is counter-productive due in part to the high abrasive nature of coral. The chemomechanical technique as used removes any residual damage from the silicon carbide stage and if pursued into the polishing mode will reveal structural information relating to the coral.

To assist with resin identification a coloured dye can be used. Figure 13.20(c) illustrates the use of dark ground to reveal the blue dyed resin. Notice how the dark ground technique has outlined the pore top surface morphology with a white line. The ability of seeing inside the sample is due to light passing through the coral and diffracting into the subsurface resin. Polarized light will also manifest the resin with the colour of the dye. Unfortunately, coloured resins are not helpful in bright field applications since they appear grey, irrespective of the dyed colour.

13.5.3 Crystalline or amorphous structures (Example 13.18)

Just as with the question of whether high density or controlled porosity affects interface bonding, so does the implant interface surface condition, the dissolution of the implant material and the tissue growth relative to crystalline or amorphous structures. This information, the microstructure from clinical trials, will be revealed.

Figure 13.21 (see Plate 7) is used to illustrate the different structural phases from the interface with a 60 μm outer layer, with its granular 10 to 20 μm particles, to the 300 μm pore surrounded by a modified enlarged structure. The technique for optically revealing this information is to use the first-order grey DIC.

Table 13.2

AUDITABLE PREPARATION PROCEDURE   No. 17

CLASSIFICATIONS:						MATERIAL: CORAL			
SECTIONING TECHNIQUE:	EQUIPMENT TYPE		BLADE			CONCENTRATION			
MOUNTING TECHNIQUE:	METHOD VACUUM IMPREGNATED EPOXY					SAMPLE/MOUNT RATIO			

	SURFACE and TYPE	ABRASIVE SIZE/TYPE LUBRICANT	FORCE (kPa)	HEAD ROTATION	Z AXIS (μm)	REFLECTIVE FACTOR	MICROGRAPH YES/NO	IMAGE ANALYSIS YES/NO	REMARKS
PLANAR GRINDING STAGE	SiC	P600 WATER	35	COMP	PLANE				

	SURFACE and TYPE	ABRASIVE SIZE/TYPE LUBRICANT	FORCE (kPa)	HEAD ROTATION	Z AXIS (μm)	REFLECTIVE FACTOR	MICROGRAPH YES/NO	IMAGE ANALYSIS YES/NO	REMARKS
SAMPLE INTEGRITY STAGE	S.1	0.06 μm COLLOIDAL SILICA	35	CONTRA	6				

	SURFACE and TYPE	ABRASIVE SIZE/TYPE LUBRICANT	FORCE (kPa)	HEAD ROTATION	TIME (s)	REFLECTIVE FACTOR	MICROGRAPH YES/NO	IMAGE ANALYSIS YES/NO	REMARKS
POLISHING STAGE	PS 7	0.06 μm COLLOIDAL SILICA	20		360		YES		13.17

14

Preparation of Minerals

The object of all these specific chapters is not just to make a collection of preparation techniques but to relate the 'new concept' of specimen preparation with its inherent advantages to each particular kind of material. Minerals, however, have traditionally been prepared using a free lapping technique and it is envisaged that this will continue to be the case. This does not mean that the new ideas are beyond the scope of the mineralogist and these will be reviewed throughout this chapter. In general the changes that could take place in the preparation technique will be those methods that manifest sample integrity in the *shortest time*. Traditional mineralogy, using lapping techniques, has in the majority of cases resulted in a faithful reproduction of the material structure, the time often being prolonged. To reduce the time factor it will be necessary to investigate the more efficient stock removal against time characteristics of grinding. As we are aware, grinding induces greater structural damage than does lapping; it will be necessary therefore to introduce the 'shock absorbing' principles if progress is to be achieved. Charged abrasive platen surfaces will undoubtedly reduce the time factor without the normally high structural damage usually associated with a grinding application. There is also a case for using fixed platen surfaces where appropriate abrasive shock absorbing takes place. Ideally with all brittle fracture materials (minerals, ceramics, etc.) the preparation technique should target for fixed abrasive application since this will be quicker, less expensive and more readily controlled and automated. The problem in the past with using fixed abrasives was the lack of any shock absorbing surfaces, in particular when using small abrasives. Today this is different; there are fixed small sized (10 to 30 μm) diamond abrasives embedded within a series of plastic type dots or similar shapes. The plastic absorbs the abrasive cutting shock; the spots *allow all* cutting swarf to be washed away. Before any surface preparation takes place, decide whether a thick reflected light sample or a thin section is required. If a thin section is required then is it for transmitted or reflected light?

14.1 Thick Sections for Reflected Light Microscopy

14.1.1 Cutting

The so-called 'thick' section is usually a small section taken from a much larger piece of rock or mineral. Diamond saws are usually employed for most, if not all, mineral sections, the initial cut being carried out on a notched rim wheel, the final cut requiring a continuous rim wheel (Figure 14.1)

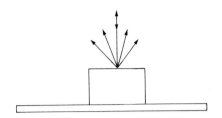

Figure 14.1 Thick sections for reflected light microscopy (1, section and mount; 2, grind—silicon carbide/diamond or 2, lap—silicon carbide/diamond or 2, fixed/charged platen surface or 2, combination of any; 3, polish)

Diamond concentration influences the cutting rate and must be taken into account, in particular at the final cut. Extremely hard minerals, for example, benefit from a low concentration blade.

14.1.2 Mounting

If the subject is very friable, immersion in a cold mounting epoxy resin should take place before cutting. Having sectioned the sample it can be mounted to the appropriate dimensions by using a cold resin, the epoxy range giving the least shrinkage and best penetration (low viscosity). Hot mounting is only possible if the sample will withstand the high pressures and temperatures; ideally the hot mounting will have a preload facility. Minerals exhibiting high levels of porosity or friable structures must be vacuum impregnated

Table 14.1 Relative structural damage

Platen surface	Structural damage number[a]	Time
Fixed diamond—metal bonded	20	1
Fixed diamond—resin bonded	17	1
Charged platen surface—grinding	10	5
Charged platen surface—lapping	5	10
Polishing surface	2	20

[a] The lower the number, the better the choice. Abrasive size identical for all surfaces.

using a low viscous resin, preferably with a warm *clean* dry specimen.

The structural damage as shown in Table 14.1 is used to illustrate the effect of platen surface relative to time.

14.1.3 Planar grinding

This step, being to achieve a uniform flatness across the sample, can best be carried out using silicon carbide papers or, for the very hard materials, fixed diamond grinding wheels. Since this operation utilizes a fixed abrasive, care must be taken not to create an intolerable level of structural damage. Should this be the case then a free lapping mode must be used. When the planar grinding creates grain pluck-out, unavoidable structural damage or exhibits high levels of porosity, it will be necessary to revacuum impregnate prior to commencing the sample integrity stage.

The designation of a planar grinding stage will only be necessary if the specimen requires some form of planarity correction. If the specimen is to be prepared by hand, or if when positioned for automatic preparation the surface remains flat and parallel to the platen surface, then no planar grinding stage is necessary. As with all previous considerations within this book, before any sample integrity stage commences it is necessary to know or estimate the extent of any residual sample structural damage. If there is no damage then avoid the sample integrity stage; if there is residual damage then it is the extent of this damage that must dictate the appropriate steps.

14.1.4 Sample integrity

On selection of the preparation procedure, attention should be paid to:

(i) Hardness of the mineral. Charged platen surface 10 is the starting point with very hard materials. When only smaller sized abrasives are found to grind, it could be necessary to use a lower numbered surface with a larger sized abrasive—platen surfaces 8/4/1 for example. If high stock removal should be necessary, fixed shock absorbing abrasive platens could be used.

(ii) Friable phases. The best selection will be a combination of lower numbered surfaces combined with smaller sized abrasives—platen surfaces 6/2/1 for example. Again, charged or fixed surfaces could be used.
(iii) Glycerol or non-aqueous lubricants have to be used with the water-soluble minerals (clay).
(iv) Brittle structures. If the induced damage is too great using charged platen surfaces it will be necessary to use a free rolling abrasive (lapping).

Experience has shown that integrity surfaces 10, 4 and 1 are quite satisfactory for most materials. Once again this balance between free and fixed abrasives must be addressed. If fixed abrasives cause high levels of damage try the nominally fixed abrasives (charged abrasives on platen surfaces). If the latter also create intolerable damage, free abrasives (lapping) must be used.

Should the mineral contain any ductile metals, as some do, be careful not to impress the cutting abrasive into the sample when using charged or loose abrasives. Impressed abrasives can occur because the platen surface is harder than the material to be cut; therefore a softer grinding surface can be the answer. Abrasives that free roll are more prone to impressing into the soft sample than abrasives that are retarded. Therefore more control over the abrasive action can also help. Finally, the use of an oil-based lubricant will reduce the tendency for abrasives to impress.

14.1.5 Polishing

Polishing (the 'new concept') is not to continue the preparation to achieve sample integrity. At the polishing stage integrity has been achieved; therefore, the object is to aesthetically improve the observed surface. Many hard materials do not require such a step, but if this step is required use a non-scratchy type of abrasive such as colloidal silica. There are occasions where it is necessary to partially destroy the sample integrity in order to resolve more clearly the structural elements. This happens when metals are etched prior to optical examination. Minerals, however, often respond to a napped cloth polishing (polishing surfaces 8 to 10), giving subject relief and hence optical resolution.

14.2 Thin Sections for Transmitted Light Microscopy

Thin sections are required for many applications, the majority being in earth sciences, but the possible applications are numerous, viz. ceramics, dentistry, civil engineering, bone implants, even frozen chocolate!!

The traditional approach is to section using a diamond saw and then to lap to the appropriate thickness. There has been much sophistication in automating the procedures but little change in the technique. In order to securely glue the sample to a glass slide it has been necessary to lap the slide top surface; some of the new epoxy resins can render this practice redundant (Figure 14.2).

14.2.1 Cutting

Using a diamond saw, on this occasion the oversized thin section must be cut subsequent to being mounted. The face to be mounted must be free from structural damage. To do this use a very thin blade on a low deformation saw and free lap the face prior to mounting onto a glass microscope slide.

14.2.2 Mounting

The mounted face will not look polished; if it has been lapped it will actually look dull. The important point is that it should be free from structural damage. For subsequent automated preparation (and better bonding) the glass slides should be of uniform size; this can be

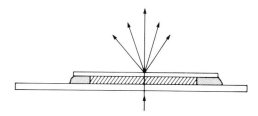

Figure 14.2 Thin sections for transmitted light (1, section; 2, lap; 3, glue; 4, section; 5, grind then lap or 5, lap)

achieved by grinding or lapping. The bond strength between the section and glass slide is also improved by this operation, ultimately achieving a thinner bond layer. Low viscosity cold setting resins are used to stick the sample to the glass slide. If glued under pressure these resins can give a near zero bond thickness.

The appearance of the bonded surface is no longer dull but bright. This is due to the similar refractive index between glass and resin.

14.2.3 Grinding

The oversized sample, now an integral part of the glass slide, requires grinding to near final thickness once the surplus has been cut off. (This could be achieved by lapping but would prolong the preparation time.) Fixed diamond grinding discs or wheels are best for this application and can on some occasions, with well-bonded non-friable minerals, actually grind the sample to the finished size. To do this care must be taken as the procedure approaches the finished size with increments of no more than 1 micrometre. As a general rule it is wise to prepare the sample just short of the finished size, and finally bring it to size by free lapping with fine silicon carbide powder. With ceramics and extremely hard materials it will be necessary to use fine diamond as the lapping agent.

The structural damage achieved by free lapping is less than grinding, so in theory this final step should always be included. The importance of having flat and parallel samples cannot be overstressed since the mineralogist would need to relate the mineral colours exhibited between crossed polars as part of the optical investigation. The optical colours observed down the microscope are a direct result of the mineral birefringence (its refractive index) and the thickness of the mineral. Since the preparation technique creates the thickness, often set for grey extinction with quartz, it can be seen why the thickness is important and why parallelism is vital.

The thickness of a thin section is often governed by the sample grain size and identification mode. If the sample, for example, is to be identified by its colour then it must be $30\,\mu m$ thick; if it is to be identified by grain morphology then ideally the section should be only one grain (or less) in thickness. With some of the fine grain ceramics ($5\,\mu m$) this is difficult; nevertheless, the sample thickness will dictate what can be optically observed. Having stated that the sample grain bond dictates the need for lapping, the sample grain hardness the grinding abrasive type and the sample grain size the sample thickness, it simply remains to define the grinding abrasive size. The general rule is to use the largest abrasive size compatible with sample integrity. The sample integrity is influenced by the type of fracture or dislocation mechanism involved in removing materials, the friable nature of the grains and the grain size. If, for example, an optimum technique were developed for a large grained sample, this would not be suitable for lapping the $10\,\mu m$ ceramic quoted above. To use the small sized abrasives necessary for the ceramic on the large grained sample would be most inefficient.

14.2.4 Mounting cover slip

Although the sample is optically observed after lapping it has a coarse upper surface which is unprotected. This can be improved by mounting to this upper surface a thin glass cover slip. The size of this cover slip is important and must be compatible with the optical corrections of the microscope objective. If immersion objectives are to be used the cover slip does not have to be fitted.

14.3 High Integrity Thin Sections

A thin section is arguably the equivalent of the surface of a thick section as prepared by the metallurgist or materialographer, adhered to a glass slide and followed by the preparation of the other surface. Surface preparation of material surfaces to the 'new concept' introduced the principle of sample integrity where the prime consideration was the residual structural damage present after each mechanical surface preparation. When any material is mechanically worked there remains some structural damage, varying from sub-micrometre depth to more than $100\,\mu m$. It is important therefore to understand and, if possible, to define the residual damage resulting from cutting, grinding, lapping and polishing. A clear definition of the abrasive functions, such as grinding and lapping, must also be appreciated. The prepared surface when completed must exhibit sample integrity; that is, it must be a faithful reproduction of the *true* microstructure and be free from artefacts or any damage induced in preparation. Although all previous references quoted related to the top surface preparation of thick sections, they apply equally to the preparation of all materials, be it for top surface reflection or for thin-section transmitted light applications. Recent work has revealed shortcomings in the classical approach to the preparation of thin sections of minerals and ceramics when a sample of *integrity* is required. The problem first came to light when inspecting some prepared neolithic thin sections. Although the thin section had been prepared with care, using traditional techniques, the grains appeared to exhibit preparation induced structural damage when observed optically. There was also a lack of clarity at the grain boundaries.

Rather than re-prepare the material again using similar methods, it was decided firstly to prepare a surface to sample integrity using reflected light methods. The technique for achieving sample integrity necessitated the use of a range of platen surfaces including composite laps and plastic/woven cloth, using diamond as the cutting abrasive (Figure 14.3, Plate 8). The final diamond preparation had to be as low as $1\,\mu$m before integrity was achieved. The prepared top surface was subsequently glued (epoxy) to a glass slide and carefully thinned to $30\,\mu$m. As the final size was approached, similar care was taken. The optical observation of the thin section was dramatically improved, the cracked grains were no longer in evidence and the clarity between grains was very strong.

The principle of high integrity thin sections is illustrated in Figure 14.4, where a comparison with the classical method is made. In recent years there has been an awareness

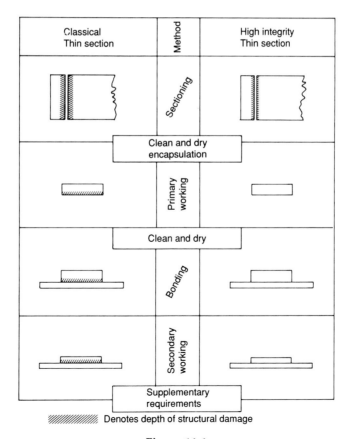

Figure 14.4

that improvements in thin section preparation can be made by adopting the so called 'double polished thin section technique'. This is accomplished by polishing both the top and underside of the chip. This method is said to produce sections of exceptional clarity and sharpness. The drawback of this approach is that sections lacking in integrity could be polished to give a sharp and clear picture of an incorrect structure. There is also the question of polishing the top surface of the chip when used in transmitted light. This has been quoted by many as being a distinct advantage in achieving image quality; fortunately the chip upper surface when accompanied by a cover slip renders any surface irregularities invisible. What is visible, however, is any structural damage and this will not necessarily be removed by polishing. This phenomenon can be demonstrated by observing a 3×1 inch cover glass slide that has been lapped. Using a $10\times$ objective with the lapped glass slide uppermost, the optical image will be matt, consisting of many irregular dark dots. If a glued cover slip is positioned to the lapped glass slide, the appearance will be totally different as the matt black image will have changed to a perfectly clear image, with the exception of very fine dots evenly dispersed. These black dots are evidence of structural damage and as such would still remain had the lapped glass slide been subsequently polished.

High integrity samples therefore require firstly, the glueing of a true and faithful planar surface onto a glass slide, followed by thinning without inducing structural damage to the finished section. Whether this can be achieved with or without polishing is secondary to the prime consideration, which is sample integrity.

Thin section preparation will be concerned with the knowledge relating to induced structural damage at each operation and will be classified by the following stages:

(i) sectioning,
(ii) encapsulation,
(iii) primary chip working,
(iv) bonding,
(v) secondary chip working,
(vi) supplementary requirements.

The 'chip working' stage is where the surface is prepared to integrity, and the identical procedure is ultimately required at 'chip sizing' in advance of the final required thickness. Supplementary requirements include polishing of the upper surface for reflectivity applications or cover slip adhesion to protect the chip. All these stages will be reviewed in detail.

14.3.1 Sectioning

In sectioning a material to a suitable shape and chip thickness from 3 to 10 mm it is often necessary to utilize more than one kind of saw. These different saws are classified by the required function, i.e.

(i) slab saws for the initial sectioning of large materials,
(ii) trim saws for routine chip sections, and
(iii) low deformation saws for extremely delicate or friable materials.

It usually follows that the associated residual damage is related to the machine type used in sectioning, the slab saw showing higher damage levels than the trim saw. This is due in part to the quality of arbor bearings and general precision of moving parts. Structural damage however is much more influenced by the wheel abrasive cutting action than by the type of machine used.

Parameters influencing efficient reduced structural damaged sectioning are:

- abrasive size,
- abrasive type,
- abrasive concentration,
- matrix bond,
- wheel speed,
- wheel size,
- wheel type,
- wheel dressing,
- lubricant type,
- lubricant flow,
- pressure,
- mechanical factors,

The chip or slice should, after sectioning, have a planar face with minimum structural damage. Material removal of minerals and ceramics causes linear elastic fracture mechanisms when stressed and can therefore, when completing a cut, leave a fractured raised portion. This must be removed in order to achieve a planar face. When removing this fractured raised portion it is important not to introduce further structural damage to the cut surface.

14.3.2 Encapsulation

Having achieved a plane surface, it is necessary to consider encapsulation. If this is to be carried out, the chip must be thoroughly cleaned in order to remove oil and abrasive residues. Ultrasonic cleaners are particularly effective but care must be taken to dry the specimens after immersion. Samples requiring encapsulation are those that are friable and/or porous (friable samples must not be ultrasonically cleaned). To optimize the encapsulating resin penetration it will be necessary to carry out vacuum impregnation. The depth of penetration is dependent upon

(i) the mercury gauge reading,
(ii) resin viscosity,
(iii) sample pore size and interconnection,
(iv) time under vacuum.

14.3.3 Primary chip working

The chip should now be in a condition suitable for mechanical working, any friable and/or porous materials being impregnated, if not totally, at least to a depth beyond the level of structural damage. The object at this stage is to achieve a planar surface, free from preparation-induced damage, which is a faithful reproduction of the material structure. Since all different materials will respond differently to any mechanical working, it follows that the technique required to achieve surface integrity will also vary. The object is also to carry out the primary chip working in the shortest time, preferably with as few steps as possible. Before any work can be undertaken, however, it is wise to have an indication of the current level of structural damage. If, as is most unlikely, the structural damage was zero, then no surface working is necessary.

The level of induced structural damage also dictates the severity of any subsequent working, i.e. if the existing damage is 100 μm deep then using a 220 grit diamond wheel which leaves 150 μm of damage after removing material is a retrograde step. It also follows that a selected abrasive surface that produces a given damage depth with a particular rock will *not* cause the same depth of damage with a different rock. The object, therefore, is to work a given damaged surface progressively to reduce the structural damage and aiming for zero damage, i.e. total surface integrity. Figure 14.5 depicts integrity curves for three different surface/abrasive combinations. Note how, with time, each combination reduces the structural damage until a point is reached where continuing operation does not reduce the residual damage. This type of curve is typical when using a non-degrading abrasive such as diamond on a lapped or ground surface.

Operational technique

The classical approach with thin-section preparation is to use a lapping technique for the primary surface. Although this is suitable for many geological materials it can be demonstrated that structural damage manifests itself when a lapped surface is used (the lapped glass slide as mentioned). It is the extent to which this damage affects the end result which dictates which additional or alternative steps are to be undertaken.

Figure 14.6 shows a typical integrity curve for two alternative methods with an associated

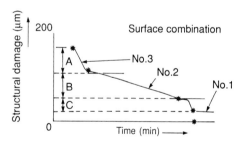

Figure 14.5 Progressing integrity

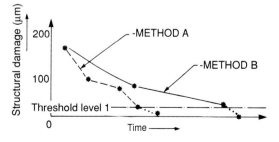

Figure 14.6 Comparison of alternative methods

threshold level. Method A is three-step, fixed diamond *grinding* to threshold level 1, total integrity being achieved using a 1 µm diamond/colloidal silica slurry. Method B is two-step *lapping* to threshold level 1, total integrity as Method A. From Figure 14.6 one can see how the desired techniques can be engineered to suit the user's needs, equipment and consumables. Take, for example, Method B using just one lapping step accompanied by the 1 µm diamond/colloidal silica slurry step, etc.

The choice of abrasives, techniques and surfaces is as follows:

(i) Abrasives
- diamond,
- cubic boron nitride,
- silicon carbide,
- alumina,
- colloidal silica.

(ii) Techniques
- grinding,
- lapping,
- polishing,
- vibratory actions.

(iii) Surfaces
- cast iron,
- glass,
- polymer/fibre,
- polymer/metal.

Optical examination

Before processing to the bonding stage it is wise to observe optically the primary surface to confirm a satisfactory achieved preparation. When dealing with semi-translucent materials it can be advantageous to use dark-ground techniques as well as bright field. Dark ground will identify areas of structural damage by reflecting the damaged area back into the optical path. This technique is also useful in identifying ingressed resin, in particular if the resin has been dyed a strong colour. Note that coloured dyes are not detectable by colour in bright field.

When what appears to be a pore is optically visible, i.e. a dark area not ingressed with resin; it is necessary to establish this dark area as being discrete or a closed porosity from induced structural damage. This again can be verified using the dark-ground techniques.

Prepared surfaces for reflected light, or as primary surfaces for transmitted light, should be prepared to the required threshold level. It can be acknowledged, however, that a lower surface reflectivity is possible when a cement of similar refractive index to glass is to be used, as in the case with primary surfaces when bonded to a cover glass.

Bonding

Bonding the chip to a microscope glass slide requires both slide and chip to be thoroughly clean, dry, and free from any grease. It may be necessary to produce a lapped surface on the bonding face of the glass slide in order to improve adhesion. This is an acceptable practice providing the structural damage to the glass slide is kept to a minimum. As has been explained previously, the matching refractive index of the glass and bonding agent will render surface irregularities invisible, but it will not disguise any structural damage. This factor is very important with thin section equipment where it is necessary to *grind* the glass slides in batches to a uniform thickness. To keep damage to a minimum, always use the smallest size abrasive, housed in the highest shock absorbing surface.

Lakeside 70 cement is the traditional bonding medium, its advantages being that it is easy to cure and, since it is a thermoplastic material, it can be re-worked. Unfortunately its bonding strength restricts its universal application.

Improved bonding strengths are achieved by using epoxy resins; these thermosetting resins can also have very low viscosities, producing the so-called 'zero bond'.

14.3.4 Secondary chip working

The object at this stage is to produce a finished thin section whose thickness will not vary and whose size will be adequate to transmit the structural information. This means that the finished dimension of a thin section will vary according to the grain size and density. There is, however, one *major* exception to this rule and that is when the material is birefringent and is required to be a specific size (30 µm) for purposes of identification by interference colours, as in geology.

In many cases the chip is too thick to commence a grinding stage and therefore requires initial thinning by re-sectioning. The finished thickness of a re-sectioned chip will be dependent upon the induced residual damage, the depth of this damage being dependent upon all the same parameters, as described in original sectioning. The proceeding stages are equally dependent upon the re-sectioned surface. There is a case for proposing that the final sizing should go through identical stages to the primary working procedure. This in fact is not always required since the bonding agent offers additional support that was not present in the original surface working.

14.3.5 Supplementary requirements

It is always wise to remove excess cement from around the chip, in particular when a protective cover glass is to be fixed. This is in part due to the stressed surface which will re-orientate polarized light and confuse the observed optical image. Those slides requiring cover glass protection should be cleaned and dried prior to applying Canada balsam to the chip surface and placing the cover glass on top. Cover glass sizes are to a specific thickness and must be adhered to when using objectives of numerical apertures in excess of 0.25, i.e. greater than 10× objective magnification. Since the geologist does most of his work using numerical apertures of 0.25 and lower, the cover glass thickness is not too critical and allows the sprayed acrylic coating method to be used.

It is a general misconception that both the cover glass and acrylic spray method will disguise a poorly prepared upper surface. If the investigator is not sure how the surface will eventually look, then this can be replicated by smearing an even layer of objective immersion oil on the chip upper surface, additionally with the cover glass in position when high NA optical observations are to be undertaken. If the specimen will not pass this test then the upper chip surface will have to be re-worked, carefully avoiding unnecessary stock removal, i.e. 0.06 μm colloidal silica or similar. Thin sections required for reflected light applications (reflectance, microindentation, structure morphology, microprobe analysis, etc.), as opposed to transmitted light observations, can be used with or without a cover glass. It will, however, be necessary to use objectives specially corrected for use with uncovered specimens. Objectives 10× and lower are insensitive to such corrections and can therefore be used on specimens with or without a cover glass. There is, however, a slightly different objective shoulder length which will result in parfocality differences. Specimens for reflected light observations very often require an additional polishing step to remove any surface scratches, but this is to improve the aesthetics since samples for both transmitted and reflected light applications are required to exhibit sample integrity.

14.3.6 Bone implants

The compatibility of various materials for implants and other biomedical applications is the subject of considerable investigation. One of the more useful analytical techniques employed is thin sectioning. With this technique, specimens of bone containing implants are carefully thinned to allow investigation using radiography or transmitted light microscopy.

This technique will gain in popularity as the trend away from metal implants increases (see Section 13.4 on 'Biomaterials'). Unfortunately, from a sample preparation point of view, metal implants will be required for many years to come, the appropriate method being very different from routine mineral thin section techniques. In many cases thin sections are not necessary since the microstructure can be revealed using a thick section, observed in reflected light.

Degeneration or regeneration of bone, penetration of tumours and other diseases, and the ingrowth of tissue into porous implants may be observed from properly prepared sections. Thin sections are also useful for routine observation of mophological and cellular details.

14.3.7 Preparation difficulties

The preparation of thin sections of bone *without* implants can be successfully carried out using a diamond slitting wheel followed by grinding on silicon carbide paper prior to a 3 μm diamond polish on the appropriate polishing cloth.

Preparation of thin sections of bone *containing* implants, however, are much more difficult to prepare due to the difference in material removal rates between the bone and the

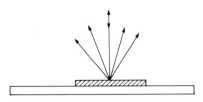

Figure 14.7 Thin section for reflected light (1, section; 2, lap; 3, glue; 4, section; 5, grind then lap or 5, lap; 6, polish)

implant. Another factor affecting successful preparation is the implant material. Metal, for example, abrades less than bone, giving rise to polishing relief—ceramic—when ground breaks away making analysis difficult if not impossible.

The bone implant when glued to the microscope slide would be first sectioned and then ground—for radiography 150 to 250 μm thick and for microscopy 50 to 70 μm. To reach the final thickness without damage the sample is thinned to size using 9 μm diamond on charged platen surface 4 (for microradiography 100 μm, for microscopy 30 μm). The metal implant will at this stage exhibit the occasional scratch. This should not be noticed in the microscope image when observing in transmitted light.

14.4 Thin Sections for Reflected Light Microscopy

The preparation for reflected light follows a similar procedure to that above for transmitted light, with minor exceptions. One of these exceptions could be the requirement for thinner specimens. If identification via birefringence is not to be used then a slightly less stringent requirement on parallelism could be tolerated; i.e. when minerals are recognized, in part, by their colour, relative to a specific thickness, then parallelism is essential. This would not be the case for thin sections observed in reflected light (Figure 14.7).

The major difference between transmitted light and reflected light is in the final surface preparation. Surfaces for transmitting light can be coarse and not smooth, since the glue, having an identical refractive index to the glass, will allow all the light to transmit through the glass/glue sample interface;

When the transmitted rays meet the top sample/air interface, they are not influenced by this surface topography. In reflected light the situation is very different and if the top surface (air/sample interface) is not *flat* and normal to the optic axis, then light will be scattered, with a resulting loss of image quality; i.e. a lapped surface is perfectly adequate for transmitted light, but totally inadequate for reflected light. To avoid surface scattering the sample top surface, when observed in the reflected light mode, must be 'smoothed off'. This is achieved by polishing with a low nap cloth (polishing surface 2 or 3) using diamond/oxides or other suitable abrasives. When these reflected light specimens are observed the cover glass must not be fitted and special objectives must be used, i.e. those corrected for use without cover glass. Once again, if oil immersion objectives are required they can be used for both cover glass and no cover glass applications (immersion oil has the same refractive index as the glass cover slip).

The microscope used for reflected light observations of thin sections will, in the majority of cases, be a transmitted light microscope with the addition of a reflected light illuminator. The problem arises as to the choice of objectives, since those used for reflected light will not necessarily be suitable for transmitted light. When making this selection, Table 14.2 should prove helpful.

Table 14.2 Choosing objectives

Objective magnification	Suitable for reflected and transmitted light	Require specially corrected
4	Yes	No
10	Yes	No
20	No	Yes
40	No	Yes
100 oil	Yes	No

14.5 Preparation Examples

The following examples illustrate procedures and results expected with thick sections for reflected light use.

Example 14.1

Family classification BB15

The material: chalcopyrite ($CuFeS_2$)

This material is one of the most widely distributed copper minerals. It is often found in veins and ore deposits with other metals. It gives a yellow colour when viewed in normal bright field reflected light. It can be mistaken for pyrite (FeS_2) which has a slightly lighter yellow appearance. Pyrrhotite ($Fe_{1-x}S$) is also similar to chalcopyrite but the veins are more bronze in appearance.

The technique

Sectioning of brittle minerals and many ceramics can best be accomplished using a resin-bonded diamond wheel. There is a tendency to always use a metal-bonded wheel; this is perfectly adequate but slower. Mounting of any porous or brittle materials must be carried out using a cold mounting technique as compression mounting will result in a damaged cracked specimen. This material, as with most vein-deposited metals, requires careful surface preparation if pluck-out is to be avoided. Preparation by traditional methods using a lapping technique is the more familiar route than the selected procedure. To utilize shock absorbing surfaces with progressively smaller abrasives, acting in a *grinding* mode will not only achieve sample integrity but achieve it in the 'shortest time'.

The method

AUDITABLE PREPARATION PROCEDURE No. 1

CLASSIFICATIONS: BB15/MINERALS	MATERIAL: CHALCOPYRITE
SECTIONING TECHNIQUE:	BLADE
MOUNTING TECHNIQUE:	CONCENTRATION

	SURFACE and TYPE	ABRASIVE SIZE/TYPE LUBRICANT	LOAD (kPa)	HEAD ROTATION	Z AXIS (μm)	REFLECTIVE FACTOR	MICROGRAPH YES/NO	IMAGE ANALYSIS YES/NO	REMARKS
PLANAR GRINDING STAGE	S.10	30 μm WATER	35	CONTRA	PLANE				

	SURFACE and TYPE	ABRASIVE SIZE/TYPE LUBRICANT	LOAD (kPa)	HEAD ROTATION	Z AXIS (μm)	REFLECTIVE FACTOR	MICROGRAPH YES/NO	IMAGE ANALYSIS YES/NO	REMARKS
SAMPLE INTEGRITY STAGE	S.4	9 μm OIL	35	CONTRA					
	S.2P	6 μm WATER	35	COMP					
	S.1	1 μm WATER	40	COMP					

	SURFACE and TYPE	ABRASIVE SIZE/TYPE LUBRICANT	LOAD (kPa)	HEAD ROTATION	TIME (s)	REFLECTIVE FACTOR	MICROGRAPH YES/NO	IMAGE ANALYSIS YES/NO	REMARKS
POLISHING STAGE	PS 7	0.06 μm SILICA	16	COMP	60		YES		14.8(Plate 8)

The micrograph Figure 14.8 (see Plate 8)

Example 14.2

Family classification BB22

The material: brake lining and friction materials

Brake linings have traditionally employed asbestos as the base material due to its exceptionally good frictional properties, thermal stability and low cost. However, due to world-wide concern, this health hazardous material is being replaced in all applications where there is any likelihood of particulate inhalation. The frictional resistance for brake linings and the operating temperature influences the combination of materials required. Braking required for a low weight motor cycle (200 kg) is very different to a large earth mover (50 tonne). These in turn are very different to braking a 747 jet aeroplane or a high speed racing car.

Non-asbestos friction materials commonly require a combination of five or more materials to replace the function of asbestos. Additionally, friction modifiers and fillers typically increase the number of raw materials to an excess of ten. The raw materials can include phenolic resin binders, rubber, metal particles, mineral fibres, abrasive particles and cheap mineral fillers. The concentration, distribution and location are all factors affecting performance and must therefore be controlled, very often by microstructural analysis.

The technique

This complex composite material with its soft, hard, friable and loosely bonded elements requires careful surface preparation to achieve a faithful reproduction. Initial sectioning is best carried out using a thin silicon carbide abrasive wheel. Although silicon carbide is suitable for the initial grinding stage, it is wise to follow with platen surface 4, to avoid preferential grinding. Once the surface is flat, the procedure can be progressed to integrity, concluding with a fine oxide or silica grind on platen surface 1.

The method

AUDITABLE PREPARATION PROCEDURE   No. 2

CLASSIFICATIONS: BB22				MATERIAL: FRICTION MATERIALS	
SECTIONING TECHNIQUE:	EQUIPMENT TYPE	BLADE THIN SiC ABRASIVE WHEEL		CONCENTRATION	
MOUNTING TECHNIQUE:	METHOD EPOXY			SAMPLE/MOUNT RATIO	

PLANAR GRINDING STAGE	SURFACE and TYPE	ABRASIVE SIZE/TYPE LUBRICANT	FORCE (kPa)	HEAD ROTATION	Z AXIS (μm)	REFLECTIVE FACTOR	MICROGRAPH YES/NO	IMAGE ANALYSIS YES/NO	REMARKS
	PAPER	P240 SiC	35	COMP	PLANE				

SAMPLE INTEGRITY STAGE	SURFACE and TYPE	ABRASIVE SIZE/TYPE LUBRICANT	FORCE (kPa)	HEAD ROTATION	Z AXIS (μm)	REFLECTIVE FACTOR	MICROGRAPH YES/NO	IMAGE ANALYSIS YES/NO	REMARKS
	S.4	9 μm OIL	35	CONTRA					
	S.1	3 μm WATER	35	COMP					
	S.1	0.06 μm COLLOIDAL SILICA	40	COMP					

POLISHING STAGE	SURFACE and TYPE	ABRASIVE SIZE/TYPE LUBRICANT	FORCE (kPa)	HEAD ROTATION	TIME (s)	REFLECTIVE FACTOR	MICROGRAPH YES/NO	IMAGE ANALYSIS YES/NO	REMARKS
	PS 8	0.06 μm COLLOIDAL SILICA	15	COMP	60		YES		14.9 (Plate 8)

The micrograph Figure 14.9 (see Plate 8)

Example 14.3

Family classification BB15

The material: graphite

Graphite is black when observed in bright field reflected light. It does, however, reward the observer with a variety of colours when orthoscopically looking at an image between crossed polars with the sensitive tint plate in position. It occurs in thin flakes and is characteristic of metamorphic rocks, often combined with limestones. In cast iron, carbon exists as free graphite and can be spheroidal or as graphite flakes. The applications for graphite are vast, viz. lubrication in metals, manufacture of iron and steel, grease and lubricants, growth of fibres, arc lamps, pencils, electric brushes, surface deposits for conduction, etc.

The technique

Graphite can very easily be broken away when grinding. Its friable structure demands the use of sharp positive rake angle cutting tools, lending itself to silicon carbide grinding. Some stages in the surface preparation process could benefit from dry cutting. Whatever route is chosen it is wise to avoid surface rubbing. Optical activity is a guide to good preparation, providing it is observed in polarized light or dark ground.

The method

AUDITABLE PREPARATION PROCEDURE   No. 3

CLASSIFICATIONS: BB15/MINERALS	MATERIAL: GRAPHITE
SECTIONING TECHNIQUE:	BLADE
MOUNTING TECHNIQUE:	CONCENTRATION

	SURFACE and TYPE	ABRASIVE SIZE/TYPE LUBRICANT	LOAD (kPa)	HEAD ROTATION	Z AXIS (μm)	REFLECTIVE FACTOR	MICROGRAPH YES/NO	IMAGE ANALYSIS YES/NO	REMARKS
PLANAR GRINDING STAGE	PAPER	240 SiC WATER	35	COMP	PLANE				

	SURFACE and TYPE	ABRASIVE SIZE/TYPE LUBRICANT	LOAD (kPa)	HEAD ROTATION	Z AXIS (μm)	REFLECTIVE FACTOR	MICROGRAPH YES/NO	IMAGE ANALYSIS YES/NO	REMARKS
SAMPLE INTEGRITY STAGE	S.4	9 μm OIL	35	CONTRA					
	S.1	0.05 μm ALUMINA	70	COMP					

	SURFACE and TYPE	ABRASIVE SIZE/TYPE LUBRICANT	LOAD (kPa)	HEAD ROTATION	TIME (s)	REFLECTIVE FACTOR	MICROGRAPH YES/NO	IMAGE ANALYSIS YES/NO	REMARKS
POLISHING STAGE	PS 7	0.06 μm SILICA	16	COMP	120		YES		14.10(Plate 8)

The micrograph Figure 14.10 (see Plate 8)

Example 14.4

Family classification BB15

The material: slag: iron oxide/quartz matrix

The waste products (gangue) formed during the manufacture of materials are called slags; their chemical composition relates to the manufactured material. Slag floats on the top of molten metals and can be separated when run off. Most slags are silicates, which, being acidic, combine with basic oxides, culminating in a lower melting point. Iron oxide gives a very fusible slag with silica and is important in non-ferrous metallurgy. In steel making the relationship between manganese and silicon affects the condition of the slag; the oxides forming as slag are mainly iron, manganese and silicon and are influenced by the condition of the hot metal. When all solid materials have been melted, samples of slag and metal are taken for analysis. From the analysis of the slag it is possible to determine the extent of carbon and phosphorus removed from the metal. The microstructure of the slag, therefore, is another fingerprint in the control of metals.

The technique

Since these materials can be brittle and friable, a diamond slitting wheel is desirable; when sectioning the most dense samples the use of a classical cut-off machine, using a resin-bonded diamond wheel, is most efficient. Encapsulating when necessary is best carried out using cold setting resin or low pressure hot mounting. The preparation of many brittle and hard materials utilizes a 10/4/1 procedure; this material is no exception. Choice of fixed or charged abrasives will depend upon the density of the sample. The final integrity stage using oxide grinding could in many cases be undertaken at double the normal pressure, in order to reduce the preparation time.

158 Surface Preparation and Microscopy of Materials

The method

AUDITABLE PREPARATION PROCEDURE									No. 4
CLASSIFICATIONS: BB15/MINERALS					MATERIAL: SLAG: IRON OXIDE/QUARTZ MATRIX				
SECTIONING TECHNIQUE:	EQUIPMENT TYPE BLADE RESIN-BONDED DIAMOND				CONCENTRATION				
MOUNTING TECHNIQUE:	METHOD COLD OR LOW PRESSURE				SAMPLE/MOUNT RATIO				

	SURFACE and TYPE	ABRASIVE SIZE/TYPE LUBRICANT	FORCE (kPa)	HEAD ROTATION	Z AXIS (μm)	REFLECTIVE FACTOR	MICROGRAPH YES/NO	IMAGE ANALYSIS YES/NO	REMARKS
PLANAR GRINDING STAGE	S.10	30 μm OIL	35	CONTRA	PLANE				

	SURFACE and TYPE	ABRASIVE SIZE/TYPE LUBRICANT	FORCE (kPa)	HEAD ROTATION	Z AXIS (μm)	REFLECTIVE FACTOR	MICROGRAPH YES/NO	IMAGE ANALYSIS YES/NO	REMARKS
SAMPLE INTEGRITY STAGE	S.4	9 μm OIL	35	CONTRA					
	S.1	1 μm WATER	35	COMP					
	S.1	COLLOIDAL SILICA	35	CONTRA	2		YES		14.11 (Plate 8)

	SURFACE and TYPE	ABRASIVE SIZE/TYPE LUBRICANT	FORCE (kPa)	HEAD ROTATION	TIME (s)	REFLECTIVE FACTOR	MICROGRAPH YES/NO	IMAGE ANALYSIS YES/NO	REMARKS
POLISHING STAGE									

The micrograph Figure 14.11 (see Plate 8)

Example   14.5

Family classification BB15

The material: cement clinker

Cement, the added ingredient to sand and stone or pebbles (coarse and fine aggregate), forms the basis of concrete. The disadvantage of this material is its low tensile strength and brittle behaviour. It also exhibits erosion problems and can with incorrect fines react with other materials, such as steel reinforcement. The stones and sand are graded in particle size to minimize the space to be filled by the cement/water matrix. This space is further reduced by compacting, resulting in increased strength. Portland cement is formed by firing a mixture of limestone and clay to a clinker and then crushing to form a powder of particle size, say, 10 to 30 μm.

Applications for cement are vast, viz. building roads, aircraft landing strips, bridges, buildings, street furniture, tunnels, etc.

Preparation of Minerals

The technique

The traditional approach would be to use a fine lapping technique. This would prolong the preparation and is not necessary, as shown by the following micrograph and method. Water is avoided throughout the preparation.

The method

AUDITABLE PREPARATION PROCEDURE		No. 5
CLASSIFICATIONS: BB15/MINERALS		MATERIAL: CEMENT CLINKER
SECTIONING TECHNIQUE:		BLADE
MOUNTING TECHNIQUE:		CONCENTRATION

	SURFACE and TYPE	ABRASIVE SIZE/TYPE LUBRICANT	LOAD (kPa)	HEAD ROTATION	Z AXIS (μm)	REFLECTIVE FACTOR	MICROGRAPH YES/NO	IMAGE ANALYSIS YES/NO	REMARKS
PLANAR GRINDING STAGE	PAPER	P600 SiC OIL	35	COMP	PLANE				

	SURFACE and TYPE	ABRASIVE SIZE/TYPE LUBRICANT	LOAD (kPa)	HEAD ROTATION	Z AXIS (μm)	REFLECTIVE FACTOR	MICROGRAPH YES/NO	IMAGE ANALYSIS YES/NO	REMARKS
SAMPLE INTEGRITY STAGE	S.2P	3 μm OIL	35	CONTRA					
	S.2P	0.05 μm ALUMINA METHYLATED SPIRITS	35	COMP	1		YES		14.12(Plate 9)

	SURFACE and TYPE	ABRASIVE SIZE/TYPE LUBRICANT	LOAD (kPa)	HEAD ROTATION	TIME (s)	REFLECTIVE FACTOR	MICROGRAPH YES/NO	IMAGE ANALYSIS YES/NO	REMARKS
POLISHING STAGE									

The micrograph Figure 14.12 (see Plate 9)

Example 14.6

Family classification BB15

The material: vitrified alumina

Vitrified alumina bonded wheels are in everyday use in the grinding industry. Vitrified wheels have a porous structure, ideal for cool and fast cutting. They have a high modulus of elasticity which makes them ideal for precision grinding. Unlike resinoid, rubber and shellac bonded wheels, the vitrified wheel is not self-dressing. The abrasive particles are held together by bridges of glass or similar vitreous material. The wheels are fired in kilns at temperatures in excess of 1000°C and are therefore unaffected by heat generated during the grinding process.

Microstructural analysis reveals much information about grain distribution and porosity, but, above all, the vitreous bond will determine the performance characteristics of the wheel. A vitreous bond is sometimes difficult to achieve due to relatively small melting temperatures. Individual bonds can therefore only be fractured by the imposed shear stress. The blue colour in Figure 14.13 is the ingress of coloured mounting resin. By using dark ground the grain morphology is depicted by a thin white line, with the grains coloured red.

The technique

Vacuum impregnation with epoxy resin can take place after any necessary diamond sectioning: fixed diamond grinding to achieve planarity; charged platen surface 4 to remove major fracture damage; followed by double pressure on platen surface 1. Note the relatively large size of diamond (3 μm) at this stage; 1 μm would be the expected choice but would take longer. Chemomechanical polishing completes the procedure.

The method

AUDITABLE PREPARATION PROCEDURE									No. 6
CLASSIFICATIONS: BB15/MINERALS					MATERIAL: VITRIFIED ALUMINA				
SECTIONING TECHNIQUE:	EQUIPMENT TYPE DIAMOND BLADE		BLADE		CONCENTRATION NORMAL				
MOUNTING TECHNIQUE:	METHOD VACUUM IMPREGNATED				SAMPLE/MOUNT RATIO				

PLANAR GRINDING STAGE	SURFACE and TYPE	ABRASIVE SIZE/TYPE LUBRICANT	FORCE (kPa)	HEAD ROTATION	Z AXIS (μm)	REFLECTIVE FACTOR	MICROGRAPH YES/NO	IMAGE ANALYSIS YES/NO	REMARKS
	RESIN BONDED	P320 FIXED DIAMOND WATER	35	CONTRA	PLANE				

SAMPLE INTEGRITY STAGE	SURFACE and TYPE	ABRASIVE SIZE/TYPE LUBRICANT	FORCE (kPa)	HEAD ROTATION	Z AXIS (μm)	REFLECTIVE FACTOR	MICROGRAPH YES/NO	IMAGE ANALYSIS YES/NO	REMARKS
	S.4	9 μm OIL	35	CONTRA					
	S.1	3 μm WATER	70	CONTRA					
	S.1	0.06 μm COLLOIDAL SILICA	70	CONTRA	1		YES		14.13 (Plate 9)

POLISHING STAGE	SURFACE and TYPE	ABRASIVE SIZE/TYPE LUBRICANT	FORCE (kPa)	HEAD ROTATION	TIME (s)	REFLECTIVE FACTOR	MICROGRAPH YES/NO	IMAGE ANALYSIS YES/NO	REMARKS

The micrograph Figure 14.13 (see Plate 9)

Example 14.7

Family classification BB15

The material: covellite

The method

AUDITABLE PREPARATION PROCEDURE			No. 7	
CLASSIFICATIONS: BB15/MINERALS			MATERIAL: COVELLITE	
SECTIONING TECHNIQUE:	EQUIPMENT TYPE DIAMOND BLADE	BLADE	CONCENTRATION	
MOUNTING TECHNIQUE:	METHOD EPOXY COLD MOUNT		SAMPLE/MOUNT RATIO	

PLANAR GRINDING STAGE	SURFACE and TYPE	ABRASIVE SIZE/TYPE LUBRICANT	FORCE (kPa)	HEAD ROTATION	Z AXIS (μm)	REFLECTIVE FACTOR	MICROGRAPH YES/NO	IMAGE ANALYSIS YES/NO	REMARKS
	PAPER	P240 SiC	35	COMP	PLANE				

SAMPLE INTEGRITY STAGE	SURFACE and TYPE	ABRASIVE SIZE/TYPE LUBRICANT	FORCE (kPa)	HEAD ROTATION	Z AXIS (μm)	REFLECTIVE FACTOR	MICROGRAPH YES/NO	IMAGE ANALYSIS YES/NO	REMARKS
	S.4	9 μm OIL	35	CONTRA	8				
	S.1	3 μm WATER	35	COMP	2				

POLISHING STAGE	SURFACE and TYPE	ABRASIVE SIZE/TYPE LUBRICANT	FORCE (kPa)	HEAD ROTATION	TIME (s)	REFLECTIVE FACTOR	MICROGRAPH YES/NO	IMAGE ANALYSIS YES/NO	REMARKS
	PS 7	0.05 μm OXIDE/ SILICA MIX	15	CONTRA	180		YES		14.14 (Plate 9)

The micrograph Figure 14.14 (see Plate 9)

Example 14.8

Family classification BB15

The material: hematite

The method

AUDITABLE PREPARATION PROCEDURE   No. 8

CLASSIFICATIONS: BB15			MATERIAL: HEMATITE	
SECTIONING TECHNIQUE:	EQUIPMENT TYPE DIAMOND	BLADE	CONCENTRATION	
MOUNTING TECHNIQUE:	METHOD EPOXY COLD MOUNT		SAMPLE/MOUNT RATIO	

*VIBRATORY POLISHING PREFERRED.

	SURFACE and TYPE	ABRASIVE SIZE/TYPE LUBRICANT	FORCE (kPa)	HEAD ROTATION	Z AXIS (μm)	REFLECTIVE FACTOR	MICROGRAPH YES/NO	IMAGE ANALYSIS YES/NO	REMARKS
PLANAR GRINDING STAGE	CAST IRON PLATE	P1200 GRIT SiC POWDER	LOW LAPPING FORCE	COMP	PLANE				

	SURFACE and TYPE	ABRASIVE SIZE/TYPE LUBRICANT	FORCE (kPa)	HEAD ROTATION	Z AXIS (μm)	REFLECTIVE FACTOR	MICROGRAPH YES/NO	IMAGE ANALYSIS YES/NO	REMARKS
SAMPLE INTEGRITY STAGE	S.2P	15 μm WATER	35	COMP					
	S.1	6 μm WATER	35	COMP					
	S.1	1 μm WATER	35	COMP					

	SURFACE and TYPE	ABRASIVE SIZE/TYPE LUBRICANT	FORCE (kPa)	HEAD ROTATION	TIME (s)	REFLECTIVE FACTOR	MICROGRAPH YES/NO	IMAGE ANALYSIS YES/NO	REMARKS
POLISHING STAGE	*PS 7	0.06 μm COLLOIDAL SILICA	15	CONTRA	600		YES		14.15 (Plate 9)

The micrograph Figure 14.15 (see Plate 9)

Example 14.9

Family classification BB15

The material: titanium oxide

The method

AUDITABLE PREPARATION PROCEDURE									No. 9
CLASSIFICATIONS: BB15 MINERALS						MATERIAL: TITANIUM OXIDE			
SECTIONING TECHNIQUE:	EQUIPMENT TYPE		BLADE			CONCENTRATION			
MOUNTING TECHNIQUE:	METHOD					SAMPLE/MOUNT RATIO			

	SURFACE and TYPE	ABRASIVE SIZE/TYPE LUBRICANT	FORCE (kPa)	HEAD ROTATION	Z AXIS (μm)	REFLECTIVE FACTOR	MICROGRAPH YES/NO	IMAGE ANALYSIS YES/NO	REMARKS
PLANAR GRINDING STAGE	S.10	30 μm OIL	35	CONTRA	PLANE				

	SURFACE and TYPE	ABRASIVE SIZE/TYPE LUBRICANT	FORCE (kPa)	HEAD ROTATION	Z AXIS (μm)	REFLECTIVE FACTOR	MICROGRAPH YES/NO	IMAGE ANALYSIS YES/NO	REMARKS
SAMPLE INTEGRITY STAGE	S.10	9 μm OIL	35	CONTRA					
	S.4	9 μm OIL	35	CONTRA					
	S.1	1 μm OIL pH ACTIVE	70	COMP	3		YES		14.16 (Plate 9)

	SURFACE and TYPE	ABRASIVE SIZE/TYPE LUBRICANT	FORCE (kPa)	HEAD ROTATION	TIME (s)	REFLECTIVE FACTOR	MICROGRAPH YES/NO	IMAGE ANALYSIS YES/NO	REMARKS
POLISHING STAGE									

The micrograph Figure 14.16 (see Plate 9)

15

Preparation of PCBs and Electronic Components

15.1 Specimen Characteristics

One associates the electronic industry with the most modern updated sector of the industrialized manufacturing industries, incorporating all the latest technology in testing. It would therefore seem surprising to see the emphasis placed on metallographic cross-sectioning as the most comprehensive in-depth method of quality evaluation. Just as the pathologist is required to section the tissue, so the materials technologist needs to section his component, be it a printed circuit board, a piece of steel or a modern engineering material.

Since cross-sectioning techniques are destructive, it would be wise to carry out all non-destructive procedures in advance, such as low power investigation ($\times 10$ to 40) using a stereoscopic microscope. Many other tests such as electrical or chemical ones can also prove fruitful.

Tests usually fall into one of two categories, i.e. failure analysis or quality control. The latter is often a closely defined specification, listing not only tests to be carried out but the acceptable tolerance of accuracy in both the procedure and the component. Failure analysis is very different; it requires care and imagination to unravel the often well-disguised cause of failure. The method must be carefully monitored and carried out to sound technical derivatives.

Before any device is subjected to a preparation technique it is wise to consider the diverse properties often associated with electronic components, viz.:

1. Components can be delicate.
2. They are often of extremely dissimilar materials.
3. They can be brittle as silicon (semi-conductive).
4. Soft as plating (gold).
5. Hard as ceramics (substrates).
6. Tough as tantalum (electrodes).

15.1.1 Delicate components

Sectioning can be the most critical operation in the preparation sequence. Use the thinnest blade possible at the lowest speed using the appropriate lubricant. Delicate components often respond to single point rather than face cutting; this is achieved by rotating the specimen against the rotating slitting wheel.

It might be necessary to 'wax' the component to the sectioning vise rather than 'clamping'. Once sectioned the component can be encapsulated to protect the geometry and allow subsequent grinding to sample integrity.

15.1.2 Dissimilar constituents

The problem with materials that are dissimilar is that their abrasion rates differ. This is further exacerbated if the grinding abrasive is allowed to rise and fall within the support backing (as it does, for example, using silicon carbide paper). This familiar condition is incorrectly referred to as polishing relief and can obscure interface analysis as well as apparently enlarge plated dimensions. The solution to this problem is to use charged platen surfaces even if it means using large sized diamonds on surface 2, possible routes being 10/4/1, SIC/4/1, SIC/2/1, SIC/8/2, SIC/6/2, etc.

15.1.3 Brittle as silicon

A grinding operation is required with maximum shock absorbing characteristics to avoid chipping the silica. If, for example, a non-encapsulated wafer is to be cross-sectioned and polished, it would be wise to lap with a 6 to 9 μm aluminium oxide slurry followed by grinding with 3 to 6 μm diamond on surface 1 or 2. Colloidal silica on PS 7 will give the desired polish.

With unmounted semiconductive chips it may be necessary to grind *without* the use of any abrasives in order to retain the surface integrity. This is carried out on a rotating frosted glass wheel and a liberal supply of water used as the chip is gently moved over the surface.

With unmounted subjects not quite so brittle (gallium arsenide) it is possible to replace the alumina lapping stage with fine grade P600 to P1200 silicon carbide paper. If the brittle subject can be encapsulated in an epoxy resin the procedure does not need to be so critical and the 10/4/1 or SIC/4/1 routes would be adopted.

15.1.4 Soft as gold plating

This has received special attention in Chapter 16. Also included in this group is the preparation of solders and other soft materials present in electronic components. If the volume of soft material is high, it would be wise to dedicate the preparation in that direction. This subject is covered by its own chapter (Chapter 17). Whenever soft elements are present in the material there is a potential attraction for every loose abrasive, particle or grinding chip and care must be taken to avoid impression. In order to satisfactorily prepare soft elements the polishing step often utilizes an abrasive slurry with a high or low pH. This will have the effect of partially etching certain alloys which can at times be an unexpected benefit.

15.1.5 Hard as ceramics

The reason for separating hardness from brittleness is in the different approach taken in their surface preparation. Although both materials remove stock by a process of cracking as opposed to plastic dislocation, their sensitivity to crack propagation varies.

Many electronic devices have a ceramic substrate or are predominantly ceramic. They often require to be sectioned within the assembly, through the centre of a vehicle spark plug for example. When these devices are in any respect delicate they must be mounted prior to sectioning. Use the drip-dry technique if they are not to be subsequently mounted in an epoxy resin. Planar grinding hard materials is a relatively easy operation; use a metal- or resin-bonded diamond wheel. When developing a procedure it is always wise to optically observe the damaged structure after planar grinding; if this is severe a charged platen surface must be used (platen surface 6/8/10). Sample integrity is achieved as the platen surfaces used are targeted for zero.

Just as there are varying brittle subjects so there are ceramics with varying hardness; ceramic capacitors could, for example, be prepared using a four-step silicon carbide stage followed by two polishing steps, i.e. P240, 600, 1200, 4000, platen surface 1 then polish.

When high levels of stock removal are required at the planar grinding stage the procedure could be:

(i) Resin-bonded 320 grit diamond wheel
(ii) P600/1200/platen surface 1
(iii) Polish

When using fully automated equipment the two silicon carbide stages would be replaced by platen surfaces 4 and 2.

15.1.6 Tough as tantalum

The use of sharp angular abrasives and the appropriate oily lubricant are required. The area fraction will dictate the degree to which the above two factors influence the preparation choice. Also affecting the results will be the presence of ceramic materials in the prepared cross-section; the latter acts like a dressing stick and tends to clean the grinding abrasives as they become 'clogged'. Whenever grinding of a soft material is undertaken, it is necessary to avoid swarf buildup around the abrasive cutting faces. This is why silicon carbide is often used for such materials, i.e. the abrasive grains break away when 'clogged'. When the soft material includes sufficient ceramic material to 'dress' the abrasive, then fixed diamond or CBN is recommended.

15.1.7 Printed circuit boards

All successful PCB manufacturers, large or small, are encouraged to utilize cross-sectioning as a quality control procedure. This is sometimes a necessity to comply with a contractor's specification. The information derived from cross-sectioning goes beyond that of just metrology, however; it can be performed at various stages of production from incoming laminates to failure analysis. Before creating a preparation procedure it is wise to have an appreciation of the possible defects since an incorrect recipe could obscure essential information. Figure 15.1 illustrates some of the artefacts that will require inspection according to the requirements of the particular specification being used.

Thickness

Copper plating is often required to satisfy a specific dimension with a limit to the minimum, such as, for example, 20 μm. The thickness of the plating can vary in size through the hole, requiring an average measurement of three places each side of the hole.

Multilayer boards require a check on the correct number of layers present with the correct dielectric spacing culminating in the correct overall thickness. Before any measurement can be made the different coatings have to be optically identified. This is achieved by using a suitable etchant such as ammonia hydroxide/hydrogen peroxide.

Plating voids and adhesion

These two effects are displayed in Figure 15.1. They are most critical to the preparation technique since overpolishing would 'lid' over

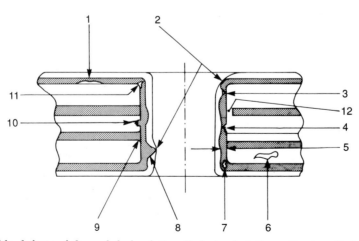

Figure 15.1 Possible defects of through-hole plating (1, lack of plating adhesion; 2, final plating faults; 3, primary plating void; 4, fibre protrusion; 5, average plating thickness; 6, laminate void; 7, plating void; 8, nodule; 9, nail heading; 10, wicking; 11, knee crack; 12, poor connection with resin smear)

the void. Lack of plating adhesion could be disguised at the grinding stage if blunt abrasives or well-used silicon carbide papers were to be used.

Quality of drilled hole

Drilling is one of the first operations the laminated board receives; the quality of this hole has an effect on future production stages. The observer will optically investigate at the copper plating stage and will ensure the hole is not rough or burred and is free from fibre protrusion. The absence of nail heading will also be investigated. It is unlikely that the preparation technique would induce or disguise such aberrations.

Desmear and wicking

With multilayers a check must be made for the effectiveness of the desmear. It should not be too aggressive, leading to wicking of the plating back along the glass fibre.

Microdelamination

Microdelamination is often caused when drilling or acid treatments attack the black oxide coating that gives bond strength. It can also be a result of using a poor encapsulating mounting material that exhibits high levels of contraction. It is important to check for registration, layer to layer and the drilled hole to pad; the hole is not permitted to break out of the pad.

Thermal cycling

This check is to identify poor ductility of the copper, resulting in cracks in the plating or the inner layer separating from the plated barrel. After thermal cycling or floating on solder at 260°C for 10 seconds the coated board is sectioned and prepared in the normal manner. When making any critical analysis for detecting cracks it is advisable to etch the sample.

The following guide-lines relate to suitable preparation techniques for through-hole plating analysis.

15.2 Preparation Techniques

15.2.1 Sectioning

(i) Hacksaws can be used, but keep well away from the required hole. Care must be exercised to avoid delamination. If there is evidence of delamination (multi-layer board) in the prepared micro, the use of hacksaws *must* be stopped.

(ii) Table top high speed diamond saws are preferred in some production environments and are extremely good providing the cut is well away from the required hole.

(iii) Table top routers perform two functions; the first is to drill two location holes for subsequent hole alignment, the other is to rout out the coupon. The router acts like a vertical milling machine and should in principle induce the *least* structural damage.

(iv) Low deformation saws could be found to match the quality of the router and are preferred for manual operation when it is required to accurately section to the edge of the through-hole without damage.

(v) Punch and die techniques are occasionally used. This is probably the quickest method for extracting coupons from within the board. Care must be taken not to cause delamination and to keep well clear of the required through-hole. This technique is not to be recommended for multilayer boards.

(vi) Test coupons can be incorporated within the board construction. They usually have the two datum holes drilled in readiness for the quantity preparation procedure.

15.2.2 Mounting

Cold mounting resins are to be favoured, the epoxy resins being the first choice. In practice there is a tendency for the user to use acrylic resins or even polyester. This is acceptable providing the resin contraction rate and viscosity are low.

It will not be the first time a through-hole has been rejected for delamination when in fact it was caused by resin contraction. Two samples were mounted from the same board, one using supplier A's acrylic, the other supplier B's acrylic—one shows delamination the other not.

15.2.3 Sample integrity

The traditional four-step silicon carbide (P320/400/600/1200) followed by two polishing steps (6 and 1 μm) can be used. There are two problems associated with the above:

(i) If the P320 paper is used until the near centre of the hole is achieved then all the subsequent steps will take you past centre.
(ii) To use two polishing steps could (will) change the actual dimension of soft layers, giving erroneous results.

To overcome these problems it is wise to use the *least* coarse silicon carbide abrasive possible in achieving near centre and then to progress with either P1200 silicon carbide paper or platen surface 2P followed by surface 1. This requires sectioning the coupon as near as possible to the edge of the through-hole.

To overcome the problem of manually grinding until 'dead centre' is achieved it is common practice to house the specimens in a fixture with diamond stops fitted. The sample is then ground until reaching the diamond stop; these stops are adjusted to give the near dead centre position. These jigs can be used manually or automatically and can incorporate many coupons in each sample. The alignment of the coupons is via the two location holes as illustrated in Figure 15.2.

The *danger* with the diamond stop technique when contacting the silicon carbide paper is that it quickly converts the optimum grains into critical grains, creating gross deformation if continued to be used. This is particularly important with automatic heads incorporating many specimens since the specimens in the holder never grind absolutely flat and have to be progressed after the first diamond stop comes into contact with the paper. It is wise under these circumstances to stop further back from dead centre and rely on surface 2P to remove stock. This can be controlled by time via the Z axis test.

15.2.4 Polishing

Short periods on platen surface 7, 8 or 9 using an oxide slurry are all that is required.

15.2.5 Etching

This can be achieved with some alloys using high pH colloidal silica as the polishing agent. Etching can be necessary beyond the purposes of just creating contrast. With every mechanical preparation there must be some even sub-micrometre surface deformation; etching can remove this. Although it is used to reveal the microstructure it can also expose finer defects sometimes obscured because of contrast.

15.2.6 P2400 and P4000 silicon carbide papers

Throughout this book there is little evidence of the use of ultra-fine silicon carbide papers. This does not mean, however, that they do not have applications. Fine silicon carbide papers principally grind (within their optimum life) and this the author favours. Providing the grains are not 'clogged' with swarf or the prepared material will not be attracted to the chipped silicon carbide particles then this must *always* be the favoured route. There are examples where the electronic material predominantly consists of laminated bonds. These and other examples would best be prepared using these small particle abrasives.

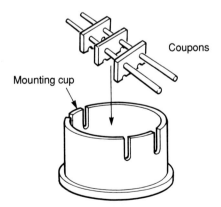

Figure 15.2 PCB for coupon alignment

15.3 Preparation Examples

Example 15.1

Family classification BB17

The material: PCBs

Printed circuit boards play a major part in our everyday life, often without the user necessarily being fully aware of it. This often multilayer board with its maze of holes, coatings and electronic devices, allows the input of signals to be stored and activated automatically. The success rate of such systems in use far surpasses the old wiring procedures. It could be said that if they are correct on installation, they will continue to be fail safe, providing they are used according to their design specification. To achieve this optimum condition, the boards have to pass a whole series of quality control tests as they are developed. These tests will inevitably include a cross-sectioning technique in order to detect any hidden defects.

Possible defects are illustrated for a through-hole plated multilayer board (Figure 15.1). A brief glimpse at the drawing gives an indication of the care that is necessary in the preparation to ensure aberrations are not artefacts of the preparation. Plating voids and knee cracks could be obscured at the polishing stage. Wicking and laminate voids could be caused by incorrect grinding. Delamination or lack of plating adhesion could be the result of internal stress, caused by incorrect mounting mediums. Average plating thickness will be influenced by contraction in the encapsulating resin and also the pressure applied at the polishing stage. With these examples, one can appreciate the need for absolute integrity in the chosen preparation technique.

The technique

The severity of the sectioning procedure dictates the Z axis stock removal at subsequent stages. Using a low deformation saw allows the board to be sectioned relatively quickly and with little induced damage. Planar grinding with P320 grit silicon carbide paper, followed by 15 μm diamond on platen surface 2P, will bring the sample to within 6 μm of the through-hole centre.

There is a tendency to follow this step with a 1 μm abrasive. The visual appearance from 1 μm would be very good but the need to avoid smearing and rubbing is paramount to preparation integrity—hence the reason for choosing 3 μm oil-based slurry prior to polishing.

170 Surface Preparation and Microscopy of Materials

The method

AUDITABLE PREPARATION PROCEDURE   No. 1

CLASSIFICATIONS: BB17/PCB	MATERIAL: MULTILAYER BOARD
SECTIONING TECHNIQUE:	BLADE
MOUNTING TECHNIQUE:	CONCENTRATION

*TWO SILICON CARBIDE STAGES COULD BE USED INSTEAD.

	SURFACE and TYPE	ABRASIVE SIZE/TYPE LUBRICANT	LOAD (kPa)	HEAD ROTATION	Z AXIS (μm)	REFLECTIVE FACTOR	MICROGRAPH YES/NO	IMAGE ANALYSIS YES/NO	REMARKS
PLANAR GRINDING STAGE	PAPER	P320 SiC WATER	35	COMP	PLANE				

	SURFACE and TYPE	ABRASIVE SIZE/TYPE LUBRICANT	LOAD (kPa)	HEAD ROTATION	Z AXIS (μm)	REFLECTIVE FACTOR	MICROGRAPH YES/NO	IMAGE ANALYSIS YES/NO	REMARKS
SAMPLE* INTEGRITY STAGE	S.2P	15 μm OIL	35	COMP					
	S.1	3 μm OIL	35	COMP					

	SURFACE and TYPE	ABRASIVE SIZE/TYPE LUBRICANT	LOAD (kPa)	HEAD ROTATION	TIME (min)	REFLECTIVE FACTOR	MICROGRAPH YES/NO	IMAGE ANALYSIS YES/NO	REMARKS
POLISHING STAGE	PS 7	0.06 μm SILICA	16	COMP	2		YES		15.3

The micrograph Figure 15.3

Figure 15.3 Through-hole PCB (×400)

Example 15.2

Family classification BB8

The material: multilayer ceramic capacitor

The quality control check requires the sectioning of the MLC to reveal, amongst other things, the quality of the alternating electrode plates on metal film. Of particular concern is the incidence of delamination, a defect whereby the layers separate at the ceramic/electrode interface. This defect can be caused by any one of several factors and is one of the reasons for the need to sample every batch.

The technique

This is an example of the softer ceramic material where the exclusive use of diamond is not necessary, planar grinding being carried out using P320 silicon carbide paper. Care has to be taken with degrading abrasives not to use them *beyond* the optimum critical period. Ceramic capacitors often require a large stock removal at the planar grinding stage. On these occasions a fixed diamond could be used, taking care not to induce a high level of structural damage. The enclosed procedure is based on the P320 planar grinding stage being used; other more aggressive steps would require more than 18 μm Z axis at the platen surface 4 stage. The last integrity stage uses double pressure. It is also interesting to note how the same abrasive (0.06 μm colloidal silica) is used at the polishing stage, this time using half the normal pressure.

The method

AUDITABLE PREPARATION PROCEDURE No. 2

CLASSIFICATIONS: BB8/ELECTRONIC COMPONENT	MATERIAL: CERAMIC CAPACITOR
SECTIONING TECHNIQUE:	BLADE
MOUNTING TECHNIQUE:	CONCENTRATION

PLANAR GRINDING STAGE	SURFACE and TYPE	ABRASIVE SIZE/TYPE LUBRICANT	LOAD (kPa)	HEAD ROTATION	Z AXIS (μm)	REFLECTIVE FACTOR	MICROGRAPH YES/NO	IMAGE ANALYSIS YES/NO	REMARKS
	PAPER	P320 SiC WATER	35	COMP	PLANE				

SAMPLE INTEGRITY STAGE	SURFACE and TYPE	ABRASIVE SIZE/TYPE LUBRICANT	LOAD (kPa)	HEAD ROTATION	Z AXIS (μm)	REFLECTIVE FACTOR	MICROGRAPH YES/NO	IMAGE ANALYSIS YES/NO	REMARKS
	S.4	9 μm OIL	35	CONTRA	18				
	S.1	3 μm OIL	35	COMP					
	S.1	0.06 μm SILICA	70	COMP					

POLISHING STAGE	SURFACE and TYPE	ABRASIVE SIZE/TYPE LUBRICANT	LOAD (kPa)	HEAD ROTATION	TIME (min)	REFLECTIVE FACTOR	MICROGRAPH YES/NO	IMAGE ANALYSIS YES/NO	REMARKS
	PS 7	0.06 μm SILICA	16	COMP	1		YES		15.4

The micrograph Figure 15.4

Figure 15.4 Ceramic capacitor (×400)

Example 15.3

Family classification BB8

The material: electronic components

Many electronic components require sectioning in order to identify the integrity of the combined unit. The components making up the unit will themselves have passed inspection prior to assembly. The hidden information can best be revealed by cross-sectioning, using very thin blades (200 μm). Figure 15.5 (see Plate 10) illustrates a multicomplex component. The central component is copper surrounded by ceramic and soft solder material. Also present are multiplatings, copper layers, silica rings, silicon rings, soft alloyed casing, etc. Figure 15.5 has tried to capture some of the hidden information available through cross-sectioning. Figure 15.5(a) and (b) is taken in dark ground and uncrossed polars, using a compound microscope.

The technique

This will vary depending on the investigation. For general component morphology it could be adequate to grind the sectioned assembly using P600 grit silicon carbide paper only. When a closer inspection is required, it will be necessary to vacuum impregnate, using an epoxy resin prior to carrying out the sample integrity stages. With plated components it can be advantageous to observe in dark ground; if so, the sample will require polishing to remove surface scratches.

The method

AUDITABLE PREPARATION PROCEDURE　　　　　　　　　　　　　　　　　　　　　　　No. 3

CLASSIFICATIONS: BB8/ELECTRONIC COMPONENT	MATERIAL: COMPLEX ASSEMBLY
SECTIONING TECHNIQUE: THIN BLADE SLITTING WHEEL	BLADE DIAMOND
MOUNTING EPOXY TECHNIQUE: VACUUM IMPREGNATED	CONCENTRATION

PLANAR GRINDING STAGE	SURFACE and TYPE	ABRASIVE SIZE/TYPE LUBRICANT	LOAD (kPa)	HEAD ROTATION	Z AXIS (μm)	REFLECTIVE FACTOR	MICROGRAPH YES/NO	IMAGE ANALYSIS YES/NO	REMARKS
	PAPER	P400 SiC WATER	35	COMP	PLANE				

SAMPLE INTEGRITY STAGE	SURFACE and TYPE	ABRASIVE SIZE/TYPE LUBRICANT	LOAD (kPa)	HEAD ROTATION	Z AXIS (μm)	REFLECTIVE FACTOR	MICROGRAPH YES/NO	IMAGE ANALYSIS YES/NO	REMARKS
	PAPER	P1000 SiC	35	COMP					
	S.2P	6 μm OIL	35	COMP					
	S.1	0.06 μm COLLOIDAL SILICA	45	CONTRA					

POLISHING STAGE	SURFACE and TYPE	ABRASIVE SIZE/TYPE LUBRICANT	LOAD (kPa)	HEAD ROTATION	TIME (min)	REFLECTIVE FACTOR	MICROGRAPH YES/NO	IMAGE ANALYSIS YES/NO	REMARKS
	PS 7	0.06 μm SILICA	16	COMP	2		YES		15.5 (Plate 10)

The micrograph Figure 15.5 (see Plate 10)

Example 15.4

Family classification BB27

The material: brazed metallized layer (Mo, Ni, Cu/Ag)

Metallizing the ceramic surface for electronic circuit applications, or simply to weld ceramics together, is proving to be a challenge to the materials technologist. Molybdenum is currently very successful with alumina and the aluminium nitride ceramics. Molybdenum has been coated with nickel, followed by gold, silver or copper. The metallized layer in Figure 15.6 (see Plate 10) is of an alumina weld. The porous looking area is the molybdenum, followed by a nickel layer, the remainder being a copper silver alloy.

The technique

Care must be exercised when dealing with a brittle ceramic (alumina), a porous interface (molybdenum) and a soft smeary sandwich (silver/copper). To use a fixed abrasive could damage the brittle elements of the material—hence the choice of a charged diamond abrasive on surface 10. The size of the diamond in the quoted example is 30 μm. With delicate components this may have to be reduced or a lower numerical platen surface used. After surface 4 with a 9 μm abrasive, the possibility of smaller sized abrasives embedding in the soft Cu/Ag sandwich becomes greater. The use of a chemomechanical technique helps to reduce the problems associated with impressed abrasives and balances the abrasion rates of the metal/ceramic. This gives a planar, well-resolved subject, as illustrated.

The method

AUDITABLE PREPARATION PROCEDURE   No. 4

CLASSIFICATIONS: BB27/BRAZED ALUMINA JOINT	MATERIAL: Mo, Ni, Cu/Ag ALUMINA JOINT
SECTIONING TECHNIQUE:	BLADE
MOUNTING TECHNIQUE:	CONCENTRATION

PLANAR GRINDING STAGE	SURFACE and TYPE	ABRASIVE SIZE/TYPE LUBRICANT	LOAD (kPa)	HEAD ROTATION	Z AXIS (μm)	REFLECTIVE FACTOR	MICROGRAPH YES/NO	IMAGE ANALYSIS YES/NO	REMARKS
	S.10	30 μm OIL	35	CONTRA	PLANE				

SAMPLE INTEGRITY STAGE	SURFACE and TYPE	ABRASIVE SIZE/TYPE LUBRICANT	LOAD (kPa)	HEAD ROTATION	Z AXIS (μm)	REFLECTIVE FACTOR	MICROGRAPH YES/NO	IMAGE ANALYSIS YES/NO	REMARKS
	S.4	9 μm OIL	35	CONTRA					
	S.1	1 μm OIL HIGH pH	70	COMP					

POLISHING STAGE	SURFACE and TYPE	ABRASIVE SIZE/TYPE LUBRICANT	LOAD (kPa)	HEAD ROTATION	TIME (s)	REFLECTIVE FACTOR	MICROGRAPH YES/NO	IMAGE ANALYSIS YES/NO	REMARKS
	PS 7	0.06 μm COLLOIDAL SILICA	17	COMP	120		YES		15.6 (Plate 10)

The micrograph Figure 15.6 (see Plate 10)

16

Thin Film Measurement

Measurement of thin plated layers is not specific to electronic components but it does have many such applications. It has been separated from the main chapter since it is not specific to the preparation techniques. The preparation technique does, however, influence the final dimension; the prepared size can be, and often is, increased through smearing or mechanical working. Measuring the prepared layer can also be subjective. This will be investigated to give a better understanding of the options reducing operator error.

16.1 Preparation Technique

With care many different sectioning techniques should be suitable. Obviously, if the coating is poorly bonded or coated onto a friable substrate low deformation saws must be used.

The area of greatest abuse is to use high contraction resins when mounting the specimen. If the coating is, say, 10 μm thick it is very unwise to use a mounting resin that contracts 25 μm. The degree of contraction permitted is compatible with the size and tolerance expected from the thin layer. Multilayer PCBs have on inspection shown a type of delamination that was caused entirely by the shrinkage of the mounting medium.

Many of these thin layers are soft ductile materials (gold, tin, zinc, etc.). This makes them very susceptible to smearing when polished with a high nap cloth (PS 8) and/or high speed and high pressure (see Figure 16.1).

It is customary for prepared samples to be highly polished and scratch free. This feature is incompatible with accurate metrology. Another polishing artefact is that of the so-called polishing relief. This occurs when the *grinding* rate between the specimen and associated mount varies, resulting in the specimen standing proud of the mount (see Figure 16.2).

The microscope image is a result of interference between the orders of refracting rays when recombined in the primary image. The zero-order rays are shown in Figure 16.2, which shows how the size will appear to be

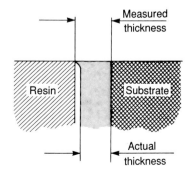

Figure 16.1 Preparation artefact—smearing

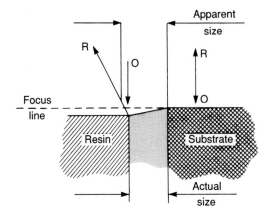

Figure 16.2 Preparation artefact—polishing relief

bigger than actual. To reduce this effect an objective exhibiting a greater depth of field would be used, but unfortunately this often results in a lower magnification and less accuracy.

16.2 Measuring Techniques

There is an increasing demand not only to measure thin film or plated areas but to increase the measured accuracy. The preferred method is to section the component, encapsulate, grind and polish prior to using the compound microscope to observe and measure the thin layer. To use the highest possible magnification gives the highest accuracy, but gives a restricted field of view. When measuring a thickness of $1\,\mu\mathrm{m}$ and less, it is wise to increase the eyepiece magnification beyond $\times 10$ to 25. If the measurement is being taken from a photomicrograph the highest camera eyepiece magnification should be employed. To increase the accuracy of measurement, systems such as image shearing could be employed (see Chapter 27).

When the accuracy and tolerance requirements are critical it is wise to utilize a taper sectioning technique. In this method the sample is ground not in the vertical mode (transverse) but inclined to preferably a given angle. The taper sectioning chart in Figure 16.3 draws a reciprocal sine curve, where it will be noticed how the sample has to be inclined to 40° from the horizontal before any appreciable benefit is derived. If the inclined samples are to be aligned by eye, without any pre-made fixture, then it is wise to use a vertical component for reference, as shown in Figure 16.4. In this example X (the plated thickness) equals AC/B.

Some suppliers offer 'angle polishing inserts' where the angle is known prior to mounting.

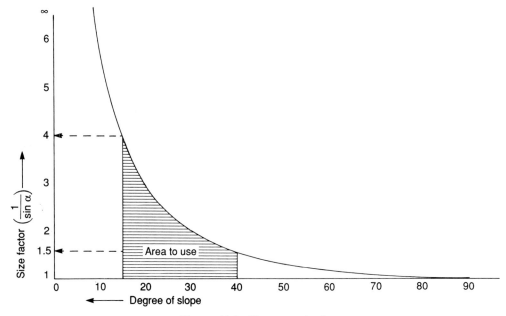

Figure 16.3 Taper sectioning

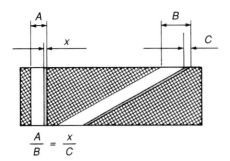

Figure 16.4 Angle ratio

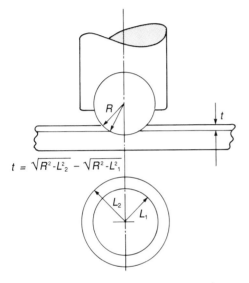

Figure 16.5 Spherical ball method—plan 2-1

The prepared sample then only requires to be multiplied by a known factor. This is a perfectly acceptable practice if working within 15° to 40° from the horizontal. When operating at 5°, as some inserts do, then it is wise to check the actual mounted angle, since slight changes of angle are very sensitive at this level, i.e. do not just multiply by 10 (actual thickness equals sine X multiplied by the measured thickness).

The method illustrated in Figure 16.5 is ideal for thin (micrometre region) plating. The idea is to have a steel ball of known dimensions fixed into a rotating head (such as a drilling machine). Lower the ball onto the plated specimen using a 3 μm diamond spray with lubricant. The greatest accuracy occurs when the ball has just penetrated the plating. When using a ball of known diameter it is easy to relate a constant to the difference of $L_2 - L_1$ from the formulae given.

The advantage of this method is speed (requires no mounting). It can also be used in a non-destructive test for on-site evaluation. Unfortunately for maximum accuracy the field of view in the microscope must be large in order to take measurements L_2 and L_1, implying a low magnification.

16.3 Graticules Measuring

Most compound microscopes depend upon a fixed eyepiece graticule or a filar micrometer system. The cause of most operator error occurs in interpreting which side of the line to take measurements from, i.e. when the micrometer line is just touching, in the middle or when there is a small light gap. This is further compounded because different operators focus at different points. To overcome all these factors the operators must become traceable to a known measured standard. They must put the micrometer lines in such a position to achieve the known standard dimension. This position will differ between operators but will overcome a major problem in subjective measuring. This subject is expanded in Chapter 27.

17

Preparation of Soft Materials

17.1 Introduction: The Problems

Technology never quite works out the way a layman would interpret it to be; as quoted in the microscope section 'anyone can take a good high magnification micrograph, but it takes an expert to repeat the quality at low magnification'. When preparing materials one would expect the hard metals to be difficult and the soft metals to be easy. This is not the case. Soft metals are extremely difficult to prepare by metallurgical means; they tend to distort (deformation), smear (polishing) and are prone to impressed abrasives. Major problems associated with the metallographic preparation of lead and lead alloys are cold working and recrystallization. Therefore methods generating the minimum mechanical strain and least generated heat are essential. Mechanical strain can be reduced by etch grinding. The stress often associated with generated heat can be reduced using lower speeds and loads. Dissolution etching (developing the microstructure by surface removal) or electrolytic polishing is often favoured with these materials. However, the following is based on mechanical techniques.

Recrystallization will always be a problem where inefficient or blunt abrasives are employed. This problem can manifest itself at the sectioning stage. Abradable cut-off wheels using sharp silicon carbide abrasives in a very soft bond and ample water-based lubricant can be successful *providing* the machine cutting parameters have been optimized. Where it is necessary to observe the microstructure quite a distance below the sample surface an alternative to sectioning is to use a sharp pointed positive rake angle tool on the traditional turning engineering lathe. The final increments would be as low as 15 μm with the cutting tool well lubricated.

When encapsulating the sample, only cold setting resins with the lowest possible exotherm should be employed (ideally use epoxy resins).

With the introduction of a deformation structural angle (see Chapter 9, Figure 9.2) it will be seen that soft materials have the greatest angle, i.e. they exhibit the greatest deformation relative to the abrasive size and surface being used. Ductile metals with high deformation structural angles are very critical to correct positive rake angle cutting and oil-based lubricants. Current grinding abrasives do not always strictly satisfy this rake angle requirement—hence the reason for using machine saws, machine lathes and microtomes, for the initial preparation. *Brittle subjects*, although also exhibiting a high structural/deformation angle, are not to be confused with *ductile metals*.

Silicon carbide is still generally recommended for grinding soft materials and in general the coarser grits create the greatest problem with impressing silicon carbide particles into the sample. If silicon carbide papers are to be used it could prove necessary to etch and ultrasonically clean the sample between every stage. Another recent recommendation was to prepare soft materials using the finest grades of silicon carbide paper to overcome the impressed particle problem. Although it reduces impressed particles it usually leads to surface rubbing and the recrystallization problem. If silicon carbide papers are to be used they must be waxed or oiled before application.

The introduction of new platen surfaces and their classification was seen by some as an essential part of the preparation of soft materials. Unfortunately this is not the case since the primary selection for platen surfaces must be PS No. 2; the problem still remains 'how do we proceed towards No. 2?' Chapter 5 gave a guide to overcoming this problem which will now be developed.

17.2 Introduction: The Solution

(i) *Use fixed abrasives*. This is always good advice but unfortunately the only non-degrading fixed abrasives tend to have negative rake angles, which if used will create high deformation and must be associated with an oil lubricant to avoid swarf buildup around each abrasive grain. Perhaps some diamond or CBN abrasives grinding supplier will provide a suitable platen surface for cutting soft materials. Using positive rake angles and reduced surface tension allowing adequate oil-based lubricant, the grinding of soft materials should be easy.

(ii) *Increase surface speed*. This can only be applied when an efficient cutting condition is being used (rake angle and lubricant). The surface speed must not be increased when poor chip removal is occurring or when the material is being polished. There is a case for reducing the speed at the polishing stage when polishing surface 7 or greater is being used.

(iii) *Reduce the load*. Deformation is relative to the abrasive size being used. Reducing the load has the effect of reducing the abrasive relative depth of penetration and is therefore a preferred option.

Currently the most efficient abrasive for soft materials is silicon carbide (sharp and pointed). Unfortunately the abrasive degrades or breaks; reducing the speed would be most essential in reducing the impact load and retaining the sharp pointed abrasive. (Do not act on the recommendations that suggest the use of a dummy to break the sharp points prior to application.) The load is also critical at the polishing stage. Although this in general should be half the grinding load there are occasions with the more softer materials when it is necessary to reduce this further to avoid surface flow and impressed abrasives. Reducing pressure can also be used to induce relief to improve particulate or phase resolution.

(iv) *Use solid lubricants*. Because of the required rake angle it is often necessary to use degrading abrasives. These abrasives as they degrade leave broken particles on the grinding surface which become attracted (impressed) to soft materials. If, however, these particles become mixed with solid lubricants (wax or solid soap) this effect is reduced.

(v) *Use strong abrasives*. The most favoured abrasive is silicon carbide because of its sharp cutting edge. Unfortunately it chips readily and can cause problems with impressed particles; on such occasions alumina would be better. Cubic boron nitride, for example, is stronger than many degrading abrasives and this could be the favoured abrasive once angular shaped particles are suitably bonded.

(vi) *Oil the abrasive surface prior to use*. This is good advice even when the surface is to be flushed with water during use. Ensure the surface is dry prior to the sprayed oil application. Much reference has been made throughout this book to the good practices of production engineering techniques; no longer is the 'bucket in the yard' mentality towards lubricants acceptable. We materialographers who are expected to perform sophisticated production engineering techniques must lose our 'flush it with water' approach.

(vii) *Use abrasive with positive rake angle.* Being aware of this can influence the preparation choice. One could, for example, not use 'blocky' diamonds which tend to be predominantly negative rake angles. Turning on a lathe is not necessarily a bad practice. Silicon carbide papers vary from different suppliers; avoid using the poorly bonded versions where the abrasives break away from the backing. When the abrasive chips or breaks, due in part to impact shock, it is wise to *absorb* the shock by using plastic, not metal, platens.

(viii) *Attack polish.* This is mainly practised towards the end of the preparation procedure to stop top surface flow.

(ix) *Avoid swarf rubbing.* Swarf when removed from the specimen lodges between the grains of silicon carbide paper. This swarf must be flushed away before it becomes trapped between the sample and the grinding paper. As the silicon carbide grains become smaller and in consequence the concentration of particles increases, there becomes less space for the swarf to lodge without specimen surface rubbing occurring. This means there is a limit with silicon carbide paper sizes where smaller grains would prove destructive and not constructive. Unfortunately this conflicts with the much documented reports which encourage the use of silicon carbide papers up to P4000 grit.

Silicon carbide can be impregnated into a cloth type paper as opposed to being stuck onto the surface as is the tradition. With this technique fine grades can be used for soft materials, unfortunately no commercial papers of this type are available beyond P1200 grit.

(x) *Use the largest size abrasive possible.* Unfortunately small sized abrasives can leave high deformation, forcing the larger sized abrasives to be used. The size difference between the abrasive used on platen surface 2 and that used at the first stage could be great. The limiting size for charged abrasives on surface 2 is dependent upon the ductility of the sample material. As ductility increases, so the charged abrasive must increase to avoid impressed abrasives. Fixed abrasives, although unsuitable, can be used successfully at this stage providing the sample is surrounded with some dressing material, i.e. ceramic ring mounts. Final grinding or polishing is achieved using small abrasives in a pH active attack technique.

(xi) *Head to platen direction.* This applies to specimens prepared in an automatic machine where the samples are *driven* in rotation across a rotating wheel. It is vitally important to observe the speed, pressure and directions provided by the equipment supplier. For soft materials it is advisable to employ a complementary direction at the polishing stage. Failure to do this will result in comet-tails and/or a directionally smeared top surface. Recent empirical tests have indicated that a much improved surface preparation of soft materials occurs when the head speed rotates slowly (33 rev/min) against a higher platen speed of 300 rev/min for sample integrity stages. This information is currently unsubstantiated; it does, however, endorse the necessity for machine optimization.

(xii) *Surface affinity.* This occurs when two similar *soft* materials are rubbed together; they 'bind'. This condition can exist if lower numbered platen surfaces are used to prepare soft materials. On such occasions these surfaces should be restricted to large sized abrasive functions or the surface changed to a non-metal type; the latter is the preferred option.

(xiii) *Polishing.* When a diamond abrasive is used on a polishing cloth it is wise to run a dummy over the surface after each diamond charge if there is any tendency for impressed abrasives. Some abrasives can be attracted to a specimen more than others: sharp fragmented abrasives (silicon carbide, chipped diamond) are more prone than the round, less sharp abrasives (silica, alumina, blocky diamond). The polishing cloth texture

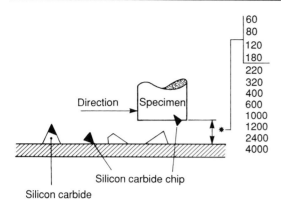

Figure 17.1 Find the limit of *

plays an important part in the final polished surface. The familiar Selvyt cloth, when used with slurry polishing, has always produced good results. Fine diamond polishing can also be successful provided that the polishing cloth has a soft fibre, low nap, high density quality. Many cloths available for polishing are soft fibre, high nap, low density types.

17.2.1 Conclusion

The following briefly reiterates some of the points raised:

(i) What is the limit of * (see Figure 17.1)?
(ii) Remember the advantages of oil.
(iii) There is a minimum tolerated size of abrasive for every grinding surface relative to the sample material.
(iv) At completion use non-angular abrasives.
(v) The pH control is important.
(vi) Complementary head direction is needed at the conclusion of the procedure.
(vii) Low pressure is needed at the planar grind and polishing stage.
(viii) Avoid metal-to-metal affinity.

17.2.2 White metals

White metals have a long history of successful operation as engine bearing linings. They have excellent compatibility, which allows them to work satisfactorily in the unlubricated condition. The object of a bearing is to reduce friction and wear which inevitably puts a great demand on the fluid lubricating film between the two sliding surfaces. Specialized bearing materials would not be necessary if this lubricating film could always be thick in relation to the mating surface roughness and dirt contamination. To maintain this oil lubricating film depends upon the working environment. It is not always possible, for example, to operate with a hydrodynamic lubricant. Grease or wick-lubrication is necessary, or alternatively, the requirements of the bearing may demand a non-lubricating environment, in which case polytetrafluoroethylene (PTFE) would have to be used. The operating temperature and load also affects the choice of bearing material. The majority of high speed internal combustion engines imposes loads on the bearings too high for white metals to withstand, resulting in fatigue cracking. White metals such as the tin-based types are, however, very suitable for engine bearings of the large slow speed diesel engines.

17.3 Preparation Examples

Example 17.1

Family classification BB26

The material: aluminium 4 to 20% tin

Tin-based white metals are soft (Hv 27 to 33) and have an excellent dirt embeddability; this might be ideal for a bearing but presents innumerable problems in mechanically working the specimen surface. Loose grit, for example, is intended to impress into the bearing surface without damage or scoring when the bearing is operational. In our preparation procedure, therefore, we must ensure that loose grinding abrasives remain on the grinding or polishing surface and are not attracted into the specimen surface. Because the specimen is soft, high preparation pressures must be avoided. Although the primary molten phase is around 240°C it is wise to keep well below this temperature; the bearing when in use is not intended to work at temperatures above 130°C.

The technique

Soft materials do not lend themselves to conventional abrasive cutting; they should be cut with sharp positive rake tools. It has been known for a hacksaw to be used; if this is so, then the planar grinding stage must be prolonged to remove the damaged layer. Oily lubricants help at every cutting stage with ductile materials. Since soft materials abrade quickly the thermosetting mounting mediums will be preferred.

Planar grinding can be carried out using silicon carbide paper provided that the SiC abrasive is large and the broken chips of SiC are not attracted to the soft materials. On this occasion 180 grit proved satisfactory. Sample integrity must be achieved using a charged abrasive large enough not to impress into the sample, yet not too large as to render the next and final step too severe. Note that 15 μm diamond is used with an oil-based lubricant ensuring that the abrasive favours the integrity surface and not the specimen.

The method

AUDITABLE PREPARATION PROCEDURE   No. 1

CLASSIFICATIONS: BB26/SOFT BEARING MATERIAL	MATERIAL: ALUMINIUM/4 to 20% Sn
SECTIONING TECHNIQUE:	BLADE
MOUNTING TECHNIQUE:	CONCENTRATION

PLANAR GRINDING STAGE	SURFACE and TYPE	ABRASIVE SIZE/TYPE LUBRICANT	LOAD (kPa)	HEAD ROTATION	Z AXIS (μm)	REFLECTIVE FACTOR	MICROGRAPH YES/NO	IMAGE ANALYSIS YES/NO	REMARKS
	PAPER OIL	180 SiC WATER	16	COMP	PLANE				

SAMPLE INTEGRITY STAGE	SURFACE and TYPE	ABRASIVE SIZE/TYPE LUBRICANT	LOAD (kPa)	HEAD ROTATION	Z AXIS (μm)	REFLECTIVE FACTOR	MICROGRAPH YES/NO	IMAGE ANALYSIS YES/NO	REMARKS
	S.2P	15 μm OIL	35	CONTRA					
	S.1	0.06 μm SILICA	16	COMP					

POLISHING STAGE	SURFACE and TYPE	ABRASIVE SIZE/TYPE LUBRICANT	LOAD (kPa)	HEAD ROTATION	TIME (s)	REFLECTIVE FACTOR	MICROGRAPH YES/NO	IMAGE ANALYSIS YES/NO	REMARKS
	PS 7	0.06 μm SILICA	16	COMP	120		YES		17.2(Plate 10)

The micrograph Figure 17.2 (see Plate 10)

Example 17.2

Family classification BB9

The material: lead

Lead with its low melting point (327°C) is easy to cast, easy to shape, easy to join and resistant to corrosion. In the past its uses ranged from drain-pipes to children's lead toys. Today, it forms the basis of car batteries, weather flashing and ornamental roofs. As an alloying agent it is also important, its uses varying from solders to free cutting steels. As a bearing alloy it is unsurpassed in tolerance to shock load and some hostile environments. Lead also finds application in sound proofing, radiation shielding and as a chemical compound in glass, giving us the often misused term 'lead glass crystal'.

The technique

Anyone who claims to be able to prepare soft materials will inevitably be asked to prepare lead. Although mechanical preparation techniques are not necessarily the best route to adopt they can be prepared with care. The method illustrated uses a fixed abrasive at the planar grinding stage (70 μm). Sample integrity is achieved with one step only, the 45 μm diamond being charged onto the platen prior to the silk cloth being fitted. This technique has limited use since the cloth can only be used as long as the abrasives remain trapped within the silk cloth. Chemomechanical techniques with this and all smeary materials are an essential ingredient of the final preparation. Vibratory polishing must also be given consideration because of its smear-free action.

The method

AUDITABLE PREPARATION PROCEDURE   No. 2

CLASSIFICATIONS: BB9/BEARING MATERIAL	MATERIAL: Pb-BASED ALLOY
SECTIONING TECHNIQUE:	BLADE
MOUNTING TECHNIQUE:	CONCENTRATION

	SURFACE and TYPE	ABRASIVE SIZE/TYPE LUBRICANT	LOAD (kPa)	HEAD ROTATION	Z AXIS (μm)	REFLECTIVE FACTOR	MICROGRAPH YES/NO	IMAGE ANALYSIS YES/NO	REMARKS
PLANAR GRINDING STAGE	FIXED DIAMOND OIL	70 μm WATER	16	COMP	PLANE				

	SURFACE and TYPE	ABRASIVE SIZE/TYPE LUBRICANT	LOAD (kPa)	HEAD ROTATION	Z AXIS (μm)	REFLECTIVE FACTOR	MICROGRAPH YES/NO	IMAGE ANALYSIS YES/NO	REMARKS
SAMPLE INTEGRITY STAGE	S.1	45 μm OIL	35	CONTRA					

	SURFACE and TYPE	ABRASIVE SIZE/TYPE LUBRICANT	LOAD (kPa)	HEAD ROTATION	TIME (s)	REFLECTIVE FACTOR	MICROGRAPH YES/NO	IMAGE ANALYSIS YES/NO	REMARKS
POLISHING STAGE	PS 7	0.06 μm SILICA	16	COMP	120		YES		17.3(Plate 10)

The micrograph Figure 17.3 (see Plate 10)

Example 17.3

Family classification BB26

The material: tin-based white metal

Planar grinding was carried out using a fixed 70 μm diamond; coarse silicon carbide could also be used with extreme care. The sample integrity stage is achieved with platen surface 2P using alcohol as the lubricant. This is because this surface is a plastic material; had the surface been a metal composite then the conditions would be different. The reader will now see a trend emerging where coarse abrasives are being used, much coarser than one would have expected. This is due to fine charged abrasives impressing and fine fixed abrasives causing swarf rubbing and surface ploughing.

The method

AUDITABLE PREPARATION PROCEDURE		No. 3
CLASSIFICATIONS: BB26/BEARING MATERIAL		MATERIAL: Sn-BASED WHITE METAL
SECTIONING TECHNIQUE:		BLADE
MOUNTING TECHNIQUE:		CONCENTRATION

	SURFACE and TYPE	ABRASIVE SIZE/TYPE LUBRICANT	LOAD (kPa)	HEAD ROTATION	Z AXIS (μm)	REFLECTIVE FACTOR	MICROGRAPH YES/NO	IMAGE ANALYSIS YES/NO	REMARKS
PLANAR GRINDING STAGE	FIXED DIAMOND OIL	70 μm WATER	16	COMP	PLANE				

	SURFACE and TYPE	ABRASIVE SIZE/TYPE LUBRICANT	LOAD (kPa)	HEAD ROTATION	Z AXIS (μm)	REFLECTIVE FACTOR	MICROGRAPH YES/NO	IMAGE ANALYSIS YES/NO	REMARKS
SAMPLE INTEGRITY STAGE	S.2P	15 μm OIL	16	CONTRA					
	S.1	0.06 μm SILICA	16	COMP					

	SURFACE and TYPE	ABRASIVE SIZE/TYPE LUBRICANT	LOAD (kPa)	HEAD ROTATION	TIME (s)	REFLECTIVE FACTOR	MICROGRAPH YES/NO	IMAGE ANALYSIS YES/NO	REMARKS
POLISHING STAGE	PS 7	0.06 μm SILICA	16	COMP	120		YES		17.4(Plate 11)

The micrograph Figure 17.4 (see Plate 11)

Example 17.4

Family classification BB26

The material: overlay plated, steel backed Al/Si

With this and all other examples using a fixed diamond at the planar grinding stage, the pressure is quoted as low. This is the minimum pressure the machine can exert that is compatible with stock removal. The sample integrity stage is similar to others with the exception of the 3 μm stage. This is due entirely to the steel backing. This example is used to illustrate how the preparation technique will change with composite changes; the material without the steel backing would have made this last step impossible (the 3 μm abrasive would impress).

The method

AUDITABLE PREPARATION PROCEDURE   No. 4

CLASSIFICATIONS: BB26/BEARING MATERIAL	MATERIAL: Al/Si OVERLAY PLATE ON STEEL
SECTIONING TECHNIQUE:	BLADE
MOUNTING TECHNIQUE:	CONCENTRATION

	SURFACE and TYPE	ABRASIVE SIZE/TYPE LUBRICANT	LOAD (kPa)	HEAD ROTATION	Z AXIS (μm)	REFLECTIVE FACTOR	MICROGRAPH YES/NO	IMAGE ANALYSIS YES/NO	REMARKS
PLANAR GRINDING STAGE	FIXED DIAMOND OIL	70 μm WATER	16	COMP	PLANE				

	SURFACE and TYPE	ABRASIVE SIZE/TYPE LUBRICANT	LOAD (kPa)	HEAD ROTATION	Z AXIS (μm)	REFLECTIVE FACTOR	MICROGRAPH YES/NO	IMAGE ANALYSIS YES/NO	REMARKS
SAMPLE INTEGRITY STAGE	S.2P	15 μm OIL	35	CONTRA					
	S.2P	3 μm OIL	16	COMP					

	SURFACE and TYPE	ABRASIVE SIZE/TYPE LUBRICANT	LOAD (kPa)	HEAD ROTATION	TIME (s)	REFLECTIVE FACTOR	MICROGRAPH YES/NO	IMAGE ANALYSIS YES/NO	REMARKS
POLISHING STAGE	PS 7	0.06 μm SILICA	16	COMP	120		YES		17.5(Plate 11)

The micrograph Figure 17.5 (see Plate 11)

Example 17.5

Family classification BB26

The material: Al/Pb plated

This example shows how simple and uncomplicated the procedure can be if it is correct, many complicated alternatives have been proposed and used. The aim for all preparation procedures is to keep them simple and aim for integrity in the shortest time.

The method

AUDITABLE PREPARATION PROCEDURE   No. 5

CLASSIFICATIONS: BB26/BEARING MATERIAL	MATERIAL: Al/Pb PLATED
SECTIONING TECHNIQUE:	BLADE
MOUNTING TECHNIQUE:	CONCENTRATION

	SURFACE and TYPE	ABRASIVE SIZE/TYPE LUBRICANT	LOAD (kPa)	HEAD ROTATION	Z AXIS (μm)	REFLECTIVE FACTOR	MICROGRAPH YES/NO	IMAGE ANALYSIS YES/NO	REMARKS
PLANAR GRINDING STAGE	FIXED DIAMOND OIL	70 μm WATER	16	COMP	PLANE				

	SURFACE and TYPE	ABRASIVE SIZE/TYPE LUBRICANT	LOAD (kPa)	HEAD ROTATION	Z AXIS (μm)	REFLECTIVE FACTOR	MICROGRAPH YES/NO	IMAGE ANALYSIS YES/NO	REMARKS
SAMPLE INTEGRITY STAGE	S.2P	15 μm OIL	35	CONTRA					

	SURFACE and TYPE	ABRASIVE SIZE/TYPE LUBRICANT	LOAD (kPa)	HEAD ROTATION	TIME (s)	REFLECTIVE FACTOR	MICROGRAPH YES/NO	IMAGE ANALYSIS YES/NO	REMARKS
POLISHING STAGE	PS 7	0.06 μm SILICA	16	COMP	120		YES		17.6(Plate 11)

The micrograph Figure 17.6 (see Plate 11)

Example 17.6

Family classification BB26

The material: steel backed Al/Cu/Si/Sn bearing

This preparation technique, with the exception of the final integrity stage, is similar to Example 17.4, which was also steel backed. The abrasive on this occasion had to be as small as 1 μm in order to clearly resolve all particles. The photomicrograph was taken in differential interference contrast (DIC) using the first order grey extinction position. This technique allows the observer to optically look *within* the subject and not just *at it*, and is the technique favoured by the writer. The uninitiated could mistake this depth as being polishing relief, but this is not the case. Examples 17.1, 17.2 and 17.3 were also photographed in DIC.

188 Surface Preparation and Microscopy of Materials

The method

AUDITABLE PREPARATION PROCEDURE   No. 6

CLASSIFICATIONS: BB26/BEARING MATERIAL	MATERIAL: Al/Cu/Si/Sn ON STEEL
SECTIONING TECHNIQUE:	BLADE
MOUNTING TECHNIQUE:	CONCENTRATION

PLANAR GRINDING STAGE	SURFACE and TYPE	ABRASIVE SIZE/TYPE LUBRICANT	LOAD (kPa)	HEAD ROTATION	Z AXIS (μm)	REFLECTIVE FACTOR	MICROGRAPH YES/NO	IMAGE ANALYSIS YES/NO	REMARKS
	FIXED DIAMOND OIL	70 μm WATER	16	COMP	PLANE				

SAMPLE INTEGRITY STAGE	SURFACE and TYPE	ABRASIVE SIZE/TYPE LUBRICANT	LOAD (kPa)	HEAD ROTATION	Z AXIS (μm)	REFLECTIVE FACTOR	MICROGRAPH YES/NO	IMAGE ANALYSIS YES/NO	REMARKS
	S.2P	15 μm OIL	35	CONTRA					
	S.2P	1 μm OIL	16	COMP					

POLISHING STAGE	SURFACE and TYPE	ABRASIVE SIZE/TYPE LUBRICANT	LOAD (kPa)	HEAD ROTATION	TIME (s)	REFLECTIVE FACTOR	MICROGRAPH YES/NO	IMAGE ANALYSIS YES/NO	REMARKS
	PS 7	0.06 μm SILICA	16	COMP	120		YES		17.7(Plate 11)

The micrograph Figure 17.7 (see Plate 11)

Example 17.7

Family classification BB6

The material: titanium

It could be argued that without titanium the development of aerospace would have been much retarded. Titanium is strong, ductile, light and corrosive resistant. It can be alloyed with many elements, such as the wrought 6% aluminium, 4% vanadium illustrated. This material finds many applications where high temperatures and/or high stress are present, such as gas turbine blades and airframe components. The decline of steel and aluminium in jet engines occurred as nickel and titanium increased. It is predicted that by the year 2010 we will see a continued reduction in steel and aluminium but also that nickel and titanium will be replaced by ceramic, metal and resin-based composites. Many parts are welded by diffusion bonding—a faithful reproduction of the joint microstructure can offer a challenge to the materialographer. Titanium is included as a biomaterial because of its considered inertness to body fluids. There is, however, a school of thought that would prefer non-metal implants for the future (coral). Cutting tools are now being coated with submicrometre layers of titanium carbide and titanium nitride to vastly improve tool life. Titanium has also made a major contribution to the paint industry, where it is used as titanium oxide to give the much desired 'whiter than white' appearance.

The technique

Provided sharp abrasives and correct lubricants are used, then titanium alloys, with the aid of a mechanochemical final stage, should offer few preparation problems. The key to the preparation of smeary materials is not to use small abrasives that are likely to cause specimen to platen surface rubbing (fixed abrasives) or impressed abrasives (charged abrasives). With pure titanium it could be necessary to oil or wax the grinding surfaces prior to use. Microtomes have been used, but prove to be extremely prone to induce deformation when the blade point is not sharp or the slice too thick. Electrolytic polishing can overcome many final integrity stage problems.

The method

AUDITABLE PREPARATION PROCEDURE — No. 7

CLASSIFICATIONS: BB6	MATERIAL: TITANIUM 6%Al - 4%V (WROUGHT)
SECTIONING TECHNIQUE: EQUIPMENT TYPE BLADE	CONCENTRATION
MOUNTING TECHNIQUE: METHOD	SAMPLE/MOUNT RATIO

*Alternatively two different grades of silicon carbide paper could be used.

	SURFACE and TYPE	ABRASIVE SIZE/TYPE LUBRICANT	FORCE (kPa)	HEAD ROTATION	Z AXIS (μm)	REFLECTIVE FACTOR	MICROGRAPH YES/NO	IMAGE ANALYSIS YES/NO	REMARKS
PLANAR GRINDING STAGE	PAPER	P240 SiC WATER	35	COMP	PLANE				

	SURFACE and TYPE	ABRASIVE SIZE/TYPE LUBRICANT	FORCE (kPa)	HEAD ROTATION	Z AXIS (μm)	REFLECTIVE FACTOR	MICROGRAPH YES/NO	IMAGE ANALYSIS YES/NO	REMARKS
SAMPLE INTEGRITY STAGE	S.4*	9 μm OIL	35	CONTRA					
	S.1	0.06 μm COLLOIDAL SILICA	70	COMP					

	SURFACE and TYPE	ABRASIVE SIZE/TYPE LUBRICANT	FORCE (kPa)	HEAD ROTATION	TIME (s)	REFLECTIVE FACTOR	MICROGRAPH YES/NO	IMAGE ANALYSIS YES/NO	REMARKS
POLISHING STAGE	PS 7	0.06 μm COLLOIDAL SILICA	17	COMP	100		YES		17.8 (Plate 11)

The micrograph Figure 17.8 (see Plate 11)

Example 17.8

Family classification BB6

The material: tantalum

The early limited use for this material was in lamp filaments; this in part was due to its high melting point, 3000°C. Its other outstanding characteristic is its excellent corrosion resistance. This had led to its current application in chemically hostile environments—none more so than when associated with heat. This material, being resistant to most acids, is also used in the pharmaceutical industry. Its application as a biomaterial for surgical implant is also known and practised. The illustration (Figure 17.9) is of capacitors, where tantalum still has an application, in particular where space is limited.

The technique

Tantalum is one of those materials that does not lend itself to mechanical preparation and, although it can be prepared by electrolytic methods, is not necessarily the preferred route. The major problem with mechanical preparation is the tendency to distort or smear the top surface. To overcome this it is necessary to avoid the use of small sized abrasives, unless associated with an appropriate high pH solution. Note how the first sample integrity stage uses an oil-based 9 μm diamond slurry. Beyond this point all abrasive actions are alkaline assisted. The convention for polishing is to reduce to half the pressure from the grinding stage; if adopted this would involve prolonged polishing times. The reduced rotational speed ensures a deformation-free top surface.

The method

AUDITABLE PREPARATION PROCEDURE No. 8

CLASSIFICATIONS: BB6			MATERIAL: TANTALUM	
SECTIONING TECHNIQUE:	EQUIPMENT TYPE	BLADE	CONCENTRATION	
MOUNTING TECHNIQUE:	METHOD		SAMPLE/MOUNT RATIO	

	SURFACE and TYPE	ABRASIVE SIZE/TYPE LUBRICANT	FORCE (kPa)	HEAD ROTATION	Z AXIS (μm)	REFLECTIVE FACTOR	MICROGRAPH YES/NO	IMAGE ANALYSIS YES/NO	REMARKS
PLANAR GRINDING STAGE	PAPER	180 SiC WATER	35	COMP	PLANE				

	SURFACE and TYPE	ABRASIVE SIZE/TYPE LUBRICANT	FORCE (kPa)	HEAD ROTATION	Z AXIS (μm)	REFLECTIVE FACTOR	MICROGRAPH YES/NO	IMAGE ANALYSIS YES/NO	REMARKS
SAMPLE * INTEGRITY STAGE	S.4	9 μm OIL	35	CONTRA	19.2				
	S.1	3 μm ALUMINA + 0.06 μm COLLOIDAL SILICA	35	CONTRA	7.4				
	S.1	0.05 μm ALUMINA + 0.06 μm COLLOIDAL SILICA	35	CONTRA	2.0				

	SURFACE and TYPE	ABRASIVE SIZE/TYPE LUBRICANT	FORCE (kPa)	HEAD ROTATION	TIME (s)	REFLECTIVE FACTOR	MICROGRAPH YES/NO	IMAGE ANALYSIS YES/NO	REMARKS
POLISHING STAGE	PS 7	0.06 μm COLLOIDAL SILICA	35	CONTRA	1200		YES		17.9

The micrograph Figure 17.9

Figure 17.9 Tantalum (×100)

18

Preparation of Ceramics

18.1 Background

In the broadest sense of the term, ceramics encompass all inorganic materials except metals. They range from polycrystalline to amorphous materials. They are brittle, sometimes extremely hard, exhibit linear fracture behaviour when stressed and in general respond adversely to incorrect stock removal techniques. Ceramics are also heat and wear resistant and find many applications in advanced materials as composites or coatings.

Ceramic components are usually produced as sintered products—hence the reason for microstructural checks on homogeneous quality. Microstructural analysis, however, is also required for the determination of material properties via the grain size, shape, distribution, grain boundary interface, impurities and phases.

Ceramics surfaces when optically observed in a reflected bright field microscope are usually of poor quality and require careful microscope aperture control or the use of alternative optical techniques. When light passes through an optical system it encounters a partial reflection at most air/glass interfaces, the remainder of the light transmitting through the optics to ultimately create the primary image. This partially internally reflected light is normally quite low in relation to the total light reflecting back from the specimen and is therefore, with higher reflecting subjects, suppressed in the final image. Ceramics are generally very low reflecting subjects, therefore making the internally scattered light more pronounced. One effect of this, beyond the blurred image, is to highlight any dust in the optical path; this is very pronounced in photomicrography. To reduce internal scattering it is important to reduce the aperture iris as per Kohler illumination; it is also wise to keep the polarizer in the optical train.

Ceramics often exhibit a glassy phase; this phase, when observed in reflected light, manifests itself as a dark area indistinguishable from porosity or preparation pull-out. Fortunately this glassy phase is bireflecting and can therefore be readily identified between crossed polars, the porosity remaining dark. The background can be introduced (rather than extinguished) by the introduction of a sensitive tint plate between the polarizer and the analyser. With high power optical observation (0.9 NA upwards) an additional aberration occurs; light reflecting back from the specimen is reflected back from the flat-faced front objective lens. This can be overcome by using an oil immersion objective.

The examples in Figure 18.1 are used to show how different optical techniques are used to assist in the analysis of ceramic

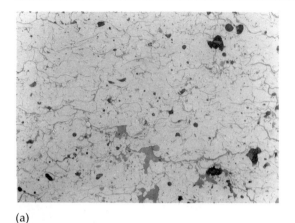

(a)

(b)

Figure 18.1 The use of crossed polars. (a) Normal bright field (×400) and (b) crossed polars (×400)

surfaces. The alumina material clearly revealed in the bright field micrograph (Figure 18.1(a)) shows the wave-like oxide boundaries and also black areas looking like pores. The micrograph of the same area (Figure 18.1(b)) shows the black areas to be in fact secondary phases or, more likely, unmelted particles. Also note from Figure 18.1(b) how the mottled background appears to include white dots; this white area is usually indicative of subsurface porosity.

The specimens used in the previous considerations were all solid blocks relying on a polished surface to reflect the image back into the microscope. Many ceramics, however, can be and are observed in a totally different manner in that light passes through the specimen, the thickness of the specimen being approximately 30 μm. The thickness of the specimen is dependent upon the grain size: if the grains are, for example, 5 μm then the thin section would have to be, say 8 μm thick; if the grains were 30 μm then the thickness would be, say, 50 μm. Any reader familiar with thin section preparation as practised by the geologist and mineralogist will find it strange that different thickness samples are needed to suit the grain size. The reason why the mineralogist keeps to the same thickness (30 μm) is because recognition of minerals is by colour; any change in different refractive index and/or thickness would change the optical colour. Techniques for preparing thin sections of ceramics are the same as those given in Chapter 14.

18.2 Preparation Techniques for Reflected Light Observation

18.2.1 Sectioning

With many ceramics this may not be necessary and is to be avoided where possible. When the specimen is small enough not to require sectioning it could require a specific stock removal amount at the planar grinding stage in order to make a representative analysis. When sectioning is required it is wise to use a low deformation diamond blade, with very hard ceramics (sialon, boron carbide) a low concentration blade and with others a high concentration blade. When sectioning proves difficult because of material hardness, choose a smaller abrasive size.

18.2.2 Mounting (encapsulation)

When it is necessary to mount the sample, care must be taken to best match the abrasive characteristics of the resin with the ceramic. This of course is difficult, the end result being relief at the specimen/resin interface. If this proves to be a problem then alumina particles should be mixed with the resin. The major problem occurring during the preparation stage is the lodging of loose abrasives at the resin/specimen interface.

This is caused by the sudden shock to the abrasive at this very much harder interface. To overcome any problems resulting from these lodged abrasives the specimen *must* be ultrasonically cleaned between stages; alternatively, avoid the use of charged abrasives.

Prolonged ultrasonic cleaning is to be avoided, in particular when the fusion between sintered grains is questionable. Ceramics should in the majority of cases be mounted using cold setting resins and only worked when the resin has set hard. Compression mounting, in *most* cases, can fracture and severely damage the ceramic. One recent example supporting this point relates to a discussion about mounting for surface preparation of three different ceramics: 99% dense alumina, aluminium titanate and aluminium nitride. It was considered *not necessary* to cold mount the subjects, so they were compression mounted in a hard phenolic resin, as a result of which:

(i) the 99% dense alumina cracked across its corners,
(ii) the aluminium titanate 'crumbled' and
(iii) the aluminium nitride cracked across the whole top surface.

18.2.3 Planar grinding

This stage takes place prior to the sample integrity stage and is required only when multiple samples are being prepared on an automatic machine. Its function is, as the name implies, to bring all samples to one common plane. The size of abrasive to be used at this stage is dependent upon (a) the amount of stock to be removed and (b) the induced Z axis structural damage from the sectioning stage. The type of abrasive is dependent upon the hardness of the ceramic. A ceramic capacitor, for example, can be quite adequately planar ground using silicon carbide paper, whereas boron carbide would require diamond. The choice of fixed or charged abrasive will vary, depending upon the amount of stock removal required and the structural damage resulting from this operation. Charged abrasive surfaces, although slower in terms of stock removal, are more shock absorbing and in consequence induce less structural damage. Because most ceramics are hard there is a tendency when choosing a charged platen surface to always select surface 10. This should not be the case where a sample with little or no induced damage requires planar grinding, i.e. use surface 8 or 6.

18.2.4 Sample integrity stage

Assuming a charged platen surface is to be used then platen surface 8 and lower are usually employed. It could be necessary to have two stages using the same platen surface, the first using $3\,\mu$m diamond, which as we know will remove material by the brittle fracture mechanism, followed by a submicrometre abrasive removing material through plastic deformation as in metals. Colloidal silica ($0.06\,\mu$m) is often used at this stage, the results being achieved by doubling the surface pressure (e.g. to -70 kPa). Material removal with colloidal silica is a complete mechanochemical mechanism.

The preparation of hard porous materials, of which ceramics must be included, can offer platen pickup problems. This is evident in the prepared sample where pores are partially filled with material from the platen surface. To overcome this problem, it will be necessary at the sample integrity stage to:

(i) increase the abrasive size,
(ii) use a higher viscosity lubricant or
(iii) avoid the use of metal composite platen surfaces when approaching the lower surface numbers, i.e. substitute platen surface 2P for 4.

Some ceramics can be extremely difficult to prepare, yet most if not all can be prepared using the correct compatible platen surface, charged with the appropriate abrasive. A recent example came to light where an investigator had used every surface with progressively smaller abrasives, noting the material removal (Z axis) required to give the least induced damage at each stage. At the end of each stage a photograph was taken for comparison. With this method it was possible to identify whether progress was taking place between stages and the rate at which it occurred. This resulted in surfaces 10/6/4/1 being selected (see * in Table 18.1). Should a databank be available for a particular material, one would simply select the appropriate stages from the initial Z axis figure.

Table 18.1 Example of procedure determined for boron carbide

Surface number[a]	Abrasive (μm)	Time	Z axis (μm)	Rate (Z/min)	
10	30	10	149	14.9	} *
8	15	10	137	13.7	} *
6	9	8	58	7.25	*
4	6	12	39	3.25	*
2	3	8	18.4	2.3	
1	1	6	6	1.0	*

[a]Surface number = charged platen surface. *Indicates preferred route.

18.2.5 Z axis material removal measurements

Measurements of the material removal in determining an appropriate Z axis can be carried out via the hardness indent method. This is a very useful and accurate way of determining the Z axis at each preparation stage. The method is to indent the sample, or a 'dummy' in the case of brittle subjects, after the planar grinding stage. The load required varies from 10 to 30 kg. The reduction in the diagonals is noted as the sample progresses or after each step. To convert the diagonal into a depth measurement simply multiply by 0.202 for pyramid indents.

18.2.6 Polishing

Since we take the view that polishing only takes place *after* the sample integrity stage is complete, i.e. after the position of the true microstructure has been reached, then polishing is often a repeat of the last integrity stage with a lower pressure and perhaps a different polishing surface. The degree to which polishing is necessary is dependent upon the hardness of the ceramic, with soft materials requiring more polishing than hard materials.

18.3 Establishing a Procedure

It is dangerous to assume that the method that has been systematically derived is the best way of preparing that particular material just because the result is good. One could, for example, be forgiven for preparing all ceramic materials by the 10/4/1 or SiC 4/1 procedure without comparing the results with different combinations. Would 8/4/1 work? Could one reduce the preparation time by using 30 μm diamond on surface 8 instead of 15 μm on surface 10? These and many more questions have to be addressed before it can be said that 'this is the best method'.

Boron carbide can be used to illustrate a live example of how a procedure is established. It was decided to use all the currently available surfaces, with 1 μm diamond on the lowest numbered platen surface and gradually increasing to 30 μm on surface 10. The two parameters of concern are the material removal rate (Z axis) and the progress towards integrity, the latter being a measure of the reduction in induced pull-out. Reflectivity measurements could be taken at this stage but were considered superfluous when taking photomicrographs.

The chart (Table 18.1) illustrates the Z axis removal at each stage. The quoted time is that which is required to achieve the best condition possible, after which there was no discernible improvement. The first surprising feature is the high stock removal rates; these are much higher than would be achieved on a metal subject. Notice how surfaces 10 and 8 both have a similar Z axis rate; the integrity from both appeared similar. It would have been interesting to see if 30 μm on surface 8 improved the Z axis rate without any loss of sample integrity. As can be seen from the chart the Z axis rate progressively reduces and the structure also showed improvement from platen surface 6 to 4 (shown in Figure 18.2). Rather than include two more stages it was considered necessary only to use platen surface 1 with 1 μm diamond supplemented by colloidal silica to improve integrity and the Z axis rate. The sample was photographed as it

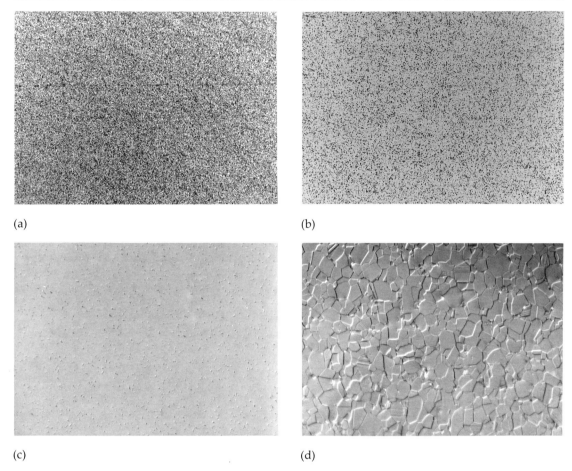

Figure 18.2 Progressing integrity of boron carbide. (a) Platen surfaces 10/8/7, (b) platen surface 4, (c) platen surface 1 and (d) etched sample

approached true microstructure; this is also illustrated. The finished sample was etched by immersing and boiling in a mixture of equal volumes of aqueous solutions containing 15% KOH and 15% $K_3Fe(CN)_6$ (Murakami's reagent). The chosen method from the chart above is platen surfaces 10/6/4/1.

18.3.1 Constants relating to the procedure

Many constants apply to the satisfactory function of a previously created procedure. The sectioning technique will, for example, influence the induced structural damage, the ratio of sample to resin will affect subsequent preparation stages and the type of encapsulating resin will influence the progress towards true microstructure. The size, type and shape of the grinding abrasive, whether fixed or charged, all influence the final result. Lubricant and head to platen direction must also be taken into account. All machine functions must be optimized.

18.3.2 Auditable visual and statistical information

Z axis

Material removal is one of the major keys to solving the problems of standard related repeatability. The use of time is to be avoided when auditing to a standard procedure. It is the actual stock removal (Z axis) that governs the rate of progress towards true microstructure and this must be controlled at every stage. Time is acceptable at the last operation provided that the stock removal, Z axis, is 1 μm or less.

Reflectivity factor

This can be used to indicate statistically the condition of the surface reflectivity. This will show a lower figure than desired if for any reason the cutting abrasive should function in an inefficient or undesirable manner.

Since reflectivity equipment can be expensive, an alternative would be to use a camera exposure device. As with all reflectivity measurements it must also be related to a standard, i.e. adjust the light intensity and/or DIN rating so that the exposure meter reads 100 with a standard coated surface (silvered mirror). Scratch pattern reflectivity readings will then give values upwards from 100 relative to their reflectivity.

Micrographs

These are obviously essential as visual comparisons with the finished sample. They are also desirable when auditing scratch patterns at the intermediate stages. Scratch patterns with ductile materials clearly show the abrasive cutting efficiency which, as indicated, can be related to reflectivity. Working with brittle materials, the rate of progress towards true microstructure is best related to the surface structural damage; this usually manifests itself as a progressive reduction in cracked-hole morphology.

Image analysis

This would generally only have applications at the completion of the preparation. It is, however, a reliable means of recording size, shape and distribution.

18.4 Preparation Examples

Example 18.1

Family classification BB10

The material: boron carbide (B_4C)

Boron carbide exhibits many desirable properties, including:

1. High abrasion resistance
2. Low density
3. High resistance to impact
4. High hardness (harder than SiC)

Its applications vary from armoured seats used in helicopters to neutron absorbers, body armour, gas bearings, seals and components which must withstand abrasion. One such application relating to wear resistance is in shot blast or slurry nozzles, wherein boron carbide outperforms conventional silicon carbide and tungsten carbide. A variety of shapes can be produced by the HIP process using specially prepared ceramic powders. As with all complex sintering operations, where high pressure and temperature are utilized, revealing the microstructure is a vital component towards understanding the material condition.

The technique

With all sintered products, the problem of grain pull-out is always prevalent unless care is taken, in particular in the early preparation stages. Care must be taken at the sectioning stage (low concentration diamond slitting wheels of minimum diameter). The sample should first be immersed in cold setting epoxy resin prior to hot mounting in the least abrading mounting resin. This is to ensure maximum adhesion at the specimen/resin interface which reduces the inevitable lodged diamond due to gross hardness (abrasion) differences. As a general rule, ceramics should not be compression mounted; on this occasion the precoated round sample showed no ill effects. The subsequent grinding stage will very much depend upon the depth of structural damage (Z axis) resulting from the sectioning technique. The proposed method will prove suitable where care has been taken during sectioning. The procedure utilizes five charged (four different) platen surfaces; this number could be reduced at the expense of extended time. The theory of removing as near as possible the initial Z axis structural damage only is not possible with this material unless a slow lapping process is used. It will be noticed by the Z axis at each stage (58, 39, 14 μm) that quite a high stock removal has taken place. This is because when grinding the diamond is propagating the already fractured structure. Incorrect preparation will result at worst with no granular appearance; hopefully the enclosed technique would never give such results. When observing the prepared sample there should be no evidence of pull-out and the grain boundaries should be clearly revealed with clear and sharply defined continuous lines; adherence to the Z axis dimensions and platen surfaces will give this result. Note how the pressure is doubled at the last integrity stage and that no polishing stage is required.

The method

AUDITABLE PREPARATION PROCEDURE   No. 1

CLASSIFICATIONS: BB10/CERAMICS	MATERIAL: BORON CARBIDE
SECTIONING TECHNIQUE:	BLADE
MOUNTING TECHNIQUE: PHENOLIC: TO BREAK OUT FOR THERMALLY ETCHING USE PRE LOAD—MINIMUM PRESSURE	CONCENTRATION

Note: Ultrasonic clean between diamond stages.

	SURFACE and TYPE	ABRASIVE SIZE/TYPE LUBRICANT	LOAD (kPa)	HEAD ROTATION	Z AXIS (μm)	REFLECTIVE FACTOR	MICROGRAPH YES/NO	IMAGE ANALYSIS YES/NO	REMARKS
PLANAR GRINDING STAGE	S.10	30 μm WATER	35	CONTRA	PLANE				

	SURFACE and TYPE	ABRASIVE SIZE/TYPE LUBRICANT	LOAD (kPa)	HEAD ROTATION	Z AXIS (μm)	REFLECTIVE FACTOR	MICROGRAPH YES/NO	IMAGE ANALYSIS YES/NO	REMARKS
SAMPLE INTEGRITY STAGE	S.6	9 μm WATER	35	CONTRA	58				
	S.4	6 μm OIL	35	CONTRA	39				
	S.1	3 μm WATER	35	COMP	14				
	S.1	1 μm OIL + 0.06 μm SILICA	70	COMP	12 MIN		YES		18.3

	SURFACE and TYPE	ABRASIVE SIZE/TYPE LUBRICANT	LOAD (kPa)	HEAD ROTATION	TIME (s)	REFLECTIVE FACTOR	MICROGRAPH YES/NO	IMAGE ANALYSIS YES/NO	REMARKS
POLISHING STAGE									

The micrograph Figure 18.3

Figure 18.3 Boron carbide DIC (×1000)   **Figure 18.4** Aluminium nitride DIC (×1000)

Example 18.2

Family classification BB10

The material: aluminium nitride

Aluminium nitride (AlN) is used as an electronic substrate, refactory, heat sink, grinding media, as a wear-resistant material, armour, ceramic or intermetallic composites, films and coatings, and fibres. It exhibits a low density, high thermal conductivity, high specific modulus and a low thermal expansion. This combination of properties makes AlN a very attractive structural material for composite heat engines and ballistic armour. It is also stable when in contact with gallium arsenide (G and As) and is therefore useful as a crucible material for single crystal growth of G and As. Since aluminium nitride is produced by a number of techniques (direct nitriding, reduction nitriding, plasma synthesis, CVD and other methods) microstructural analysis is required to analyse the AlN quality.

The technique

Unlike the boron carbide procedure, where a whole series of surfaces were evaluated to identify an optimized route, on this occasion the traditional 10/4/1 using a fixed abrasive on surface 10 followed by charged abrasives on surface 4 and 1 was progressed. Notice how the mechanochemical grinding with high pressure is used towards the completion of the stages. With reference to Figure 18.4, etching is not necessary when the surface is analysed in differential interference contrast with first-order grey position.

The method

AUDITABLE PREPARATION PROCEDURE   No. 2

CLASSIFICATIONS: BB10/CERAMICS		MATERIAL: ALUMINIUM NITRIDE	
SECTIONING TECHNIQUE:	EQUIPMENT TYPE BLADE	CONCENTRATION	
MOUNTING TECHNIQUE:	METHOD	SAMPLE/MOUNT RATIO	

	SURFACE and TYPE	ABRASIVE SIZE/TYPE LUBRICANT	FORCE (kPa)	HEAD ROTATION	Z AXIS (μm)	REFLECTIVE FACTOR	MICROGRAPH YES/NO	IMAGE ANALYSIS YES/NO	REMARKS
PLANAR GRINDING STAGE	RESIN BONDED DIAMOND WHEEL	P320 GRIT	35	CONTRA	PLANE				

	SURFACE and TYPE	ABRASIVE SIZE/TYPE LUBRICANT	FORCE (kPa)	HEAD ROTATION	Z AXIS (μm)	REFLECTIVE FACTOR	MICROGRAPH YES/NO	IMAGE ANALYSIS YES/NO	REMARKS
SAMPLE INTEGRITY STAGE	S.4	9 μm OIL	35	COMP	46				
	S.1	3 μm WATER	35	COMP	9				
	S.1	1 μm HIGH pH	70	CONTRA	5				
	S.1	0.06 μm COLLOIDAL SILICA	70	CONTRA	10 MIN		YES		18.4

	SURFACE and TYPE	ABRASIVE SIZE/TYPE LUBRICANT	FORCE (kPa)	HEAD ROTATION	TIME (s)	REFLECTIVE FACTOR	MICROGRAPH YES/NO	IMAGE ANALYSIS YES/NO	REMARKS
POLISHING STAGE									

The micrograph Figure 18.4.

Example 18.3

Family classification BB10

The material: sialon

This ceramic (silicon, aluminium, oxygen, nitrogen) has been introduced to meet some of the demands of machine tool wear when operating at high speed. High speed cutting tools operate at elevated temperatures up to, say, 1000°C. Diamond, although excellent for non-ferrous applications, is expensive and is not so suitable for ferrous metals.

Alumina is susceptible to mechanical and thermal shock. Cemented carbides offer many attractions but chip dissolution of the tool top face occurs at elevated temperatures caused by high speed cutting. Thin coatings, such as TiC and TiN, are used to overcome tool dissolution and flank wear. It is hoped, with the introduction of ceramics like β-sialons, to overcome some of the problems of tool dissolution caused by the chip, therefore increasing tool life and operating speeds.

Microstructural features of interest include the Vf of the glassy phase and the β-sialon grain size. With this material unetched observations in DIC were unfruitful. Chemical etching is slow and the glassy phase is selectively removed. In contrast to chemical etching 'plasma etching' was used; this technique etches principally the β-sialon phase, clearly delineating the two phases.

The technique

The specimen being prepared was not a coating but the more difficult solid section of material. Due to the volume of very hard sample, it was not possible to use a charged platen surface such as 10 (the 30 μm abrasives were knocked away and smaller abrasives would not remove sufficient material in the relatively short time). Planar grinding took place using the 320 grit resin bonded diamond grinding wheel (metal/resin bonded would also be suitable). Note the use of high pressure and a mechanochemical final stage.

The method

AUDITABLE PREPARATION PROCEDURE   No. 3

CLASSIFICATIONS: BB10		MATERIAL: SIALON 101	
SECTIONING TECHNIQUE:	LOW DEFORMATION MACHINE	BLADE	DIAMOND LOW CONCENTRATION 100rev/min 5in DIAMETER
MOUNTING TECHNIQUE:	EPOXY/PHENOLIC	CONCENTRATION	20%

	SURFACE and TYPE	ABRASIVE SIZE/TYPE LUBRICANT	LOAD (kPa)	HEAD ROTATION	Z AXIS (μm)	REFLECTIVE FACTOR	MICROGRAPH YES/NO	IMAGE ANALYSIS YES/NO	REMARKS
PLANAR GRINDING STAGE	RESIN BONDED DIAMOND WHEEL	320 GRIT WATER	35	CONTRA	164		NO	NO	

	SURFACE and TYPE	ABRASIVE SIZE/TYPE LUBRICANT	LOAD (kPa)	HEAD ROTATION	Z AXIS (μm)	REFLECTIVE FACTOR	MICROGRAPH YES/NO	IMAGE ANALYSIS YES/NO	REMARKS
SAMPLE INTEGRITY STAGE	S.6	9 μm OIL	35	CONTRA	44		NO	NO	
	S.2P	3 μm OIL	35	COMP	19		NO	NO	
	S.2P	1 μm WATER	35	COMP	8		NO	NO	
	S.1	1 μm HIGH pH	70	COMP	10 MIN		YES		18.5

	SURFACE and TYPE	ABRASIVE SIZE/TYPE LUBRICANT	LOAD (kPa)	HEAD ROTATION	TIME (s)	REFLECTIVE FACTOR	MICROGRAPH YES/NO	IMAGE ANALYSIS YES/NO	REMARKS
POLISHING STAGE									

The micrograph Figure 18.5

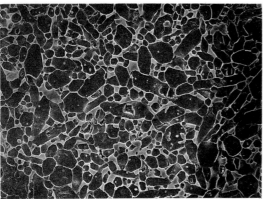

Figure 18.5 Sialon 201—plasma etched SEM (×2000)

Example 18.4

Family classification BB13

The material: silicon carbide particulate/silicon nitride matrix (CMC)

This β-Si_3N_4 with 5% Y_2O_3 is a typical example of the new breed of ceramic matrix composites being developed to overcome some of the low toughness qualities with current ceramics. Ceramic matrix composites also offer improved thermal and wear characteristics. Reinforcing media can be added to the glass ceramic and engineering ceramic matrix composites, including continuous fibre, whiskers, platelets and particles. The benefits of reinforcement to the material characteristics are (a) load transfer where good interface bond is necessary, (b) continuous line tension where the reinforcing fibre must exhibit the higher ductility and (c) prestressing where the difference in thermal expansion coefficients puts the matrix into compression to inhibit crack growth. The technology of crack deflection is also exploited where propagation is restricted by energy absorbing interfaces. Applications for these materials range from the space shuttle reentry tiles to the aeroengine manufacturers' need to operate under continuous stress at temperatures above 1400°C.

The technique

Previous experience with the GMC Nikalon (Chapter 13, Example 13.8) had led to the belief that this would be a difficult material to prepare. These fears were unfounded and the material responded to a 10/4/1 technique. As with most ceramic materials the use of 1 μm diamond, along with a high pH solution, proved advantageous at the final integrity stage.

The method

AUDITABLE PREPARATION PROCEDURE — No. 4

CLASSIFICATIONS: BB13/CMC		MATERIAL: SILICON NITRIDE/PARTICULATE SiC	
SECTIONING TECHNIQUE:	EQUIPMENT TYPE BLADE LOW DEFORMATION—DIAMOND	CONCENTRATION	
MOUNTING TECHNIQUE:	METHOD NON-COMPRESSION	SAMPLE/MOUNT RATIO	

	SURFACE and TYPE	ABRASIVE SIZE/TYPE LUBRICANT	FORCE (kPa)	HEAD ROTATION	Z AXIS (μm)	REFLECTIVE FACTOR	MICROGRAPH YES/NO	IMAGE ANALYSIS YES/NO	REMARKS
PLANAR GRINDING STAGE	S.10	30 μm OIL	35	CONTRA	PLANE				

	SURFACE and TYPE	ABRASIVE SIZE/TYPE LUBRICANT	FORCE (kPa)	HEAD ROTATION	Z AXIS (μm)	REFLECTIVE FACTOR	MICROGRAPH YES/NO	IMAGE ANALYSIS YES/NO	REMARKS
SAMPLE INTEGRITY STAGE	S.4	9 μm WATER	35	CONTRA					
	S.1	3 μm WATER	70	COMP					
	S.1	1 μm WATER	70	COMP					
	S.1	1 μm + HIGH pH COLLOIDAL SILICA	70	COMP	10 MIN		YES		18.6

	SURFACE and TYPE	ABRASIVE SIZE/TYPE LUBRICANT	FORCE (kPa)	HEAD ROTATION	TIME (s)	REFLECTIVE FACTOR	MICROGRAPH YES/NO	IMAGE ANALYSIS YES/NO	REMARKS
POLISHING STAGE									

The micrograph Figure 18.6

Figure 18.6 Silicon nitride matrix DIC (×200)

19

Hardness

To give an absolute definition or measurement of hardness has proved elusive yet an appreciation of hardness is fundamental to everyone. If a subject resists scratching or does not abrade, cut or wear we say it is hard. In the English language it is used in a similar way to the word 'difficult'. In the field of engineering it is the degree to which a subject is 'hard' that requires classification. With metals, hardness is related to the plastic resistance—hence the correlation between hardness and tensile strength. With ceramic materials or brittle minerals other connotations apply.

With ductile material, hardness increases with yield and tensile strength and reduces with plasticity and ductility.

19.1 Hardness by Indentation

If a hard object is forced into a softer material it will impress an indentation. The size of this indentation will be a measure of the material hardness. In order to have some hardness relationship with other materials the operating conditions must be standardized so that hard materials manifest small indents and softer materials larger indents.

Features affecting the size of the indent are:

(i) *Applied load*. The greater the load the larger the indent. Large loads (kg) are usually expressed as *macro* hardness testing, low loads (g) as *micro* hardness testing. *Submicro* hardness testing also takes place (less than 1 g) using ultra-light loads.

The application of the applied load very much depends upon the size of the load. With large loads (macro) the lever arm mechanism can be applied since this will overcome the friction taking place at the fulcrum point. With low loads it is advisable to use the dead-weight mechanism, i.e. the load is applied, without leverage, as an integral part of the indentor.

As a general rule it is always wise to use the largest sized indent to minimize errors of dimensional interpretation. By implication this would suggest the use of the larger rather than the smaller loads when indenting. The selected load will also depend upon the observed field of view. With heterogeneous materials it is wise to select a load such that the size of the indent adequately encompasses all the constituent phases. Thus, the macro style of indents gives a consistent value which could be described as a combination of all constituent hardnesses and is representative of no individual phase.

To carry out indents on individual constituents the load needs to be much lower and the magnification higher (microhardness testing). It still follows that this load needs to be as high as possible, compatible with the field of view and the indenting constituent size.

With microindentation the applied load must always be quoted since the hardness value changes with changes in load. The low loads become increasingly inaccurate due to operator error and surface/subsurface tension and distortion. It has been well recognized for many years that there is a need to relate recommended loads to particular materials when designating microhardness values. This has to a certain extent been carried out in the field of mineralogy but to date is completely lacking in the materials field.

(ii) *Time*. Time can influence the indent size if the load is removed prior to reaching equilibrium. The cycle time should therefore be adjusted to allow the indent to impress and additionally a 10 second 'dwell' time should be allowed before withdrawing the indentor. The recorded cycle and dwell time is very important when there is a possibility of 'creep' occurring (low melting point metals at ambient temperature). With the majority of materials ambient hardness testing is adequate since creep is negligible.

The indent time, although not critical to theoretical hardness values, can influence the hardness value when affected by vibration levels. This is the reason for quoting a 10 second dwell time, which could be increased, if desired, under vibration stable conditions and can be reduced to *no less* than 5 seconds when susceptible to vibration.

(iii) *Rate of load*. The rate of speed at which the indentor approaches the subject influences the indent size and in consequence the hardness value. The size of the indent will be smaller with a slow rate and larger with a higher speed. There is a similarity in the ratio of speed to size until the velocity is very low, when it no longer remains arithmetical. As a consequence the indent rate must be carefully controlled. With brittle subjects the speed of the indent has to be reduced to avoid specimen cracking. With this type of equipment constant operator calibration to a known standard is required. Using an indentor approach velocity range of between 20 and 70 μm/s shows little variation with most materials. The approach velocity should nevertheless be quoted when compiling a reference or standard.

(iv) *Work hardening materials*. When carrying out any form of indentation we are effectively cold working the subject. When considering, for example, the Vickers diamond hardness indent, work hardening will vary across the indent relative to the variation in cold working via the pyramid impression. The hardness values from such tests must take this factor into consideration.

(v) *Indent morphology*. The hardness values using the indent methods are related to load and cross-sectional area of the indent. This being so, you would expect image analysis of the area to be the preferred method of measurement, particularly since most errors in hardness testing result from operator error. Unfortunately due to elastic recovery (with ductile materials) the observed impression exhibits a different shape to the indentor and this shape will change with the degree of elastic recovery.

Indentation also causes a change in the top surface flow around the indent; high ductility results in a raised surface and any degree of work hardening can cause a lowering of the surface. This phenomenon can readily be demonstrated using differential interference contrast which is an optical technique extremely sensitive to variation in surface *slope*. It is for the reason given above that only the diagonals of a Vickers indent are recognized as dimensions for calculating the area, since they are more stable than a picture point image analysis system, even though the latter would be a more faithful representation of cross-sectional area.

(vi) *Surface roughness and friction*. The condition of the top surface requiring macro and micro indenting can influence the final result. The basic requirement is the surface brightness, which must be adequate to reflect the surface and indented image back into the microscope. The surface requirements for macro indenting are not as critical as those for micro indenting and can be satisfactorily performed on a ground surface. This ground surface would be the result of P600 grit silicon carbide paper, silicon carbide or alumina 120 grit stone ground and on some occasions a lathe-turned condition. As with all mechanically worked surfaces the structure and condition of the material will change; this does not have a dramatic effect on the *macro* hardness value but *must* be taken into account with micro indentations.

The surface required for micro indenting must be compatible with the results achieved from 1 to 3 μm diamond on platen surface 1 or 2. There is a tendency for microhardness testing to be carried out on highly polished samples. This should only be the case when platen surface scratches impede the clarity of the diagonals in the indent. If we wish to eliminate the work hardening effect caused by mechanical working of the prepared surface, the sample where possible should be electrolytically polished. Mechanical/chemical polishing, if part of the polishing procedure, must be used with caution since this can give false hardness values. If we are to accept the point that with any form of mechanical working there must be some structural change, call it deformation or structural damage, then the depth of that damage will influence the hardness value. We must therefore prepare the sample to integrity, thereby accepting the surface aberrations as being to a minimum (they could be submicrometre). The indent is also required to be as large as possible with error in true hardness readings increasing as the indent load decreases.

Surface friction between the indentor surface and the sample material will influence the hardness value. Take, for example, a machined component having an oiled surface; the friction will be less and more energy will be channelled to the indent resulting in a greater size, i.e. a lower hardness value. This increase in the indent size is not sufficiently significant to cause concern; a recent microhardness test (MHT) on aluminium and iron gave results as shown in Table 19.1. These values should be taken as relative and not absolute since the testpieces were not 'certified' blocks.

(vii) *Vibration*. Vibration is not so critical with the more robust macro hardness testing machines. High frequency vibrations associated with everyday life, such as fans and motors, should not create a natural affinity with the machine components. Low frequency ground-borne vibrations should, however, be avoided. Do not site the equipment next to a

Table 19.1 Microhardness test results

Loads (gm)	Material	MHT diagonals (μm)	VHN
200	Aluminium alloy	105 dry 107 oiled	33.6 32.4
1000	Iron	94 dry 97 oiled	210 197

high load press or shearing machine and only indent at intervals between low frequency ground-borne vibrations.

Microhardness testing equipment is critical to some high frequency air-borne and low frequency ground-borne vibrations. Equipment should be selected having low frequency vibration absorbers as an integral part. High frequency vibrations will not prove to be a problem with robust well-dimensioned internal components. High frequency can be dampened by sitting the instrument on an antivibration platform; this would be more desirable as the necessity for low loads increases. Air-borne vibrations can also be minimized by enclosing the tester in a soundproof compartment.

(viii) *Traceability of hardness value*. The UK national standards for hardness, previously held by the National Physical Laboratory, were devolved to IMGC of Italy (1989) who maintain the currently monitored scales. IMGC therefore satisfies the traceability requirements for the UK NAMAS accreditation. These scales and standards do not apply to micro indentations, only macro indentations. Although standard MHT test blocks are supplied by many companies they currently are not traceable to any UK national standard.

Microhardness testing is being increasingly used and industrial inspection and certification of all standards is closely monitored. Suppliers of test blocks will issue a test certificate with each block indicating:

(a) Serial number
(b) The approving organization or specification from which all tests are traceable to
(c) Calibration conditions such as indentor approach speed, dwell time and load, and information about the measuring condition, primary magnification, numerical aperture and final magnification

(d) Test results, including the mean diagonal length of the test indentations taken at five points, mean value and relative span, as well as the hardness deviation and thickness of the comparison reference plate.

19.2 Microhardness (Micro Loads)

If we take the word 'micro' as being small then we would literally be confined to 'small hardness'. This is not the case, the word 'microhardness' applies to micro loads resulting in micro indentation.

Microhardness testing has in the past been a very subjective art where, due to poor equipment and/or the inability of operators to relate accurate and consistent results, this very important test lost prominence. Today the situation is different. Equipment is good and reliable, standards are traceable as references and operators are being made aware of the need to produce accurate results from these standards (in the past the standard could be wrong or the machine out of calibration—today this would be very rare).

The weakness of MHT is it is still operator subjective and attempts to eliminate this by scanning the impression have met with much resistance; i.e. MHT is based on a formula related to area derived from measurement of the diagonal lengths and not from a direct area of measurement. To use the morphological area would be a more accurate value of actual area but would introduce new factors beyond those that are presently used in defining hardness.

One way of improving the operator subjective function is to use a double thin line for each graticule line; the edge is reached when the double line just shows a central point. Additionally, CCTV can be employed, the filar lines being superimposed onto the screen along with the indentation. Where there is a clear amplitude difference between the background and indent, the intensity curve system could be employed.

Tests carried out in 1980 using reference standards with over 50 different people using the same standard on different pieces of equipment resulted in readings varying by 20% from the true value. In one case a

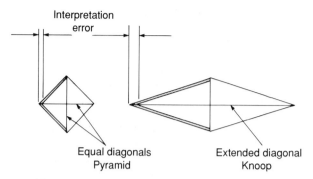

Figure 19.1 Hardness indentation shapes

manufacturer of MHT equipment came up with a 4% error due entirely to the operator reading high, not his equipment. The above serves to illustrate the need for care, particularly since MHT is becoming more and more an everyday requirement. It is used not just for actual hardness values but for hardness traverses and for a combined hardness/reflectivity formula in mineral identification.

Micro indentation consists of forcing a diamond indentor into the subject and measuring the diagonals remaining after removing the indentor. The hardness value (HV) is derived from this mean measurement using one of the following equations:

$$\text{Pyramid HV} = \frac{1854.4f}{\text{mean length of diagonals}^2} \quad \text{using kgf}$$

$$\text{Pyramid HV} = \frac{189\,090F}{\text{mean length of diagonals}^2} \quad \text{using N}$$

or

$$\text{Knoop HV} = \frac{14\,229f}{\text{longest length}^2} \quad \text{using kgf}$$

The hardness value would also indicate the test force. In other words, 463 HV 0.5 indicates a Vickers hardness value of 463 using a test force of 0.5 kgf, or 4.903 N. There are two different shaped diamond microindentors: the pyramid and the knoop.

The indentation shapes are shown in Figure 19.1: the pyramid with its equal diagonals, the knoop with a much extended single measurement diagonal. The depth of penetration of the pyramid indent is one-fifth of the diagonal length. The knoop penetration is much reduced and the diagonal increased. There are, therefore, two advantages with the knoop, one being the small penetration which is essential when indenting thin components or thin platings, the other being the extended diagonal. This extended diagonal is approximately three times longer than the pyramid, and hence, in theory, should result in a factor of three improvement in accuracy. This, however, is where the problems can arise. The following two examples are used to illustrate possible pitfalls:

(i) *Mechanically worked top surface.* When an indent is carried out on ductile materials that have been mechanically worked, even when prepared for microstructural analysis, this top surface can have a stressed or work hardened upper layer. The greater the indent preparation beyond this layer, the less influenced will be the results; this therefore could favour the pyramid shape indentor.

(ii) *Interpretation error.* The idea that an increased length will result in increased measured accuracy does look attractive. Figure 19.1 illustrates how the increased interpretation error, due to the proximity of two lines, can in fact work in reverse.

19.3 Possible Operating Errors

(i) The indentor must not be damaged; it must have sharp edges and be free from cracks. The axis of the moving indentor must be 90° to the plane axis of the equipment. This can be checked by rotating a planar subject between indents to ensure that the diagonals do not show a bias irrespective of the subject rotation.

(ii) The selected load should ideally result in 50 μm diagonal lengths or greater. This is not always possible, particularly with ultra-light loads, and extra care is necessary in the calibration of the equipment and the interpretation of the operator. As the indent becomes less than 10 μm the indentor point radius becomes most critical and governs the accuracy of the reading. It is essential with such low indents to calibrate the operation to a known standard and in some cases construct a special formula for that particular equipment and operator.

Table 19.2 shows a range of *suggested loads* relative to a range of material hardness in order to achieve an adequate indent size. The mechanism within each piece of equipment activating the selected force must be periodically checked and if necessary rectified and recalibrated. The dead-weight system, favoured by the writer, is the least critical, provided that the weights are independently machine activated. Dead-weight systems requiring operator activation can be dropped or abraded, necessitating recalibration. Any system depending upon a fulcrum lever arm or mechanical activation operating pneumatics must be regularly inspected.

(iii) *The test block and testing machine* should not be subjected to large temperature variations and all testing should take place at 20 ± 3°C. The test blocks must always be kept in a dry atmosphere, clean and *free from grease* and surface contact with human hands.

(iv) *Planarity of specimen.* The specimen, if not plane, will result in an anisotropic effect where one leg of the diagonal is longer than the other. Lack of planarity can be confirmed if the 'long leg diagonal' rotates in the direction of the rotated specimen, observed after subsequent indents.

(v) *Visual acuity.* Having taken care to ensure that the equipment is in calibration, the correct load is selected and the microscope illumination sharply displays the diagonals, what about the operator? Since different operators see different sizes, they have a different focus; they also take the position of the filar line at various positions, i.e. superimposed on the diagonal curves, to the left or to the right. Because of this aberration it is necessary for the operators to confirm readings taken from indents

Table 19.2 Hardness of a range of loads

Material hardness (HV)	Suggested load (g)
240	300
300	500
400	500
540	1000
620	1000
720	1000
840	1000

taken from the standard test block, i.e. using the standard test block as a 'reading standard' *without* making an impression.

Each individual operator must learn the exact position required in positioning the filar line in order to achieve correct and consistent results.

(vi) *Magnified image.* The observed size of the indent can change with a change in lens resolving power. The resolving power is usually expressed as the numerical aperture and is engraved on all lenses. It would be unwise therefore to supplement a ×40/0.65 objective for a ×40/0.6 objective. To a lesser extent the wavelength of the illumination will also affect the final measured dimension. Dedicated microhardness testing machines usually have a fixed aperture setting with a built-in green filter in the illumination; this must not be changed.

Microhardness indentors are sometimes offered as accessories to routine compound microscopes without any reference to the necessary aperture iris position or the effect of filters. One such supplier goes so far as to say 'no special training is required to achieve accurate and reproducible results'. The author acknowledges the manufacturer's need for good product marketing but on this occasion the weakest aspect of the design has been misquoted as a benefit. Without recourse to fixed apertures, filters and fixed indent centring positions, the results could be *consistently inaccurate*. To neglect the training so necessary with any MHT equipment is not the recommended norm.

(vii) *Interaction between indents.* This occurs as subsequent indents are too close to each other and are being affected by the stress caused around the indent. If this situation requires clarification a non-destructive differential interference contrast technique could be employed. The cross-section preparation to observe the microstructure will also reveal the required information.

19.4 Vickers Hardness Under Load

A new non-destructive microhardness test method has been developed (see W. W. Weiler, *A New Microhardness Test Method*, The American Society for Testing and Materials) using the conventional Vickers indentor, which is coupled to a displacement measuring device. An HVL (hardness Vickers under load) is computed from the indentor displacement when still under load. The instant advantage of such systems is that they are not operator subjective, as is the case when optical measurements of the diagonals are employed. The results, although theoretically correct, could differ from the conventional diagonal system of measuring, and this is to be explored along with other possible beneficial features.

19.4.1 Shortcomings of diagonal measuring system

It should be clear to all of us that macro indentation will never yield the same hardness results as micro indentation, unless the subject is a totally uniaxial crystal. One would expect when making successive indents on identical materials to achieve similar values; unfortunately this is not the case. Microhardness numbers obtained are dependent upon specimen surface preparation, test load and operator interpretation of diagonals. The surface preparation has been covered earlier, as has the interpretation of diagonals. The use of low load knoop indents on mechanically worked surfaces must also be in question. The suggested test load has been based on an adequate reading dimension and not on the material surface condition. Attempts to overcome some of the operator errors by digitizing the optical image failed because it recognized areas that proved different to a calculation based on diagonals. To overcome this problem the digitized image would have to be based on an accurate black and white separation between the end of a diagonal and the specimen matrix, which brings back into the equation the subjective element, particularly with knoop.

19.4.2 Benefits with displacement measuring

Since this method does not take into account plastic recovery or material displacement, the results are arguably more accurate. Figure 19.2

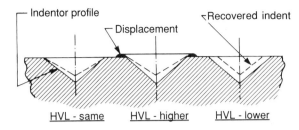

Figure 19.2 HVL compared with HV

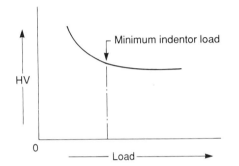

Figure 19.3 Optimum test load

compares the results under three different conditions; as can be seen, two of these examples could give conflicting results, the HVL method being more accurate if displacement and material recovery are to be excluded.

The testing of rubber and plastics could also be carried out using HVL. Displacement measuring would, with an anisotropic material, give results relative to the harder of the two values, which could be considered a disadvantage. Without some visual check, this condition would not be recognized.

19.4.3 Selected load

Instead of using one fixed load, the load could be applied by increments, the hardness value changing from the low load until a stabilized point is reached. Figure 19.3 shows the type of results one would expect from incremental load testing, the higher values being attributed to the top surface condition. The 'hardness profile' capability indicates the mechanical worked surface condition, but of equal importance is the hardness testing of thin coatings. To compile the hardness profile from the load versus indentation would reveal the thickness of the worked surface or the thin coating.

19.5 Certified Test Blocks

These are the standards supplied to a whole range of hardness, the standards chosen corresponding to the hardness of interest. Two different types of blocks are used, the 'reference standard' and the 'reading standard'.

19.5.1 Reference standard

Each block is certified to at least three values of load, viz. 100, 300 and 1000 g. There are five indentations per load. The results are certified by the national standardizing authority and recertified periodically.

These test blocks are to be used in the calibration and verification of the indenting instrument, a test which should be carried out annually. The blocks are certified to a given hardness value. It is wise to use a traceable graduated stage micrometer for the measurement verification, reserving the test block for verifying the indented hardness calibration.

If calibration is based on the standard test block results only, then it is possible for an incorrect measuring system combined with a faulty indentor to indicate the instrument as being within calibration (two wrongs don't make a right).

19.5.2 Reading standards

These are not necessarily used to verify the hardness of the test blocks. They are used to calibrate the measuring system. They are manufactured from a stable but not necessarily totally homogeneous material and are impressed with indentations of specific sizes. This does not necessarily mean that they are capable of accepting new indentations; they are intended for operator use, enabling each operator to repeat the measured diagonals in accordance with the stated value. It is as important to ensure that the operator is traceable in measurements to standards as it is to ensure that the instrument is traceable to a standard. Reading standards will also be traceable to the national standardizing authority. The use of graduated stage micrometers to confirm operator measurement is sometimes recommended, but this is a poor substitute for real life tests when using reading test blocks.

20

Training in Metallography

20.1 Introduction

Although academia more than adequately covers the development and analysis of metals and materials, it is difficult to pin-point areas of training specifically orientated to the preparation of surfaces for structural analysis. By 'specifically orientated' it is intended to imply a structural training programme with some type of examination both written and also in the practical sense. The practical work must include the ability to demonstrate 'elements of competence' and to conform to standards that are traceable. Because of this shortfall in training the author has found a high proportion of materialographers, when subjected to a basic written examination, fail to achieve a satisfactory pass mark. (This examination is part of a structured course directed by the author.)

In an attempt to find solutions to this problem a list of 'elements of competence' was compiled which formed the basis of an assessment procedure. The Engineering Training Authority (ETA) are currently soliciting major industrial organizations in Britain with a view to finalizing this draft proposal towards microstructural determination of traceable standards. Because of the complexity of training, and also the recognition that industry requires different levels of expertise, the proposal is based on two grades, i.e. technician level IV and technician level II. In order to evaluate the credibility of such a training requirement a pilot test run was carried out involving delegates with different levels of experience; all were practising metallographers. As a result, all delegates passed the elements of competence for technician level II, no pass being given at level IV. It was also concluded that with more training 30% would pass at level IV providing a greater awareness was given to traceability.

The European Society for Microstructural Traceability is another example of the move towards industrial standards; this along with the ETA will raise awareness of the need for structural training in metallography.

The following 'elements of competence' are intended as an illustration of the requirements at technician level IV and are not necessarily totally representative of the ETA final draft.

20.2 Elements of Competence

Microstructural Determination to Traceable Standards

Technical Level IV

1. Specify cut-off systems.
(a) Performance criteria
   1.1 The appropriate wheel type and dimension are specified correctly.

1.1 Machine speeds, feeds, type of increments and load are defined.
1.2 The appropriate lubricant is specified correctly.
1.3 The correct clamping system for the sample is specified.
1.4 The common faults and their likely causes are advised to the course tutor.

(b) Competence tasks
To be carried out using low deformation and/or high speed abrasive cutters or other more suitable means. Materials used are spray coating, B_4C, saffil, carbon–carbon, steel, titanium, white metal, aluminium silicon.

2. Specify mounting system.
(a) Performance criteria
   2.1 The appropriate hot or cold mounting medium is specified (appropriate to abrasion and contraction rates).
   2.2 Cold mixes, if required, are specified to the correct proportion, giving centipoise figure where necessary.
   2.3 Hot mixes are specified to the correct proportions and the appropriate mounting press details are specified.
   2.4 Vacuum impregnation is selected if needed.
   2.5 Common mounting faults and their likely causes and remedy are correctly described to the course tutor.
   2.6 Ratio of sample to resin is defined.
   2.7 Need for preload when necessary.

(b) Competence tasks
To be carried out using the previously sectioned material. Written report to be submitted on the hot mounting technique to be used when mounting a U-shaped plated component.

3 Define stock removal required (Z axis).
(a) Performance criteria
   3.1* A technique appropriate to the specimen for establishing the stock removal depth needed to reach a true and faithful surface (Z value) is selected from the appropriate established databank.
   3.2 A section of a similar specimen is cut parallel to the Z axis.
   3.3* This specimen is prepared for examination using the selected techniques.
   3.4 The specimen is prepared for structural examination by etching or other appropriate means.
   3.5 The Z value required is defined by optical examination of the specimen.
   3.6 If required, sampling frequency is defined.
   3.7 Sample orientation when relevant must be noted.
   3.8 The need or frequency for ultrasonic cleaning must be recognized.
   3.9 The number of samples per head is to be recorded when using automatic preparation systems.
   3.10 Specific optical techniques are to be defined when this is necessary to monitor preparation progress.

(b) Competence tasks
Use one of the samples in element 2 above.

4. Choose appropriate grinding system.
(a) Performance criteria
   4.1 Select abrasive type and lubricant for grinding to a traceable standard.
   4.2 Define auditing frequency requiring surface hardness indentation (Z axis).
   4.3 Define need for reflectivity or image analysis.
   4.4 Select grinding equipment to traceable standard.
   4.5 Select grinding system to traceable standard.
   4.6 Grind to traceable criterion using intermediate surface examination and note conditions and times at each stage.
   4.7 Define if preparation satisfactory by using appropriate optical examination routine.
   4.8 Proceed with grinding system according to the appropriate traceable route to at least the Z depth defined by performance criteria 3.
   4.9 Ability to carry out microstructural analysis on prepared sample.

---

*When this is not available it must be created from first principles and logged.

(b) Competence tasks
Use one of the samples mounted in element 3 above. The written procedure is required for 0.37% C steel and aluminium bronze using traditional techniques.

5. Choose appropriate polishing system.
(a) Performance criteria
   5.1 Select an appropriate polishing cloth to meet traceability criterion.
   5.2 Select a polishing abrasive to meet traceability criterion.
   5.3 Select a carrier to meet traceability standards.
   5.4 Select suitable polishing speed.
   5.5 Test polish to confirm suitability of polishing parameters.
(b) Competence tasks
Use sample from element 4 above.

6. Prepare preparation sheet.
(a) Performance criteria
   6.1 Be able to complete a preparation sheet, having established required traceability route.
   6.2 Prepare a visual display of the Z axis–time scale ratio, in relation to total sample integrity.
   6.3 Be able to database all information in accordance with the Z axis curve.

7. Recognize preparation faults.
(a) Performance criteria
   7.1 Can recognize preparation faults in prepared specimen.
   7.2 Can modify preparation route to traceability standard to correct preparation faults.
   7.3 Can demonstrate an understanding of how to technically and statistically compare alternative preparation methods to reduce subjective procedures.

20.3 Conclusion

So much for surface preparation but what about microscopy? There is still a serious shortfall in the structural training for industrial microscopy, in particular techniques relating to structure identification. There are various courses already available and structured for microscopy but these tend to be either life science orientated or reflected light techniques only. Industrial microscopists are often aware of the different optical techniques available but without the knowledge of when to use such methods, what information can be derived, how best to interpret such information and how to display such information enabling others to appreciate it. Experience of short courses directed by the author has revealed this shortcoming in the materialographer's training; to date this has not been rectified. This lack of training means that when information is published displaying micrographs in an unfamiliar mode, it often results in incorrect interpretation.

21

Supplementary Materials, Techniques and Methods

21.1 A Summary

Example number	Family classification	Material
1	BB4	Copper
2	BB21	Yttria barium copper oxide
3	BB14	Tungsten carbide/cobalt with titanium nitride coating
4	BB18	Corrosion of copper
5	BB3	Pure aluminium
6	BB5	Nickel
7	BB9	Tin
8	BB3	Magnesium
9	BB3	Zinc
10	BB24	Roman glass

21.2 Preparation Examples

Example 21.1

Family classification BB4

The material: copper

This history of this material is from pre-bronze age weapons to today's conductors of electricity and heat. Its uses have been vast and it is still one of the most commonly used metals (steel, aluminium, copper, zinc). A high proportion of the pure copper used today is for carrying electric current. One of the problems of using long span overhead cables is the lack of strength, due to density and extra loads subjected by wind and/or snow. To increase the strength by alloying reduces conductivity; this restricts the available compromise choice to cadmium. Copper is losing out to steel reinforced aluminium for long span applications. An example of heat conductors can be found in the kitchen, i.e. copper-bottom pans.

The technique

Pure copper is very soft and ductile and is therefore prone to smearing and impressed abrasives. The method given includes only one silicon carbide grinding stage; copper alloys, on the other hand, could include more. Polishing without deformation is best carried out using vibratory equipment or alternatively electrolytic methods.

The method

AUDITABLE PREPARATION PROCEDURE   No. 1

CLASSIFICATIONS: BB4/BB15		MATERIAL: COPPER ORE	
SECTIONING TECHNIQUE:	EQUIPMENT TYPE BLADE SiC ABRADABLE WHEEL	CONCENTRATION	
MOUNTING TECHNIQUE:	METHOD	SAMPLE/MOUNT RATIO	

*Vibratory polish.

	SURFACE and TYPE	ABRASIVE SIZE/TYPE LUBRICANT	FORCE (kPa)	HEAD ROTATION	Z AXIS (μm)	REFLECTIVE FACTOR	MICROGRAPH YES/NO	IMAGE ANALYSIS YES/NO	REMARKS
PLANAR GRINDING STAGE	PAPER WAXED	SILICON CARBIDE P180 WATER	35	COMP	PLANE				

	SURFACE and TYPE	ABRASIVE SIZE/TYPE LUBRICANT	FORCE (kPa)	HEAD ROTATION	Z AXIS (μm)	REFLECTIVE FACTOR	MICROGRAPH YES/NO	IMAGE ANALYSIS YES/NO	REMARKS
SAMPLE INTEGRITY STAGE	S.2P	30 μm OIL	35	COMP					
	S.1	3 μm OIL	30	COMP					

	SURFACE and TYPE	ABRASIVE SIZE/TYPE LUBRICANT	FORCE (kPa)	HEAD ROTATION	TIME (s)	REFLECTIVE FACTOR	MICROGRAPH YES/NO	IMAGE ANALYSIS YES/NO	REMARKS
POLISHING STAGE*	PS 8	0.05 μm OXIDE/SILICA	15	COMP	500		YES		21.1 (Plate 12)

The micrograph Figure 21.1 (see Plate 12)

Example 21.2

Family classification BB21

The material: yttrium barium copper oxide ($YBa_2Cu_3O_7$)

The recent discovery of high temperature superconducting ceramics is said to have potentially revolutionized the electric power industry. Superconductivity is usually defined as the complete loss of all electrical resistance. The major application for superconductors is to be found in magnets, the advantage being the reduction in size over conventional machines. Applications in electronic circuits and devices have not as yet fulfilled the original expectations. This is not to say that the level of interest or R&D has reduced. The powder preparation and the sintering conditions are major factors influencing the microstructure and therefore the mechanical and superconducting properties. Sintering temperatures between 875 and 925°C give a dramatic structure change, the so-called 1, 2, 3 phase changes from small rounded grains (5 μm) to sharp edged, twin structured columnar grains.

The technique

The sintered materials usually exhibit high levels of porosity (18%); therefore, vacuum impregnation is often suggested. This is often to no avail since porosity is usually in the 1 μm range with interconnecting holes considerably less. Because of the submicrometre interconnections it is unlikely that any resin ingress will occur. Superconductors do not need high pressure when carrying out any surface preparation. It is therefore quite easy to carry out specimen preparation by hand, using classical methods. To avoid any chemical reaction to the specimen, water is to be avoided. One such method would be to use waxed silicon carbide paper to P4000, if available, followed by 6, 3 and 1 μm diamond on platen surfaces 2 and 1. The method illustrated utilizes platen surface 4 in the preliminary stages, the time or Z axis required at this stage being dependent upon the coarseness of the planar grinding stage.

The method

AUDITABLE PREPARATION PROCEDURE									No. 2	
CLASSIFICATIONS: BB21/SUPERCONDUCTORS						MATERIAL: YTTRIUM BARIUM COPPER OXIDE				
SECTIONING TECHNIQUE:	EQUIPMENT TYPE ABRASIVE CUTTER		BLADE SILICON CARBIDE (OIL)			CONCENTRATION				
MOUNTING TECHNIQUE:	METHOD					SAMPLE/MOUNT RATIO				

*P4000 silicon carbide paper as alternative.

	SURFACE and TYPE	ABRASIVE SIZE/TYPE LUBRICANT	FORCE (kPa)	HEAD ROTATION	Z AXIS (μm)	REFLECTIVE FACTOR	MICROGRAPH YES/NO	IMAGE ANALYSIS YES/NO	REMARKS
PLANAR GRINDING STAGE	PAPER OIL	P500 GRIT SILICON CARBIDE	30	COMP	PLANE				

	SURFACE and TYPE	ABRASIVE SIZE/TYPE LUBRICANT	FORCE (kPa)	HEAD ROTATION	Z AXIS (μm)	REFLECTIVE FACTOR	MICROGRAPH YES/NO	IMAGE ANALYSIS YES/NO	REMARKS
SAMPLE INTEGRITY STAGE	PAPER OIL	P1200 GRIT SiC	30	COMP					
	S.4*	9 μm OIL	30	CONTRA					
	S.1	1 μm OIL	30	COMP					

	SURFACE and TYPE	ABRASIVE SIZE/TYPE LUBRICANT	FORCE (kPa)	HEAD ROTATION	TIME (s)	REFLECTIVE FACTOR	MICROGRAPH YES/NO	IMAGE ANALYSIS YES/NO	REMARKS
POLISHING STAGE	PS 8	0.25 μm ALCOHOL	16	COMP	600		YES		21.2 (Plate 12)

The micrograph Figure 21.2 (see Plate 12)

Example 21.3

Family classification BB14

The material: tungsten carbide/cobalt with titanium nitride coating

Sintered carbide cutting tools have for many years been successfully operated at high speed, replacing high speed steel (HSS) cutting tools for many applications. Tungsten carbide is sintered at around 1500°C, using about 6% cobalt as the binder to produce pore-free tools. Throwaway indexable tool tips with more than one cutting edge, along with improved life via thin film coatings, are proving advantageous. Figure 21.3 (Plate 12) is one such example. Single or multicoatings of sizes from submicrometres upwards are currently in use, increasing the tool life and operating temperature. Coating examples could be from substrate outwards—TiC 4 μm layer, Al_2O_3 4 μm layer and TiN < 1 μm layer.

The technique

Tungsten carbide materials can be prepared using high speed wheels (500 rev/min with 300 mm diameter) with a fixed metal bonded diamond abrasive, followed by 1 to 3 μm diamond on a hard cloth (surface 1); card or paper has also been used at this final step. The above route is used to illustrate the ease with which many hard materials can be prepared (please do not take this as a recommendation). Generally there are two points to be addressed, viz. structural damage to the coatings or matrix and induced and/or enlarged porosity. If the sample was totally dense, the procedure would be less critical than if porosity had been present. The same can be said of the grain size; fine grains require a different approach to coarse grain size. If the hard coating is submicrometre, the progress towards integrity must be more carefully controlled—the most suitable mounting resin must be hard and without contraction.

The simple three-step method illustrated will produce faithful results for the majority of tungsten carbide materials. When preparing coated materials the addition of a 3 μm surface 1 could be necessary. Figure 21.3 (Plate 12) illustrates a well-prepared coated layer, the differential interference contrast techniques showing the different layers and the damaged radial cracks. The high magnification shows a poor resin/coating interface, which has caused the breakup of the final coating. This sample was prepared by hand and was hot mounted in phenolic resin. Possible faults, therefore, are:

(i) Phenolic resins incur a poor surface bond, even though they are shrunk onto the sample. This could be overcome by using an epoxy resin.
(ii) The radial cracks could be induced through high pressure during mounting. It would be necessary to eliminate this possibility before settling for a hot mountng technique.

The method

AUDITABLE PREPARATION PROCEDURE   No. 3

CLASSIFICATIONS: BB14/SINTERED CARBIDE		MATERIAL: TUNGSTEN CARBIDE/COBALT
SECTIONING TECHNIQUE:	EQUIPMENT TYPE BLADE	CONCENTRATION
MOUNTING TECHNIQUE:	METHOD	SAMPLE/MOUNT RATIO

*Additional stage for thin coatings.

	SURFACE and TYPE	ABRASIVE SIZE/TYPE LUBRICANT	FORCE (kPa)	HEAD ROTATION	Z AXIS (μm)	REFLECTIVE FACTOR	MICROGRAPH YES/NO	IMAGE ANALYSIS YES/NO	REMARKS
PLANAR GRINDING STAGE	S.10	30 μm WATER	35	CONTRA	PLANE				

	SURFACE and TYPE	ABRASIVE SIZE/TYPE LUBRICANT	FORCE (kPa)	HEAD ROTATION	Z AXIS (μm)	REFLECTIVE FACTOR	MICROGRAPH YES/NO	IMAGE ANALYSIS YES/NO	REMARKS
SAMPLE INTEGRITY STAGE	S.4	9 μm OIL	35	CONTRA					
	S.1*	1 μm WATER	70	COMP	3		YES		21.3 (Plate 12)

	SURFACE and TYPE	ABRASIVE SIZE/TYPE LUBRICANT	FORCE (kPa)	HEAD ROTATION	TIME (s)	REFLECTIVE FACTOR	MICROGRAPH YES/NO	IMAGE ANALYSIS YES/NO	REMARKS
POLISHING STAGE									

The micrograph Figure 21.3 (see Plate 12)

Example 21.4

Family classification BB18

The material: corrosion of copper

The word 'corrosion' conjures up images of rust on steel; this we are all familiar with in everyday products. We are also aware of the green appearance (verdigris) of copper plate when subjected to the atmosphere. Corrosion comes in many forms, sometimes to protect the material, more often to degrade. Corrosion can be localized (pitting), layered or intercrystalline. It can be caused by a chemical reaction, contact, friction or stress. Intercrystalline corrosion is to be found along the grain boundaries emanating from an outer surface. This is the reason for a plated protection layer of the same pure material being used to combat this effect. Layer corrosion is signified by segregation lines running parallel to the outer surface. Acid rain is a new problem associated with corrosion; high sulphur compounds released from burning coal have already caused damage to the stonework of historic buildings. The marine environment is very corrosive and has to be carefully controlled in areas such as off-shore structures. Corrosion protection is a vast industry and is becoming very scientific, demanding that protection be just good enough; i.e. protection on automotive bodies is balanced to match a given manufacturer's car life.

The technique

The preparation of a copper material when concentrating on corrosion aspects is very different to a normal copper route. Copper is soft and ductile; the corrosive elements are often brittle and friable. The specimen will ideally be encapsulated in a low viscous epoxy type resin. Grinding can take place using silicon carbide paper, provided that it is used within the optimum life period. As the sample progresses to integrity it will be necessary to use a charged platen surface of about 4 with 9 μm diamond abrasive, using an oil lubricant. This step must be progressed until the corroded layer is visible, complete and unbroken. Completion of the integrity stage will then reveal the friable elements in a matrix that could be scratched. Careful polishing should remove these scratches. If, however, they degrade the corrosion layer, it could be necessary to partially compromise the polish (never compromise integrity). Figure 21.4 (Plate 12) shows the massive corrosion ingress.

The method

AUDITABLE PREPARATION PROCEDURE   No. 4

CLASSIFICATIONS: BB18						MATERIAL: CORROSION OF COPPER			
SECTIONING TECHNIQUE:	EQUIPMENT TYPE		BLADE			CONCENTRATION			
MOUNTING TECHNIQUE:	METHOD					SAMPLE/MOUNT RATIO			

	SURFACE and TYPE	ABRASIVE SIZE/TYPE LUBRICANT	FORCE (kPa)	HEAD ROTATION	Z AXIS (μm)	REFLECTIVE FACTOR	MICROGRAPH YES/NO	IMAGE ANALYSIS YES/NO	REMARKS
PLANAR GRINDING STAGE	SiC PAPER	P320 WATER	35	CONTRA	PLANE				

	SURFACE and TYPE	ABRASIVE SIZE/TYPE LUBRICANT	FORCE (kPa)	HEAD ROTATION	Z AXIS (μm)	REFLECTIVE FACTOR	MICROGRAPH YES/NO	IMAGE ANALYSIS YES/NO	REMARKS
SAMPLE INTEGRITY STAGE	SiC	P600 WATER	35	COMP	20				
	S.4	9 μm OIL	35	CONTRA	10				
	S.1	1 μm OIL	35	COMP	6				

	SURFACE and TYPE	ABRASIVE SIZE/TYPE LUBRICANT	FORCE (kPa)	HEAD ROTATION	TIME (s)	REFLECTIVE FACTOR	MICROGRAPH YES/NO	IMAGE ANALYSIS YES/NO	REMARKS
POLISHING STAGE	PS 7	0.06 μm COLLOIDAL SILICA	17	CONTRA	120		YES		21.4 (Plate 12)

The micrograph Figure 21.4 (see Plate 12)

Example 21.5

Family classification BB3

The material: pure aluminium

Pure aluminium is soft and ductile, lending itself to cold rolling into very thin aluminium foil (silver paper). The end product finds many applications in our everyday life from cigarettes to baking foil. Recent applications have been in the peel-off lidding used on drink cans. Aluminium, when polished or as an aluminized layer, is extremely reflective, finding application in microscopy and astronomy. No longer is there the necessity for highly polished steel or silvered reflecting mirrors. To look into a good front surface mirror can be an unbelievable experience; all the facial defects are clearly resolved and visible. Aluminium is also a good reflector of heat and is used to advantage in reflecting away the sun's rays, when protecting food or explosive flashpoint products.

The technique

This material must be treated totally differently to an aluminium alloy. Its softness and ductility make it extremely vulnerable to impressed abrasives and surface smearing. From the method given the jump from 15 to 0.06 μm is very large—hence the reason for increased pressure. Great care must be exercised when using high pressure with soft smeary materials if structural changes are to be avoided. These materials are often more suitably prepared by electrolytic polishing or, alternatively, are finally polished on vibratory polishers.

The method

AUDITABLE PREPARATION PROCEDURE   No. 5

CLASSIFICATIONS: BB3		MATERIAL: PURE ALUMINIUM	
SECTIONING TECHNIQUE:	EQUIPMENT TYPE BLADE	CONCENTRATION	
MOUNTING TECHNIQUE:	METHOD	SAMPLE/MOUNT RATIO	

* If impressed abrasives occur use fixed diamond wheel

	SURFACE and TYPE	ABRASIVE SIZE/TYPE LUBRICANT	FORCE (kPa)	HEAD ROTATION	Z AXIS (μm)	REFLECTIVE FACTOR	MICROGRAPH YES/NO	IMAGE ANALYSIS YES/NO	REMARKS
PLANAR GRINDING STAGE	PAPER OIL	P320 SiC WATER	35	COMP					

	SURFACE and TYPE	ABRASIVE SIZE/TYPE LUBRICANT	FORCE (kPa)	HEAD ROTATION	Z AXIS (μm)	REFLECTIVE FACTOR	MICROGRAPH YES/NO	IMAGE ANALYSIS YES/NO	REMARKS
SAMPLE INTEGRITY STAGE	S.2P	15 μm OIL	35	COMP					
	SILK	0.06 μm high pH COLLOIDAL SILICA	45	COMP					

	SURFACE and TYPE	ABRASIVE SIZE/TYPE LUBRICANT	FORCE (kPa)	HEAD ROTATION	TIME (s)	REFLECTIVE FACTOR	MICROGRAPH YES/NO	IMAGE ANALYSIS YES/NO	REMARKS
POLISHING STAGE	PS 7	0.06 μm COLLOIDAL SILICA	17	COMP	120		NO		

Example 21.6

Family classification BB5

The material: nickel

Nickel has many similarities to iron, with the added advantage of being corrosion resistant. It was used in tableware as a thin electroplated coat and without the familiar letters EPNS (electroplated nickel silver) would, and did, fool many as being silver plated. It forms the corrosion-resistant layer prior to chrome plating on modern-day utensils. It can also be applied to plastic materials to give that polished metal appearance—yet another act of public deception. Electroplating a thick layer of nickel onto plastic has led to the perfect replica for mould manufacture. The manufacture of records is one such example. Nickel as an alloying agent accounts for most of its current usage, such as austenitic stainless steels, nickel-based super alloys, austenitic cast iron (nickel-resist irons). Heat-resistant alloys employing high levels of nickel (80% Ni–20% chromium) are employed as heating elements (electric fires). Nickel-based nimonics, Inconel and Incoloy, are further examples where high strength, corrosion resistance and elevated temperatures are important (in aeroengine parts).

The technique

Nickel blocks can be successfully prepared by classical methods if care is taken at the grinding stages. This usually involves slightly lower surface speeds and optimum abrasive cutting. The method given includes charged abrasive surface 4 using an oil lubricant.

The method

AUDITABLE PREPARATION PROCEDURE — No. 6

CLASSIFICATIONS: BB5		MATERIAL: NICKEL	
SECTIONING TECHNIQUE:	EQUIPMENT TYPE BLADE	CONCENTRATION	
MOUNTING TECHNIQUE:	METHOD	SAMPLE/MOUNT RATIO	

* Could be substituted with two stages of SiC grinding.

	SURFACE and TYPE	ABRASIVE SIZE/TYPE LUBRICANT	FORCE (kPa)	HEAD ROTATION	Z AXIS (µm)	REFLECTIVE FACTOR	MICROGRAPH YES/NO	IMAGE ANALYSIS YES/NO	REMARKS
PLANAR GRINDING STAGE	PAPER	P180 SiC	35	COMP	PLANE				

	SURFACE and TYPE	ABRASIVE SIZE/TYPE LUBRICANT	FORCE (kPa)	HEAD ROTATION	Z AXIS (µm)	REFLECTIVE FACTOR	MICROGRAPH YES/NO	IMAGE ANALYSIS YES/NO	REMARKS
SAMPLE INTEGRITY STAGE	S.4*	9 µm OIL	35	CONTRA					
	S.1	3 µm OIL	35	COMP					
	S.1	0.06 µm COLLOIDAL SILICA	35	CONTRA					

	SURFACE and TYPE	ABRASIVE SIZE/TYPE LUBRICANT	FORCE (kPa)	HEAD ROTATION	TIME (s)	REFLECTIVE FACTOR	MICROGRAPH YES/NO	IMAGE ANALYSIS YES/NO	REMARKS
POLISHING STAGE	PS 7	0.06 µm COLLOIDAL SILICA	15	COMP	300	/	NO	/	

Example 21.7

Family classification BB9

The material: tin

The majority of uses for this soft low melting point (230°C) material will be as a coating or alloying element. It is non-toxic and extremely corrosive resistant—hence its use as a coating on steel for food storage (tin cans). Tin plate is made from rolled low carbon steel with a thin electrodeposited tin layer. Pewter is a tin–copper–antimony alloy and has been used in early kitchen ware and candlesticks. Although pewter is fashionably connected with German beer tankards, it was originally used in Britain as a standard volume half pint and full pint mug. The name 'pewter' is also a misnomer for some lesser desirable ornaments made from cheaper materials. Tin, as with lead, makes good bearing alloys; it is also used with aluminium as a bearing alloy. Bronze, as we understand the material, exists with a high proportion of tin. Lead–tin alloys not only make good bearings but also metal joining solders. Its low melting point and good alloying characteristics make it ideal for this use.

Example 21.8

Family classification BB3

The material: magnesium

Low density and relatively high alloy strength make this material extremely useful. In this modern age where weight savings can result in reduced fuel and therefore emission levels, magnesium is increasingly an attractive option. Alloying magnesium with up to 10% aluminium and very low values of zinc and manganese produces a good casting material, with a greatly improved strength (better than twice that of pure magnesium). Magnesium can be considered as a potential replacement candidate for aluminium where weight savings are important. There is, however, a pay-back consideration of extra material cost and extra protection against corrosion. The 'new materials' applications for metal matrix composites of fibres or particulates has stimulated more interest in magnesium.

Example 21.9

Family classification BB3

The material: zinc

Brass being a copper–zinc alloy accounts for the greatest volume of zinc usage, although much is still used in zinc die casting. Protecting steel from rust is another major function, where the zinc coating is applied by dip galvanizing—hence the term 'galvanized steel'. Zinc and aluminium can be metal sprayed to steel structures, such as bridges, to extend corrosion protection.

The technique

Zinc, magnesium and tin have been linked together since the problems associated with surface preparation are similar, viz. a deformation layer easily forms when mechanically working the surface. Sharp silicon carbide abrasives using an oil or wax lubricant are best for the initial integrity stage. Avoid rubbing contact which could occur with fine grained silicon carbide papers. When using charged or loose abrasives, ensure that they are applied in an oil slurry. Final integrity stages will require careful pH control. To avoid the analysis of false structures, heat and mechanical strain must be kept to an absolute minimum. Alumina or magnesium oxide is often chosen at the polishing stage, along with an etch-polish mechanochemical polish or the technically more desirable electrolytic polish. Microtomes using a tungsten carbide blade are sometimes recommended for the integrity stage. Extreme care must, however, be exercised to reduce specimen residual stress.

The method

AUDITABLE PREPARATION PROCEDURE   Nos. 7 to 9

CLASSIFICATIONS:		MATERIAL: ZINC, MAGNESIUM and TIN	
SECTIONING TECHNIQUE:	EQUIPMENT TYPE	BLADE	CONCENTRATION
MOUNTING TECHNIQUE:	METHOD		SAMPLE/MOUNT RATIO

* Vibratory polishing is often advantageous with all soft, smeary materials.

	SURFACE and TYPE	ABRASIVE SIZE/TYPE LUBRICANT	FORCE (kPa)	HEAD ROTATION	Z AXIS (μm)	REFLECTIVE FACTOR	MICROGRAPH YES/NO	IMAGE ANALYSIS YES/NO	REMARKS
PLANAR GRINDING STAGE	PAPER OIL	P180 SiC	35	COMP	PLANE				

	SURFACE and TYPE	ABRASIVE SIZE/TYPE LUBRICANT	FORCE (kPa)	HEAD ROTATION	Z AXIS (μm)	REFLECTIVE FACTOR	MICROGRAPH YES/NO	IMAGE ANALYSIS YES/NO	REMARKS
SAMPLE INTEGRITY STAGE	S.2P	30 μm OIL	35	COMP					
	S.1	3 μm OIL	35	COMP					

	SURFACE and TYPE	ABRASIVE SIZE/TYPE LUBRICANT	FORCE (kPa)	HEAD ROTATION	TIME (s)	REFLECTIVE FACTOR	MICROGRAPH YES/NO	IMAGE ANALYSIS YES/NO	REMARKS
POLISHING STAGE*	PS 7	0.05 μm OXIDE/SILICA	15	COMP	300		NO		

Example 21.10

Family classification BB24

The material: Roman glass

Glass is usually relatively easy to prepare. However, Roman glass is more difficult since it contains areas of manganese ingress and interfaces where the glass is extremely friable. The sample illustrated (Figure 34.3) had to be resin impregnated prior to sectioning on a low deformation diamond blade saw. After sectioning, the sample was vacuum impregnated using a low viscous resin. The traditional approach would have been to free lap the surface, but this was not possible since the rolling abrasives would have impressed into the resin. Fixed abrasives were considered too drastic; therefore charged abrasives using platen surfaces 4, 2P and 1 were favoured. Although iron oxide powder (jeweller's rouge) is commonly used for the polishing of glass, on this occasion cerium oxide was used.

The method

AUDITABLE PREPARATION PROCEDURE   No. 10

CLASSIFICATIONS: BB24			MATERIAL: ROMAN GLASS	
SECTIONING TECHNIQUE:	EQUIPMENT TYPE	BLADE	CONCENTRATION	
MOUNTING TECHNIQUE:	METHOD		SAMPLE/MOUNT RATIO	

PLANAR GRINDING STAGE	SURFACE and TYPE	ABRASIVE SIZE/TYPE LUBRICANT	FORCE (kPa)	HEAD ROTATION	Z AXIS (μm)	REFLECTIVE FACTOR	MICROGRAPH YES/NO	IMAGE ANALYSIS YES/NO	REMARKS
	S.10	30 μm WATER	35	CONTRA	PLANE				

SAMPLE INTEGRITY STAGE	SURFACE and TYPE	ABRASIVE SIZE/TYPE LUBRICANT	FORCE (kPa)	HEAD ROTATION	Z AXIS (μm)	REFLECTIVE FACTOR	MICROGRAPH YES/NO	IMAGE ANALYSIS YES/NO	REMARKS
	S.4	9 μm OIL	35	CONTRA					
	S.2P	6 μm OIL	35	CONTRA					
	S.1	3 μm WATER	35	CONTRA					
	S.1	1 μm OIL pH ACTIVE	45	COMP					

POLISHING STAGE	SURFACE and TYPE	ABRASIVE SIZE/TYPE LUBRICANT	FORCE (kPa)	HEAD ROTATION	TIME (s)	REFLECTIVE FACTOR	MICROGRAPH YES/NO	IMAGE ANALYSIS YES/NO	REMARKS
	PS 7	1 μm CERIUM OXIDE	17	COMP	120		YES		34.3

The micrograph Figure 34.12 (see Plate 14)

List of Preparation Examples

Family classification	Material group	Material	Example number	Page
BB1	Ferrous, except cast iron	Cast alloy steel	5.5	53
		Sintered iron	5.6	54
		Tool steel	5.9	57
		0.37% steel	11.1	91
BB2	Cast iron	White cast iron	5.3	51
		SG iron	5.7	55
		SG iron (nodular)	11.6	96
BB3	Aluminium, magnesium and zinc alloys	Aluminium bronze	5.2	50
		Aluminium–silicon alloy	11.4	94
		Pure aluminium	21.5	220
		Magnesium	21.8	223
		Zinc	21.9	223
BB4	Copper and alloys	Copper phosphorus	5.1	49
		Brass	11.2	92
		Copper	21.1	214
BB5	Nickel alloys cobalt super	Nimonic 105	5.4	52
		Nickel	21.6	221
BB6	Titanium alloys	Titanium 6/4	17.7	189
		Tantalum	17.8	190
BB7	Refractory metals			
BB8	Battery and electronic components	Electronic component	11.5	95
		Ceramic capacitor	15.2	171
		Electronic components	15.3	173
BB9	Soft and noble metals	Lead base alloy	17.2	184
		Tin	21.7	223
BB10	Ceramics	Boron carbide	18.1	198
		Aluminium nitride	18.2	199
		Sialon 201	18.3	200
BB11	MMC	Alumina fibres in aluminium	13.5	123
		(L) Silicon carbide particulate/aluminium	13.6	125
		Borsic fibres in aluminium	13.7	126
		Alumina fibres in aluminium/copper	13.11	132
		Silicon carbide fibres/graphite in aluminium	13.12	133

Family classification	Material group	Material	Example number	Page
		Ceramic particles in steel	13.13	134
		Titanium carbide in titanium	13.14	135
		(S) Silicon carbide particles in aluminium/silicon	13.15	136
BB12	PMC	Carbon fibre	5.8	56
		Carbon fibre (PEEK)	13.4	122
BB13	GMC and CMC	Silicon carbide fibres in glass	13.8	128
		Silicon carbide particulate/silicon nitride	18.4	201
BB14	Sintered carbide	WC/cobalt with titanium nitride coating	21.3	217
BB15	Rocks and minerals	Chalcopyrite	14.1	154
		Graphite	14.3	156
		Slag—iron oxide/quartz	14.4	158
		Cement clinker	14.5	159
		Vitrified alumina	14.6	160
		Covellite	14.7	161
		Hematite	14.8	162
		Titanium oxide	14.9	163
BB16	Spray coatings	Zirconia/yttria	12.1	107
		Alumina thin	12.2	108
		Alumina thick	12.3	110
		Tungsten carbide	12.4	111
		Copper, nickel, indium	12.5	113
		Complex nickel	12.6	114
		Aluminium, silicon, polyester	12.7	114
		Nickel, graphite	12.8	115
BB17	Thin platings including PCBs	PCB	11.3	93
		Through-hole PCB	15.1	170
BB18	Corrosion products	Copper	21.4	219
BB19	Paint	SiC/graphite/paint	13.10	131
BB20	Polymers			
BB21	Magnetic materials and superconductors	Yttria barium copper oxide	21.2	216
BB22	Composites	Carbon–carbon	13.9	130
		Brake lining	14.2	155
BB23	Refractories			
BB24	Glass	Roman glass	21.10	224
BB25	Biomaterials	Hydroxyapatite	13.16	139
		Coral (porites and acropora)	13.17	142
BB26	Bearing materials	Aluminium/tin	17.1	183
		Tin-based white metal	17.3	185
		Al/Si overlay plate	17.4	186
		Al/lead plate	17.5	187
		Al/Cu/Si/Sn	17.6	188
BB27	Solder, brazed, welded	Brazed metallized layer	15.4	174
BB28	Thick coatings			

PART II

APPLIED MICROSCOPY

Unlike many authors on microscopy I have no claim to an early childhood involvement; nor was I immersed in the subject by parental influences. There was no microscope at primary school and the only one worth using in secondary school was too good to be let loose with students. This then was the background prior to further education and apprenticeship as a mechanical engineer with by chance a company making microscopes whose history spans 300 years (Cooke, Troughton and Simms changed to Vickers Instruments, currently called BIORAD).

I therefore became involved in microscopes by accident rather than design. Little did I know the effect microscopy would have as my career moved into designing instruments for reflected and transmitted light applications. To design any product it is essential not just to understand the technicalities of that product but how it is to be used in its working environment. To list the applications of microscopy would be an endless task but to be involved, be it only peripherally, with so many faculties has enriched a period of my life described by others as work.

Some two decades ago circumstances necessitated a closer involvement in industrial microscopy and its associated preparation equipment. This broad-based involvement went from conceptual design, through manufacture to technical marketing. It was during this period an awareness was made of the lack of training in both the technologies of preparation and the use of optical microscopy. Part I of this book demonstrates a completely new approach to surface preparation; the question is 'what's new in microscopy?' The following chapters will illustrate with practical examples the use of microscopic techniques as applied to the materials technologist. Kohler illumination will get a new look and photomicrography will be biased towards application. The book will conclude with photomicrography in practice showing the 'beauty of materials' and illustrating not just the techniques but where best to use them.

If the best this volume can achieve is an awareness by the microscope user to look *into the subject* and not *at it*, then the observer is starting on the path to a lifetime of fascination.

22

The Microscope—A Résumé

Unfortunately history does not record the first use of the simple single lens microscope but it is conceivable that the ancient Egyptians who were skilled in the art of lapidary would have noticed that objects viewed through a lens-shaped glass appeared enlarged or diminished in size. Greek and Roman writers made reference to the magnifying effect of a glass sphere filled with water.

The early history of the compound microscope is equally vague, and is said to begin in the late sixteenth century. Principal influences must come from Galileo in the early seventeenth century with his design for the telescope and later in the century from the Dutch scientist Huygens who developed the eyepiece which still carries his name today.

In that century Robert Hooke published what was called his *Micrographia*, combining a description of a compound microscope utilizing an eyepiece to magnify the primary image created from the single lens objective. The first compound microscope consisted of a small single lens objective and an eyepiece with a large field and eyelens. This is how it remained optically for over 100 years. It did, however, vary considerably in its mechanics and this allowed the instrument making art to thrive, producing aesthetically beautiful pieces of equipment.

The major disadvantage with the Hooke type of compound microscope was the gross chromatic aberration, resulting in resolved lines looking like reflections of the rainbow. Chromatic aberration is the result of unequal refraction taking place with the components of white light; i.e. different colours having different wave lengths refract at a different angle. Chromatic aberration is absent from the reflecting telescope and was the reason for much research into a possible reflecting microscope in an attempt to overcome this problem.

The middle of the eighteenth century saw the introduction of the achromatic lens for astronomy; this was later (early nineteenth century) developed for microscopy. Later in that century Abbe produced the first apochromatic objective using the mineral called fluorite. With all our present-day technology these lenses have scarcely been surpassed in optical quality.

Metallurgists have to thank Sorby for opening up a new world to the microscopist; in the second half of the nineteenth century he looked at metals as never before and revealed grains and structures. Utilizing reflected light techniques the microscope was to be an industrial tool as industry required materials to function in many different ways; the microscope was the key—the fingerprint of metallography. It is now more than just a research tool; it is used for the quality control of components to ensure there are no weak links in the chain of manufacture of articles as diverse as motor cars and musical instruments.

Every item we touch or see has at some time been investigated by the microscopist: nearly every subject has a microstructure and this is what the microscopist is looking for. The condition of this structure tells all: why the bolt failed, why the bridge collapsed, why the paint peeled off.

Microscopy has become a vital science in the quality control of materials and components, without which our confidence in travelling on aeroplanes, in motor cars, on trains, etc., would be greatly reduced. New engineering materials have brought a new dimension in our microscope studies since we are required not just to observe the material grain structure but also the relationship of one composite with another: viz.

(i) Has it fused together?
(ii) Has it stuck?
(iii) Is good contact taking place?
(iv) Has it caused a secondary phase action?
(v) Has it been stressed?
(vi) What is the distribution?

These and many other aspects have brought great demands upon the materials technologist, requiring the maximum use of all optical microscope techniques.

The previous part in this book deals with the essential care necessary to achieve a faithful structure, one of integrity. It is the microscope, however, that is necessary to *reveal* the true microstructure and this is not always apparent from a normal bright field observation.

One very important aspect of microscopy is the ability of every lens to give a three-dimensional picture of what appears to be a two-dimensional subject (within its depth of field). Metallurgists have been concerned generally with preparing the flattest specimens possible, then making an analysis from the sometimes etched subject across this flat plane. This technique is adequate when making an analysis of the microstructure but leaves much to be desired when trying to interpret physical relationships with composites. It is necessary therefore to continue to prepare planar surfaces but also to explore the different optical characteristics giving the best three-dimensional (or Z axis) information.

Any move from qualitative analysis towards quantitative analysis is a forward approach. The use of three-dimensional analysis is the basis of quantitative stereology (exploration in three dimensions from a two-dimensional subject). This introduces the microscope via its optical techniques to the twentieth century computer analysis of grey level information. Quantitative values for volume fraction as well as two-dimensional metrology can be statistically derived and stored.

The fully automatic robot operated microscope can be (and is being) used when the revealed image is portrayed in a manner that can be statistically defined. Take, for example, a hardness indent; the eye can readily differentiate the impression from other visually observed information.

The eye recognizes this subject primarily by its shape or morphology; software packages could be written allowing the scanned image to recognize and record these features. This is the basis of digital image analysis and is applied to many aspects of both life and material science. Once the field of view has been analysed an incremental series of fields can be programmed for assessment. An extension of this is to carry out other functions beyond just the analysis at each incremental station; microhardness is an obvious example of this where the subject can, for example, be a toothed gear where it is necessary to indent, analyse and increment along, say, the involute curve.

The microscope in the future will play an integral part in the software-generated laboratory management plan. The analysis could be software driven and will in any event have the analysis requirements, conditions and acceptable tolerances dictated via the laboratory management plan. The progress of the component or material is subjected to quality clearance at various stages, the images and statistics (qualitative and quantitative) being archived ready for instant recall

The optical microscope has over the last 50 years, with minor exceptions such as fluorescence and reflectance techniques, remained stable in its design. Modern aesthetics and ergonomics have updated the appearance and improved the functions without overcoming the increased demand for improved resolution and depth of field. This has been satisfied by the transmission electron and the scanning electron microscope; an extension of the electron microscope is the statistical analysis of elements.

Having looked at the past and present, what future has the optical microscope in materials analysis? To answer this we need to look at the analytical requirements of materials and components and future development. The demand for new materials—composites of existing material—to reduce weight, increase useful life, improve wearability, operate at very high temperatures, be degradable when redundant, be stronger, withstand shock, be made from a plentiful resource, etc., means that structural analysis and hence the optical microscope will be in great demand. The optical microscope may represent an old technology but it still represents the most important tool in macro- and microstructural analysis. Other forms of microscopes are better at submicrostructural analysis and as such will only complement the information derived from the optical image.

The future then is good providing we, the materials scientists, make good use of all available techniques. Although there are many reference books on microscopy they tend to be either theoretical or biased towards life science. The following chapters will redress the balance and present the subject in an applied manner, biased towards materialography.

23

Microscope Types and Nomenclature

Putting a name to objects is an important part of successful communications; very often, particularly in microscopy, dialogue becomes confused through a lack of nomenclature. The author for example always names the graduations observed orthoscopically in the microscope eyepiece as graduations on a graticule, yet many others use the word reticule. Within this last sentence is the word 'eyepiece'; many call it an ocular and so the confusion grows.

23.1 Upright Compound Microscope

The microscope shown in Figure 23.1 is an upright compound microscope constructed for both reflected and transmitted light applications and having facilities for taking photographs (micrographs). The earliest form of microscope, the 'simple microscope', consists of a single lens or cluster which magnifies an erect image which is predominantly of low magnification, i.e. magnifying glass or eyepiece. The compound microscope is an extension of the simple microscope in that it further increases the primary magnification by utilizing another lens or cluster forming a secondary image. This combination we know

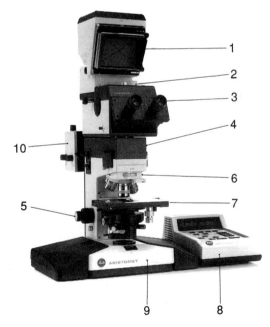

Figure 23.1 Upright compound microscope (Photo: Leica)

as the 'objective' and the 'eyepiece'. The objective creates the initial magnified primary image; the eyepiece positioned to focus onto the magnified primary image further magnifies, giving the secondary image as seen by the eye, projection screen, television camera or photographic plate.

The upright compound microscope can be built for applications in (a) reflected light, (b) transmitted light or (c) a combination of both, as illustrated in Figure 23.1.

The following is a list of major microscope parts and the options available when constructing a specification.

(1) Photographic screen

The system illustrated is to take large format cameras made to an MPP fitting. The screen is used for viewing the area of interest and can be used for specimen focusing prior to taking a photograph, but it is much wiser to focus through the eyepiece system, provided that the primary image plane and camera plane are par-focal.

The ergonomics and rigidity of this type of design are excellent but it involves a series of extra mirrors, prisms and relay lenses. This ultimately must compromise the photographic quality when compared with direct non-reflected systems.

One positive advantage of having the screen and camera inclined towards the operator is so that the dirt caused when sliding cameras in and out does not drop on the camera relay lens, as is the case with horizontal systems (vertical optics). Compound microscopes are not confined to the large format camera; 35 mm can also be accommodated.

Camera focusing methods vary, the through the lens reflex style can be used or alternatively via (a) the eyepiece tube, (b) a special focusing camera tube or (c) on a camera projection screen. If the latter is chosen it is wise to use the aerial image method of focusing via a magnifying telescope.

(2) Magnification changer

The magnification of a compound microscope is usually based on the objective times the eyepiece (times any tube length factor). This can be supplemented by further magnifying the primary magnification using stepped increments or incorporating a zoom relay. The magnification changer sometimes incorporates a Bertrand lens which allows the lamp filament in the objective back focal plane to be instantly imaged. Magnification changers have the advantage of being able to fill the fields of view with just the desired area, but they tend to allow the operator to work with fields of 'empty' magnification. The writer has been called to task about the poor quality of a particular objective only to find that the operator when using the magnification changer worked at magnification beyond the capability of the lens numerical aperture.

(3) Eyepiece and eyepiece tube

The eyepiece enlarges the primary image and is available with many different magnifications from $\times 6.3$ to $\times 25$. Different types are available such as positive, negative, high eyepoint, wide angle, non-focusing, focusing, normal field, wide field, ultra wide field, etc. Most require correction for chromatic differences of magnification; one manufacturer corrects for this within the objective (Nikon).

Eyepiece tubes are available as monocular, binocular, trinocular and quadrocular. Some monocular versions incorporate a Bertrand lens. It is important that the optical tube length remains constant; therefore the binocular tubes have to compensate for the variations created in accommodating different user interocular positions. If the binocular head should not have this facility then the observer with narrow interocular separation will use the microscope at a reduced tube length. Those people with wide separation would operate at increased tube length. The result of these two conditions manifests itself as:

(a) woolly images with high power objectives (high NA),
(b) loss of objective par-focality (the ability to go from one objective to the next and remain in focus) and
(c) out of focus photographs if focused through the binocular eyepiece.

Since the action of surface reflections within a prism system can unequally polarize the light passing through the binocular head it is advisable to have a depolarizing element built into the system. If this is not present, different levels of image intensity can be noticed between the two tubes. This phenomenon if not corrected can cause directional glare from such features as pearlite.

The magnification factor is kept where possible to unity, i.e. does not increase the magnification beyond the objective times eyepiece. Where this is not the case it should be engraved on the eyepiece tube body, viz. ×1.2. This case is more likely to occur with finite tube length systems and by implication means there will be a tube length correction lens built into the tube or microscope body. This lens will correct for any objective aberrations that would take place; it can, however, reduce the photographic potential of the microscope. When a trinocular or a quadrocular design is incorporated there will be a facility to direct all or part of the image into specific tubes or directions. It is, for example, an advantage to have all the available light entering the camera to reduce exposure times but for CCTV it would be better if the group observing the television monitor could see what the microscopist sees. One very attractive feature of the illustrated microscope is the ability to pivot the eyepiece tube to locate the most comfortable viewing angle. Also incorporated in the design is an auxiliary lens system to aid low power photomicrography; thankfully many manufacturers offer this near-essential feature. Discussion devices for simultaneous viewing of the microscope image by two or three people along with an externally movable internal arrow can be accommodated on many microscopes. Additional drawing devices for easy tracing of the microscope image can also be fitted (sometimes called camera lucida).

(4) Microscope limb

The strength and shape of the limb limits the 'add-ons' before vibrations become a problem. We have all seen microscopes looking very like Christmas trees. The limb, often a die cast component, is bolted to the microscope base.

(5) The concentric coarse and fine focusing controls

These controls are housed in the limb and are usually graduated in micrometres. They can be used for depth measurements (Z axis). It is not wise to use these controls for depth measurements far exceeding, say, 100 μm without checking that the Z axis is linear throughout its run. Dynamic laser autofocusing or intensity autofocusing systems are available. In reflected light the upper limb will incorporate the lamp fitting and in many cases the reflected light relay lens, aperture and field diaphragms.

(6) Nosepiece and objectives

The nosepiece or objective changer is predominantly of the rotatable type as opposed to a sliding design. The task is to accurately locate different objectives to give par-centricity and par-focality.

They can be from triple to a sextuple and usually incorporate the familiar RMS (Royal Microscopical Society) thread. Because of the demands for wider apertures with dark field objectives some universal objective changers have a much bigger thread size; this restricts the number of objectives available to a maximum of five. The nosepiece can be inclined in such a manner that the objectives face away from the observer when not in use. The advantage of this system is that it allows unobstructed vision to the specimen and objective in use along with a much clearer access for introducing and removing specimens. Its advantages are obvious. It does, however, create more design problems and hence the reason for its infrequent use (it is included in Figure 23.1). When the objectives are facing the observer, obscuring vision to the engraved magnification, it is helpful to have the objective magnification engraved on the nosepiece.

The objectives usually have a 45 mm shoulder length (thread shoulder to object) and can be technique specific and/or chromatic correction specific. They can also be selected to be strain free. Magnifications vary from ×1 to ×100. Many objectives are manufactured with magnifications above ×100 but are specific to techniques and must not be considered better or necessarily as good as objectives with equivalent numerical apertures. Reflected light objectives have a slightly different shoulder length to those used for transmitted light. More importantly, objectives with medium and high numerical apertures (above ×10 magnification) are specially corrected for use with or without a specimen cover glass. The objectives must not be mixed, which is a particular problem for users wishing to work in transmitted and reflected light modes. There is a full explanation of this in Chapter 25.

(7) Microscope stage and substage

The microscope stage is used to position the specimen in the X and Y directions. Many different versions are available, viz.

(a) plain stage,
(b) attachable mechanical stage for fitting to a plain or rotating stage,
(c) mechanical stage with NS and EW graduated movements up to 6 in × 6 in for wafer inspection,
(d) automatic scanning stages with minimum scan points as low as 0.5 μm,
(e) micrometer drum stage,
(f) levelling stage, which avoids the necessity of levelling specimens for reflecting light,
(g) tilting stage, used to vary the fringes in surface interference,
(h) circular stage, essential to the mineralogist,
(i) warm, hot and furnace stage for specimen investigation at elevated temperatures and recrystallization investigations—special long working distance objectives are necessary as the temperature rises,
(j) 3 to 5 axis stage used in crystallography,
(k) counting stage, which moves at pre-selected intervals,
(l) remote control stages used in hot cells,
(m) gliding stage, etc.

The substage (often focusing) is the name given to the bracket housed below the stage. The substage incorporates the substage condenser, of which there are many types. The substage also includes the aperture iris for transmitted light and also the polarizer when required.

(8) Photographic exposure

The more sophisticated instruments include some form of automatic exposure measuring device which electronically activates the electromagnetic shutters when required. Some microscopes have this device built in; others as illustrated are free standing. Exposure devices can be from the simple built-in integrated style of the SLR camera to the automatic spot metering favoured by microscopists operating in extreme variations of subject intensity.

(9) Microscope base

This is sometimes called the illumination base since it can house the illumination system including the transformer. The field diaphragm is located in the microscope base along with the illumination relay lens system. The microscope base when supplied for use in reflected light only is often an empty shell and tends to lose some stability, in particular when taking photomicrographs. Under such circumstances some suppliers fit supplementary bases to improve rigidity.

(10) Reflected light illumination

This is sometimes called the epi-illuminator; the style shown in Figure 23.1 is a sophisticated design with all but the lamp and filter station built into the microscope limb. It also has an additional illumination relay system for superimposing graticules in the image and camera planes. Many microscopes have the illuminator as an 'add-on' to the microscope limb. There are various types available from simple low voltage critical or poor man's illumination to the fully corrected Köhler. Bulbs can be tungsten filament, quartz halogen, mercury vapour or xenon. The field diaphragm and aperture diaphragm will be built into all epi-illuminators, some models having the facility to 'trip-out' the semi-reflecting mirror for clearer transmitted light observations.

The microscope illustrated incorporates an infinite tube length optical system and therefore an extension to the tube length by means of an epi-illuminator within the parallel optics is possible. Those manufacturers with a finite tube length optical system have to compensate for this increase in length. This will be discussed further in Chapter 28.

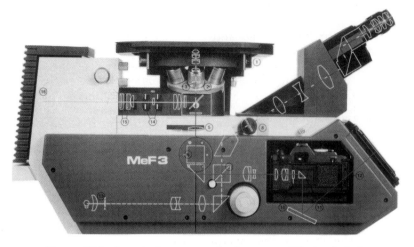

Figure 23.2 Inverted compound microscope (Photo: Leica)

23.2 Inverted Compound Microscope

The inverted metallograph has over the years been the metallurgist's work-horse—starting in the 1930s with the British Vickers projection microscope (VPM), unfortunately no longer manufactured. The metallograph shown in Figure 23.2 is a very much improved instrument compared with the old VPM. Today's microscopes have a large flat field image, modular accessories and ergonomic design. The author recalls the time when ergonomics in microscope design (1960) was so newsworthy as to warrant a presentation in a British design journal. Just prior to publication an artist's impression was solicited depicting how not to ergonomically design a microscope showing the microscopist exerting contortions; unfortunately the resemblance to the VPM (M55) was too close. The drawings in Figure 23.2 give an excellent appreciation of the different accessories and optical path ramifications. Note how the illuminated image is supplemented by projected luminous lines in both the eyepiece and the camera. Follow the light path from the lamp (16) through the relay lens past the aperture and field iris, reflected through the objective to the specimen, then down again through the reflector, this time on its way to a fully reflecting prism. This prism (7) diverts the image into the eyepiece and to the appropriate camera. All metallographs of this type use

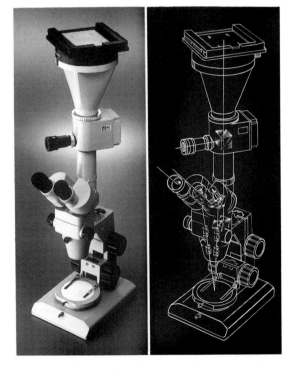

Figure 23.3 Stereoscopic microscope (Photo: Nikon)

infinite tube length optics, in part because of the devious optical route but also to allow prisms or accessories to be introduced without the need for constant optical tube length corrections.

Inverted microscopes are also made for transmitted light applications where the illumination comes from above the microscope stage. These instruments are usually more suited to life science (cell culture) applications.

23.3 Stereoscopic Microscope

Figure 23.3 shows the optical train within the zoom microscope. The stereoscopic microscope can be considered as two microscopes built into one. Following the optical path from the stage plate where the specimen is positioned, upwards, the three-lens (with six elements) twin objectives create a primary image. The next cluster of lenses is the zoom mechanism. The right-hand optical path shows a prism positioned in the optical train; this deflects the light through into the camera and bypasses the right-hand ocular. This prism when tripped out allows normal binocular vision.

The front lenses of the objectives can be designed to allow one large lens element covering both optic axes instead of two separate smaller lenses. This is why the observer could be confused when a visual exterior inspection is made. The single large format lens is extremely popular, even with the most expensive instrument. It is an optical compromise but will not be found lacking unless taken to the limit of stereoscopic vision, i.e. ×10 objectives.

The photographic equipment shown in Figure 23.3 is of the large format design; a 35 mm camera would also fit this unit. Notice how the focusing on this occasion is by a special camera focusing eyepiece; the prism which allows the observer to focus on the specimen is tripped out of the optical path during camera exposure. When observing the optical train from the camera down through the zoom system and objectives it becomes apparent that a flat field sample would not be normal to the incident ray. This is why macrographs taken on stereoscopic microscopes can show an out-of-focus effect in the east/west direction.

Photography on a stereoscopic microscope that takes an image from one optical axis only must be considered as an optical compromise. Critical macrographs must include both axes or alternatively a macro microscope should be used. Since any macro photograph by definition implies low magnification it also means a large depth of field, and the human eye has difficulty coping with this. Whenever possible reduce the visual depth of focus by the use of eye lens magnifiers or optical accessories supplied by the instrument manufacturer.

To give true stereoscopic effects the view from each eye must have an exact conjugate focus at the specimen. This requires careful diopter adjustments which are different for every user. Extreme care must be exercised with zoom stereoscopic microscopes to retain not just the three-dimensional vision but also the focus throughout the zoom range. Very briefly one would set both diopter positions marked on the binocular tubes to the zero position; with the zoom at its maximum setting, focus the subject and then zoom down to the lowest magnification. If the image remains in focus that is your best setting; if not the eyetubes must be rotated individually, without recourse to the focusing mechanism, to bring the subject back into focus. It should now remain in focus throughout the zoom range and the microscope is now calibrated for that individual observer only. The student having carried out this calibration could find when a subject is focused at a low zoom setting that further adjustment to the instrument focusing could be necessary when zooming up in magnification. This is because it is difficult with the lower magnifications to focus in the mid focus position.

The included angle for the two optic axes remains constant from one manufacturer to another and is based on the average interocular distance when focused at 10 inches, i.e. normal reading distance.

This accounts for the difficulty microscopists have with narrow interocular separations when fusing a stereoscopic image (seeing the image in true three dimensions). Microscopes that have true separate axes that conform to this angle are said to be Greenough systems.

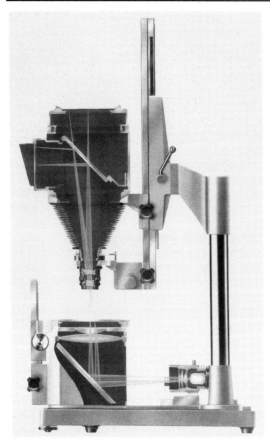

Figure 23.4 Macro microscope (Photo: Nikon)

Many designs use a single large front lens as previously mentioned. These instruments could have a central optical train where the optics run parallel; this is a design function facilitating easier zoom, camera ports or extra accessory accommodation such as coaxial illumination.

Stereoscopic microscopes come in many different styles from fixed magnification, dual magnification and multistepped magnification. They can, as illustrated, incorporate a zoom mechanism and also accommodate cameras within the construction. Magnifications vary from, say, ×5 to ×100 and different eyepieces are also available from below ×10 to as high as ×20.

Illumination is usually an add-on of the high intensity oblique type, single or dual fibre optic or ring illuminators. Built-in illuminators are commonplace with the transmitted light base, which would be connected to the under side of the flat microscope base. If true incident down the optics illumination is required the coaxial illuminator would fit below the binocular head. The coaxial illuminator is often used to exploit anisotropy (see Chapter 34) in the specimen; this entails encompassing a polarizer in both ray paths and in consequence requires a quarter lambda plate fitted below the objective to convert plane polarized light into circular polarized and vice versa. When observing an isotropic subject it will be necessary to rotate the quarter lambda plate into a position of maximum light transmission; this is usually 45° away from extinction.

23.4 Macro Microscope

The name given to types of photography is based on the reproduction ratio of image size divided by object size. A standard focal length lens when focused at its shortest distance gives a ratio of 1 divided by 10. 'Close up' photography is the name describing reproductions of, say, 1 divided by 10 to 1 to 1, photomacrography 1 divided by 2 to 25 divided by 1 and photomicrography 10 divided by 1 to 1500 divided by 1. The difference between photo*micro*graphy and photo*macro*graphy at the same magnification is the size of the field of view.

As the word implies, 'macro' is for large field, low power observation. It uses a camera-type lens as the objective (see Figure 23.4). The primary image is focused onto a screen for observation and then tripped into the camera position for 35 mm or large format photography. Some macro systems are simply 35 mm cameras fitted to a large stand. Macro observations and macro photography require a comprehensive understanding of the effects of different illuminations. For example, a large flat macro welded subject can best be photographed using large floodlights yet the crack or defect within that subject would require a pencil light to highlight the defect by shadows.

The quality of the photomicrograph is rarely if ever related to the quality of the camera. With macro photography expensive exposure devices would have to be fitted to the camera plane (due to the use of bellows). As this is impracticable much simpler systems are employed.

Macrophotography is different to microphotography in that the object to lens distance will vary with every bellows extension that takes place. In microphotography the object to image distance remains fixed. It is important to remember that the depth of field changes with object distance changes. Depth of field is also influenced by a given lens and a particular 'f' setting. To increase the depth of field:

(a) increase the distance between the lens and object,

(b) reduce the lens aperture of, say, f4.5 to f32.

In general the advice for maximum image quality is 'use the longest focal length accompanied by the *shortest bellows distance*'. The first thing one tends to do when setting up a macro system is to wrongly open out the bellows.

The quality of the macro photograph is also dependent on the correct use of the aperture diaphragm, *but* above all it relies on the intelligent use of subject illumination.

24

Creating the Microscope Image

24.1 Lens Functions—Objectives

The chief function of a microscope objective is to produce the primary generally magnified image, which is in turn viewed by the eyepiece in a compound microscope. To simply magnify the object and then further magnify this image is scarcely sufficient; there are many factors to be considered such as aberrations, resolutions, depth of focus, depth of field, etc.

The apparent size of an object as seen with the naked eye depends on the angle subtended by the object to the eye. Hence, the nearer an object is brought to the eye the larger it will appear. At half the distance it appears twice as large, at a quarter of the distance, four times and so on. It then follows that to magnify a subject one simply has to get closer to it. The shortcoming to this analogy is that there is a limit of focus for the normal eye. This distance or the limit of distinct visions for the human eye is said to be *ten inches*. So what can the eye see at ten inches and when would we need to assist the human eye to see smaller subjects?

From Figure 24.1 the limit of resolution for the human eye is given as 100 μm. This means that the eye could resolve subjects separated by this amount as two distinct separate objects; at a separation of less than 100 μm the eye would see the two objects as only one. If we

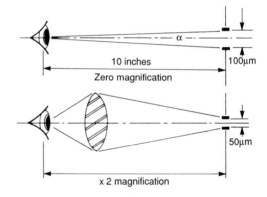

Figure 24.1 Man's limit of resolution

were therefore to look at two subjects separated by 50 μm it would be necessary to magnify the gap to at least 100 μm, i.e. a ×2 lens would be necessary.

We can therefore conclude that magnification and resolution have a relationship, and the ability to see the subject clearly depends on both features which must be taken into account in lens design. Since a nominal value for lens magnification should not be based on man's limit of resolution the figure for comfortable resolution is often based on 200 μm. In other words,

Zero magnification at 10 in = 200 μm

$$100 \, \mu m \text{ at } 10 \text{ in} = \frac{200}{100} = 2\times \text{ lens}$$

and

$$50\,\mu m \text{ at } 10\,in = \frac{200}{50} = 4\times \text{ lens, etc.}$$

In the compound microscope the objective lens magnification is multiplied by the eyepiece magnification. To find the limit of resolution using, say, a ×100 objective and ×10 eyepiece the figure would be

$$\frac{200}{\text{Magnification}} = \frac{200}{1000} = 0.2\,\mu m$$

Resolution is much more complex than this simple analogy but it does illustrate how the object must appear nearer in order to see clearly and resolve the subject; this is achieved as shown by magnification. Therefore if we place a lens within this ten-inch distance, we help the eye to focus an object at a much shorter distance. We are in fact not looking at the object but an image of the object created by the lens.

Figure 24.2 shows that when light is parallel to the central axis of the lens the rays are imaged at one focal point. This gives the focal length of the lens and the principal focus. If we now consider an object placed at a point outside the principal focus of the lens, a real inverted image is formed on the opposite side of the lens. This image can be cast upon a screen or photographed.

Figure 24.3 shows the principal focus (PF) for a lens, illustrating how the image of the

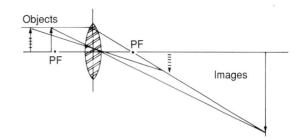

Figure 24.4 Magnifying the real image

object is inverted. This image is said to be in the primary image plane (PIP). If one were to use a lamp filament, simple hand lens and projection screen (ground glass) the hand lens would magnify a real, enlarged, inverted image of the lamp filament onto the screen. If the object filament were now to be moved towards the hand lens, the screen would have to be moved outward to focus the *new* primary image, which would show an increase in size (magnification) (see Figure 24.4).

To continue this experiment by bringing the object still closer, a point occurs when the object is placed within the principal focus of the lens. At this stage it is impossible to focus an image of the object onto the ground glass screen, i.e. no real image is formed, but the eye, if placed in the emerging beam, will see a *virtual, erect* and magnified image of the filament, apparently projected at a distance of distinct vision, i.e. ten inches. Figure 24.5 illustrates this point.

If the eye is placed in the plane of the real image nothing will be seen, but if the eye is withdrawn to a distance from the image of, say, ten inches (the distance of distinct vision) then an inverted image, suspended in air, will be observed. The position of the object relative to the principal focus therefore dictates whether the image is real or virtual; there must therefore be a point where it is 'in between'.

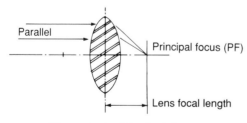

Figure 24.2 Principal focus

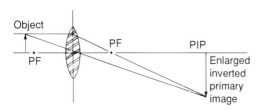

Figure 24.3 Inverted primary real image

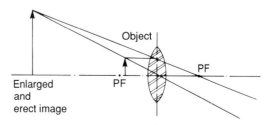

Figure 24.5 Virtual erect image

This condition occurs exactly at the lens principal focus when minor adjustments in both directions instantly change the image from real to virtual and vice versa; this phenomenon is utilized in the compound microscope.

When looking at an object through the eyepiece of a compound microscope we are looking at a virtual image; this is said to be *orthoscopic* observation. When we take a photograph of that object through a microscope we are looking at a *real* image. This is to be developed later.

24.2 The Compound Microscope

Instead of comparing resolution with magnifications as was previously done it is wiser to relate the normal distance of distinct vision against the subtending angle in Figure 24.1. The normal distance of distinct vision is a reference distance against which the subtending angle of the virtual image can be projected and compared with that of the object itself when positioned ten inches from the eye.

If we were to examine an object at, say, one inch from the eye, the angle will be ten times greater but the object out of focus. In order to magnify the object ten times and obtain a sharply focused image, we will have to interpose a lens giving a virtual image of the object which will subtend the same angle as the object subtends when placed one inch from the eye. This will be found to be a lens of one-inch focus.

It follows therefore that the magnification will be given by the expression:

$$\text{Magnification} = \frac{\text{Distance of distinct vision}}{\text{Focal length of lens}} = \frac{Dv}{f}$$

Another way of expressing magnification is to relate the subtending angles, i.e.:

$$\text{Magnification} = \frac{\text{visual angle of image}}{\text{visual angle of object}}$$
$$= \frac{\text{tangent image}}{\text{tangent object}}$$

There is obviously a limit to the degree of magnification one can achieve with a single lens. A lens can, however, be arranged to give a real magnification image of the object. Since the image is real, it can be further magnified by means of a second lens. This secondary image can be real or virtual. This is shown in a simplified form in Figure 24.6, where the first lens is the objective, the second lens the eyepiece. Figure 24.6 shows:

(A) Formation of a real, inverted, primary image in the focal plane of the eyepiece,

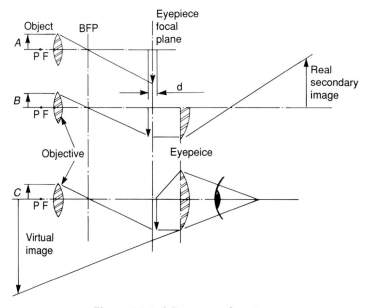

Figure 24.6 Microscope functions

(B) formation of a real, erect, secondary image with the primary image just outside the focal plane of the eyepiece,
(C) formation of a virtual, inverted secondary image with the primary image just within the focal plane of the eyepiece.

It can be seen that with the real image two inversions have taken place resulting in the erect image; the virtual image has only one inversion and the object appears inverted.

Note: If an object is focused orthoscopically (through the eyepiece and objective) the instrument gives a virtual image. Therefore the PIP must lie within the principal focus of the eyepiece. Should a ground glass screen now be placed, say, ten inches away from the eyepiece it will be noticed that a real erect image is obtained (without making any adjustment to the microscope). This confuses most students since it implies that the primary image is both within and outside the principal focus of the eyepiece at the same time. Because the object has depth, the objective lens has a depth of focus. The space image of the primary image also has depth, i.e. both within and outside the eyepiece's principal focus at the same time.

In practice the eyepiece would never be refocused from virtual to a real image; this would be carried out by adjusting the microscope fine focusing and changing the object from *within* the objective principal focus to outside the principal focus.

Figure 24.7 is an exaggerated scale view illustrating principles of the compound microscope. The mechanical tube length is fixed at 160 mm for finite optical tube lengths and greater for infinite optical tube lengths. The infinite system is fixed, however, and not usually variable. The object to the primary image distance is 195 mm for finite optical tube lengths—again it is greater but fixed for infinite systems. The object to the Royal Microscopical Society (RMS) objective thread shoulder distance is 45 mm in air (45.06 mm with 0.17 mm cover glass); this is called the objective par-focalizing distance. The eyepiece par-focalizing distance is 10 mm as shown; the diameter of the eyepiece can be 23.2 mm for normal tube and 30 mm for wide tube.

The general construction of modern microscopes is to keep both the finite and infinite optical systems to a fixed mechanical tube length. Objectives of finite correction will not function correctly if they are used at any but their subscribed distance (excepting those below 0.25 NA). Objectives of infinite tube length have a distance in the optical train where light runs parallel. It is only this distance that can be changed; it must not be varied within converging rays. In practice the tube lengths are not changed. The reason for infinite tube length is not to make it variable but to accommodate accessories within the parallel train without necessitating the use of correction lenses. This is why the tube length

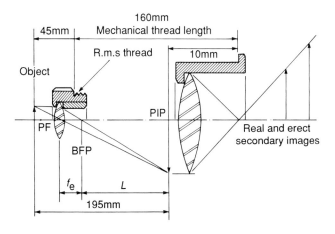

Figure 24.7 Microscope standards

of infinite systems is longer than finite. The optical tube length (L) varies with different focal lengths of objectives.

From Figure 24.6 objective magnification can be shown to be L/f_e. Take, for example, a 2 mm lens and consider the optical tube length to be 160 mm. Then

$$\text{Magnification} = \frac{160}{2} = \times 80$$

If we were now to consider a 25 mm lens the optical tube length would be much less, being approximately 125 mm. Then

$$\text{Magnification} = \frac{125}{25} = \times 5$$

The implication of this last point is that the back focal plane (BFP) of objectives varies, as does the optical tube length. This factor is very important when discussing Köhler illumination in Chapter 28.

25

Objective Aberrations

The optical designer has many considerations to take into account when designing the objective to suit specimen needs. As well as the obvious points such as magnification, field of view and resolution account has to be taken of optical aberrations. If all aberrations are removed the objective would be too expensive to purchase and too sensitive to use. It is therefore not a compromise but a calculated assessment of the specific user needs. The geologist, for example, has a different requirement to the biologist, as is also the case with the metallurgist or the bacteriologist. It is important therefore to know what to look for when selecting optical systems.

25.1 Spherical Aberration

The method of manufacture of lenses dictates that the lens be generated, not formed, and as such can only be produced as true spheres. This in turn leads to spherical aberration; an aspherized curve would eliminate this problem but would result in moulded lenses of poor quality, sometimes acceptable as low quality condensers.

Figure 25.1 shows rays passing through the periphery of the lens focus on the axis at a shorter distance than those passing through the more central or near-axis rays. This effect could be corrected by using a non-spherical lens but since this is impracticable an alternative method is required. The method used is to cement a negative meniscus lens to a double convex element having a different marginal thickness and refractive index. By bringing the marginal and near-axis rays (and all others) to a common focus the lens is said to be *aplanatic*. The consideration below (Figure 25.2) shows how this can be done for one specific wavelength. Since this is a result of refraction and different colours have different degrees of refraction, it would be necessary to correct for at least the three primary colours to be totally corrected for spherical aberrations. This being

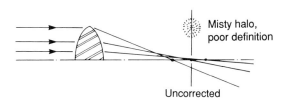

Figure 25.1 Spherical aberration—uncorrected

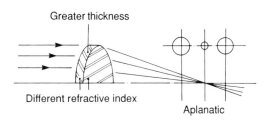

Figure 25.2 Spherical abberration—corrected

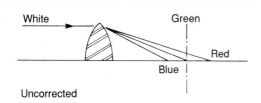

Figure 25.3 Chromatic aberration—uncorrected

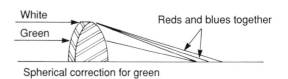

Figure 25.4 Chromatic aberration—partial correction

impracticable only the primary colour green is totally corrected for one focal point. Highly corrected objectives (*apochromatic*) are additionally corrected for blue.

Spherical aberration exhibiting poor definition can sometimes be confused with (a) an oil smear on the front surface of a dry lens or (b) an objective corrected for use in air being used with a cover slip or vice versa.

These points will be developed later.

25.2 Chromatic Aberration (Axial)

The refractive index (the degree to which light is bent when passing through a subject) is not the same for all frequencies of light, the refractive index for the blue end of the spectrum being higher than that of the red. This is known as the dispersion characteristic and varies from one substance to another.

When white light is refracted by the lens as in Figure 25.3 it will be noticed how the light is dispersed, the blue focusing short. This figure also illustrates the position of the primary colours. This effect can be demonstrated by looking down the microscope (orthoscopically) at a subject with pin-holes; they exhibit colour fringes. This can be minimized by using monochromatic light (preferably green). This type of aberration is what one sees when viewing a subject through a prism or better still light dispersed through a rain cloud, i.e. the rainbow.

Fortunately different types of glass are more dispersive at the blue end of the spectrum and others at the red.

25.3 Types of Objectives

25.3.1 Achromatic objective

By combining glasses of varying refractive index and marginal thickness we have shown (Figure 25.2) how spherical aberration can be controlled. Likewise by combining different degrees of dispersion characteristics we can reduce the colour or chromatic aberration (Figure 25.4). Therefore an achromatic lens is one which:

(a) brings the central and marginal rays for the primary spectrum (green) to the same focus and
(b) causes rays of the secondary spectrum to concentrate near this point.

25.3.2 Fluorite objective (semi-apochromatic)

The residual or secondary spectrum can be considerably reduced by the use of fluorite glass in the lens manufacture. Because of this a more faithful colour rendition of the object can be achieved. This has particular advantages when observing small coloured objects or fine structures and gives the subject a more defined boundary when observing, say, fine lines. Although this lens is better corrected for chromatic aberration it is still only corrected spherically for the primary colour green.

Fluorite glass exhibits small strain birefringence and is therefore not an ideal combination within objectives used for critical polarized or fluorescence applications.

25.3.3 Apochromatic objectives

This type of objective (Figure 25.5) has been fully corrected for the three primary colours blue, green and red, practically eliminating any residual chromatic error. Spherical aberration is corrected for two colours (green and blue).

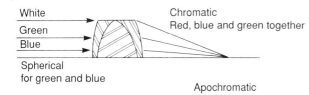

Figure 25.5 Chromatic aberration—fully corrected

Objectives of this kind are used in very exacting applications where no trace of residual colour can be tolerated or when the subject structure is small and ill defined. The bacteriologist needs to identify by means of very subtle colour differences with small subjects. The study of chromosomes or very finely resolved particles, lines or grain boundaries in materialography are all examples of the need for such lenses.

(a)

(b)

Figure 25.6 High carbon steel—bright field: (a) ×40 achromatic objective and (b) ×40 apochromatic objective

Figure 25.6 of high carbon steel is used to illustrate the clarity of line manifest through total axial chromatic correction. The clarity of grain boundaries, e.g. in Figure 25.6(b), far exceeds the strength and clarity of line achieved through partial chromatic correction, shown in Figure 25.6(a). Although this looks obvious in the photomicrographic comparison, this is not the case when observing an orthoscopic image down the microscope. During a recent (1991) microscopy course only two out of twenty delegates said they could observe a difference.

Apochromatic lenses are normally associated with a higher than normal resolving power (numerical aperture). As with the fluorite lens, one must exercise extreme caution using these objectives in polarized or fluorescent light but it does not mean they should never be used.

Because of all the chromatic and spherical corrections built into the apochromatic lens it is very expensive (say three times the price of an achromatic lens), but it does not follow that this is a better lens for every application.

The apochromatic lens has two major disadvantages:

1. Because of improved resolution it has a *reduced* working clearance (distance from the bottom of the lens to the object plane).
2. The increased numerical aperture causes increased glare and has to be corrected via the microscope illumination diaphragm controls. Take, for example, the ferrous metallurgists looking at grain boundaries. The achromatic lens will show a strong black boundary line; the apochromatic could show a more faithful grey line built up from dots. The latter is not so easy to see and could give information beyond the terms of reference of the observer.

25.3.4 Summary

Table 25.1 indicates some of the characteristics of the three different objective lenses as they apply to the field of materials.

25.4 Chromatic Difference of Magnification (Transverse)

A second form of chromatic defect is due to the fact that different colours undergo different

Table 25.1 Objective lens characteristics

Objectives	Applications	Disadvantages	Optical corrections
Achromatic	General purpose; the lack of total corrections can sometimes be an advantage. When subjects are optically stained for contrast it also works well. All the micrographs in this book were confined to this range of objective	Not the best for critical resolution (pearlite)	Partial chromatic. Spherical green only
Fluorite	Very often the best all-round choice for improved orthoscopic applications, but its lack of spherical corrections for blue could be evident in the most critical micrograph	Requires more care in utilizing the good practice of Köhler illumination in order to benefit	Improved chromatic. Spherical green only
Apochromatic	When subject requires improved resolution. When subject lines or grains are not sharp, usually when they are very thin. Improves sharpness in micrographs with small or thin lines	Shorter working distance, possible glare, and requires optimization of illumination and diaphragm stops to achieve benefits. Expensive	Chromatic blue/green/red. Spherical green/blue

degrees of magnification, resulting in unequally magnified images. Blue magnifies greater than red. The effect of this is to superimpose a number of images of different colours and of varying sizes, resulting in a poorly defined resultant image. This can be demonstrated by looking orthoscopically at an object exhibiting small dark spots. If these dark spots or objects are placed near the edge of the field their outer side will appear red, the inner blue. A bright object on a dark slide will show the reverse.

25.4.1 Correction

In the case of low power objectives this defect can be corrected *out* in the optical design stage. With medium to high power objectives from 0.5 NA upwards this is not possible; therefore an equal and opposite error must be built into the eyepiece, which gives the name 'compensating eyepiece'. To be correct one should change eyepieces as objectives are changed from low to high. This being totally impractical has prompted manufacturers to provide correction or compensation at the eyepiece throughout the range where objectives are uncorrected.

One manufacturer has, however, managed to correct for chromatic difference of magnification throughout the range of his objectives, therefore utilizing non-compensating eyepieces. The reader will appreciate the reasons for not switching optics from one manufacturer's microscope to another unless these and many other aspects are investigated. It has been suggested that objectives corrected for chromatic difference of magnification at the objective are 'chromatic free'. This only applies to transverse corrections and not chromatic axial corrections. Since the majority of compound microscopes utilize the compensating eyepiece technique the following argument is often propounded.

The lens designer could be restricted in compensating for field curvature, astigmatism, coma and other aberrations that exert an influence upon the image around the field periphery if transverse (lateral) chromatic corrections are included in the objective lens design. The point is that transverse chromatic corrections can easily be compensated-out in the eyepiece without having an adverse effect on other aberrations. The compensating eyepiece method is, however, more complex when graticules are fitted in the eyepiece.

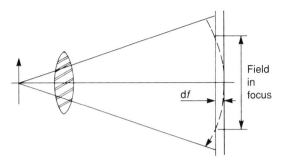

Figure 25.7 Field curvature

25.5 Curvature, Coma, Astigmatism and Distortion

25.5.1 Field curvature

Many objectives produce a curved image of a flat object; more complex lens systems can produce an almost flat image of a flat object. This is more apparent when photographing the magnified object since the eye can be very accommodating when observing an object orthoscopically; however, the camera is not.

Figure 25.7 illustrates field curvature. Because of objective depth of focus the image observed shows an out-of-focus effect at the field extremity only. This condition is more apparent with high numerical aperture objectives where the depth of field is less.

It is possible for the optical designer to produce eyepieces that make this out-of-focus effect less apparent, but this cannot be disguised when taking photomicrographs. Because of this, special flat field objectives are manufactured. When corrected for maximum flatness of field objective they are generally called *plan* objectives. These objectives, although much more expensive, are not necessarily better quality lenses. If field curvature can be accommodated one could have a higher performance objective.

Depth of focus—depth of field

In the explanation on field curvature the words depth of focus/field were used. This has in the past been a source of confusion. Every lens has a position where there is a focal point for the object resulting in a primary image. Both at the *object* and the *image* there is an axial distance where the subject remains in focus, but this distance is often different. One good example of this is in photomicrography, where the axial distance of the object remaining in focus when using low power objectives is large in relation to the axial distance of the *image* of the subject remaining in focus in the camera plane. To overcome any confusion the expression 'depth of field' is related to the *object* space and 'depth of focus' to the *image* space.

25.5.2 Coma

Coma is the aberration which when observed in transmitted light causes a point object, such as a pin-hole in a silver film, to be imaged with a flare, like the tail of a comet. It radiates from the centre of the field—more often a peripheral effect (see Figure 25.8).

This type of aberration is marginally evident in many objectives, even the more expensive apochromatics. Its confinement to the periphery of the lens often causes it to go unnoticed. The pin-hole test is a very searching test for objective aberrations, much more critical than the everyday specimen; students of microscopy will be fascinated with its revelations. Start with a large pin-hole in the axial plane looking for chromatic aberrations and use a small pin-hole to look for damaged

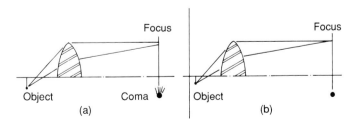

Figure 25.8 Optical defect of coma. (a) sine law—unfulfilled and (b) sine law—fulfilled

lens defects (astigmatic or coma). Move the pin-holes to the periphery and observe the comet tail with small holes and chromatic difference of magnification with slightly larger holes. When there is no observed chromatic difference of magnification change the compensating eyepiece for the non-compensating one or vice versa, and see the introduction of this aberration. It does not follow that the objective with the least pin-hole aberrations is necessarily better than those of another manufacturer with less; i.e. this test, as good as it is, must not be used in isolation in judging what is best for particular specimens. Objectives used for cover glass transmitted light are identical in optical construction to those used for non-cover glass reflected light. High NA objectives have slightly different lens separations but a good transmitted light objective could be poor in reflected light. A good transmitted light bacteriology objective is not necessarily the best cytological objective, etc.

Coma is a defect confined entirely to rays coming from points lying outside the optic axis of the lens; therefore evidence of axial coma would undoubtedly suggest a damaged objective. Coma is a defect of design computation rather than a manufacturing fault. Apochromatic lenses do appear to suffer from this aberration more than achromats. Coma is a non-fulfilment of the sine law and is illustrated in Figure 25.9.

25.5.3 Astigmatism

This occurs when a point is imaged as a line. There will be another image at a different focus orientated at 90°. When light is reflected (or refracted) obliquely from a concave surface it shows the cone of light transformed into two transverse focal lines. This defect when evident in an objective is indicative of an

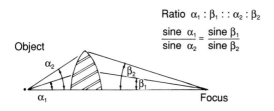

Figure 25.9 Sine law fulfilled

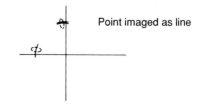

Figure 25.10 Optical defect of astigmatism

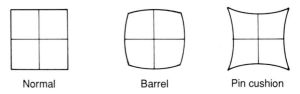

Figure 25.11 Optical defect of distortion

off-axis component lens, i.e. a damaged objective (see Figure 25.10).

25.5.4 Distortion

This is a defect found in the majority of cases with inexpensive stereoscopic microscopes (see Figure 25.11). A good test for this is to use graph paper as the object and observe the degree of barrel or pincushion that takes place. It is important to know to what extent distortion is present if the microscope is used with an eyepiece graticule for measurement purposes.

25.6 Contrast

One of the major advances over the last 20 years in objective design has been in lens surface coatings. The 'blooming' of today is very sophisticated and often involves multi-layer coatings, the object being to achieve maximum transmission and thus reduce reflected glare-causing rays. All smooth surfaces have a reflecting characteristic. If, however, a coating of one-quarter wavelength should be used then the optical path difference between the top and bottom surface reflection will be a half lambda, i.e. destructive interference, allowing total transmission of the rays. In practice it is not possible to eliminate all wavelengths. The original blooming was to allow green transmission, reflected green being extinguished, and hence the reason for the purple colour of bloomed films. This is yet another reason for using a green filter in black and white photomicrography or measurement applications. Multilayer coatings are for other wavelengths of light. Some eyepieces can be found with a green cast; this is to reduce the effect of external light scattering. Object contrast is often lost because the peripheral rays where most reflection or light scattering occurs are obscuring the general image. This is one of the reasons for stopping down the aperture diaphragm. Contrast therefore is one of the most important characteristics of modern objective design. There is little point in resolving the subject and ensuring chromatic correction if the contrast in the image is poor. Defining contrast is not too easy since it tends to be a physiological phenomenon, but it is easy when making comparisons. The eye can readily detect the general sharpness of subject detail as a result of improved coating technology which is confirmed on a micrograph. A good, well coated achromatic gives better results than an apochromatic with a lower quality coating. Specimen contrast is also influenced by chemical and mechanical features; the remarks above are confined to optical considerations.

26

Improving the Image

26.1 Numerical Aperture (NA)

This is one of the most important factors in any optical system. The numerical aperture sets a limit to the size of the subject that can be viewed or resolved by a particular lens. This figure cannot be increased—only reduced by incorrect use of the microscope.

The ability of an objective to separate fine detail does not depend on magnification, but on the resolving power (NA) of the lens. Resolution is dependent on two factors:

(i) wavelength of the illuminated source and
(ii) numerical aperture of the lens

The second factor (NA) is itself dependent on two factors:

(i) the maximum cone of light the lens can collect and
(ii) the refractive index of the medium from which the light enters the lens

The numerical aperture of an objective is defined as NA = $n\sin\alpha$, where n is the refractive index and α is the axial angle made by the most oblique ray (see Figure 26.1).

In practice this will *not* be a total lens aperture but one taking into account any constraints made in mounting the lens. This number is engraved on every objective outer barrel

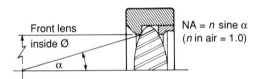

Figure 26.1 Numerical aperture

and is a theoretical value based on the inside diameter of the front or other lens cell elements.

Variations from the engraved value will be greater with the higher NA objectives. An objective displaying ×40/0.65 would mean it had a primary magnification of power 40 with a numerical aperture of 0.65.

26.2 Refractive Index

Since some lenses are intended to be used not just in air but immersed in oil, water or glycerol, thus bending the light, this must be taken into account.

With reference Figure 26.2, consider a beam of light along axis AXA normal to the surface of the liquid. The beam passes (though retarded) without deviation through the liquid. When the beam is incident at any other position, say BX, the beam will be refracted to XC. The ratio of BB1 to CC1 gives a constant regardless of where the

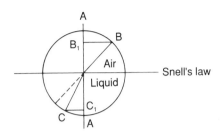

Figure 26.2 Refractive index

26.3 Resolution and Magnification

The ability to see the object as it really is has been defined as the ability of the lens to resolve all the object details. This could be restated as the ability of the lens to separate two small points close together and not to recognize them as one. Each objective has its own limit of resolution. Therefore, one objective just recognizing two close points as two would, if the points could be positioned closer, recognize only one sausage-shaped subject; i.e. the subject would not be resolved.

Resolution is a function of:

(i) illuminating wavelength of light,
(ii) the cone of light entering the front objective lens (numerical aperture),
(iii) the refractive index of the medium between the object and the objective front lens.

26.3.1 Empty magnification

Take two points; gradually reduce the distance between them until they can no longer be defined as separate points; now magnify the image. This, it will be found, has no effect on resolving the single sausage shape into two points. Thus we can say there is a limit whereby extra magnification would be 'empty magnification'. If one was now to change to a higher numerical aperture objective then the two points would once again be resolved.

26.3.2 Useful magnification

It follows that if a point is reached where the object is just resolved there must be a useful point where further magnification proves empty. As a general rule this is said to be

incident beam commences. This constant is called the refractive index.

Taking air in a vacuum as unity (since air in air will not refract), pure water was found to be 1.33, glass say 1.5, microscope immersion oil 1.518, etc. Since the sine of $90° = 1.00$ it follows that dry objectives will have a practical limiting NA of less than one—say 0.95. This explains why oil immersion objectives used with oil can have an NA in excess of one. Objectives for use in oil (or other liquids) have this information engraved on the outer barrel. For example 100/1.32 would signify a primary magnification of ×100 with a numerical aperture of 1.32; also engraved on the objective body should be 'oil'. Although the refractive index has been standardized at 1.518 some lenses use other values and the designed values must be used if a woolly image is to be avoided. Lenses are supplied with corrections for glycerol or water; the appropriate medium must always be used.

Beware: do not oil a ×100 objective with an NA of, say, 0.95 unless it clearly states 'oil' on the objective barrel. The author would like to confirm the story about the microscopist who thought oiling the objective meant filling the objective with oil—this anonymous person did get a free replacement.

Figure 26.3 shows how light can be both refracted and reflected. In this case light when incident to the 1.52 refractive index glass at 56° to the vertical causes relected light to vibrate in one direction only. Knowledge of this so-called Brewster angle is put to good use by the fisherman using polaroid glasses to extinguish the reflected glare from the water. It can also be put to good use in the microscope in extinguishing specific unwanted rays, a subject to be developed later.

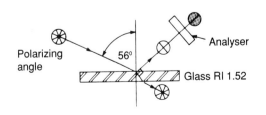

Figure 26.3 Polarized reflections

×1000 NA for comfortable vision. There are occasions when this rule is broken for good reasons. Take the case of measuring a very small subject; it can be further magnified to make measurement easier even though no extra detail is resolved. If we look at useful magnification as the minimum magnification required then an interesting point arises.

Take the case of the human eye. This can just resolve two points (at a distance of distinct vision 10 inches) whose distance apart is 100 μm. From this it follows that the smallest object visible down the microscope when magnified must be 100 μm. The smallest interval that can be resolved with the light microscope is 0.2 μm. To magnify this dimension to 100 μm would mean multiplying by 500. The theoretical minimum magnification for distinct vision is therefore ×500 (so much for the ×3000 magnification *superior* microscopes can achieve). As interesting (or provocative) as this last statement is, we must not lose sight of the fact under consideration was the resolution of *space* between objects. Submicrometre objects below the limit of resolution are visible down the microscope requiring magnification greater than ×1000 for comfortable vision.

To use the factor 1000 times the objective NA is good advice for othoscopic observations. When reproducing the orthoscopic image, as in micrography, as high as 2000 times the NA is necessary to overcome the loss of reproductive quality yet still retain comfortable vision.

26.3.3 Diffraction disc

Consider pin-hole points in a silvered plate used as the specimen in transmitted light. Unfortunately a true point in the object plane cannot be reproduced as a true point in the image plane. This is due to the wave nature of light which reproduces the image as a bright dot surrounded by diffraction rings. If the interval between the two points in the object plane becomes too small, the two smallest diffraction rings merge and the points are no longer resolved as two but one (further magnification will have no effect).

The intensity curve is shown in Figure 26.4. It was found (empirically) when two close points were just resolved that the intensity curve showed a reduction between the two peaks as Figure 26.4(c). At this point the interval between the two peaks was found to be a half lambda of light. It follows that the shorter the wavelength of light, the smaller the interval resolved.

26.3.4 Angle of light

It has been shown that wavelength affects the separation of the smallest diffraction ring which in turn affects resolution. This is also the case with the angle of illumination: the wider this angle becomes, the smaller the first-order diffraction ring. The practical conclusion from all this is that we can improve resolution by reducing the wavelength, using the maximum numerical aperture, and can also ensure (in transmitted light) that the illuminating aperture matches the objective aperture. The resultant numerical aperture of an optical system is the objective working NA (not necessarily that engraved on the body) times the illuminating NA divided by two. This applies only if the illuminating NA is less than the objective NA.

26.3.5 Limit of resolution

Substituting from Figure 26.4,

$$\text{The limit of resolution } R = \frac{½ \text{ lambda}}{\text{NA}}$$

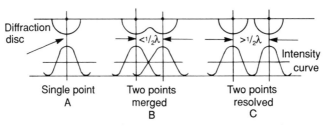

Figure 26.4 Resolution by diffraction

In practice this has been found to be

$$R = \frac{0.61 \text{ lambda}}{\text{NA}}$$

From this we can find the actual limit of resolution for the light microscope, taking blue light as 490 nm and NA as 1.4:

$$R = \frac{0.61 \times 490}{1.4} = 200 = 0.2 \, \mu m$$

Rather then remember the formula, it is better to remember the rule:

To improve resolution, reduce lambda and/or increase NA.

26.3.6 Reduce wavelength

This is not always possible, since although to use the short wavelength (blue–ultraviolet) part of the emitted spectrum would in theory improve resolution it would also adversely affect the image produced by the objective. Only apochromatic lenses are corrected in the blue region for spherical aberration.

Having over the years built up a collection of colour photomicrographs it was decided to have them displayed as large size (20 in × 16 in) framed pictures. It was then noticed that there was an absence of red colours, the tendency being towards blue. From a balanced colour scheme some reddy-type micrographs were produced. The conclusion to this was that, when given an option of infinitely variable colours (as in the case with optical staining), reds are for aesthetics and blues for contrasting resolution. It is also wise to note that, when using optical staining techniques, the type of illuminating source no longer influences the image resolution.

26.3.7 Increase numerical aperture

This seems the most likely approach. To achieve maximum numerical aperture the user must go to an oil immersion objective, with the oil having a high refractive index (objectives are corrected for specific refractive index oils and must not be used beyond their correct value). As has already been mentioned an increase in NA associated with an increase in glare is to be avoided, oil immersion objectives often reducing glare; i.e. a dry objective NA of 0.65 could with certain specimens exhibit a lack of clarity not apparent with an oil immersion objective of identical NA.

26.3.8 Resolution

It will now be evident that the true criterion for the performance of an objective is the resolution and not the magnification. There are limitations in practice to just increasing the numerical aperture for every objective magnification. Not least of these is the reduction in working clearance (distance from the object to the front lens) and the reduction of the objective depth of field (inversely related to numerical aperture). Table 26.1 will help to illustrate this point.

26.3.9 Working clearance

This figure is often quoted in manufacturers' publications and relates to the optical working clearance which is often a greater value than the physical working distance (since the lens is fitted into its mounting). The working distance of low power objectives can be

Table 26.1

Objective magnification (160 mm TL)	df[a] (μm)	NA	Resolution (μm) (green light)	Working clearance (mm)	Useful magnification
×5	55	0.10	2.75	20	×100
×10	8.7	0.25	1.18	8	×250
×20	2.0	0.50	0.55	2	×500
×40	1.4	0.65	0.42	0.5	×650
×60	0.58	0.85	0.33	0.3	×850
×100 dry	0.49	0.90	0.31	0.2	×900
×100 oil	0.47	1.25	0.22	0.18	×1250

[a] df = depth of field

affected by the state of accommodation of the observer's eyes, especially in young people. This results in an apparent lack of par-focality (the ability to focus all objectives at the same plane) where the observer has to refocus when changing from low power to high power. This situation is exaggerated if the tube length should vary with interocular separation in the binocular head.

26.4 Depth of Field (df)

This is the depth remaining in focus without recourse to the microscope focusing mechanism. Depth of field is inversely proportional to the objective NA; i.e. the greater the df the lower the NA and vice versa.

From Figure 26.5 assume an object focused at B and then moved to B1. The central and marginal rays will have changed; if this distance does not exceed a quarter lambda the subject will remain in focus. In the statement 'lower the NA and increase the focus space' we can see how the angle of light (cone) affects this dimension. Notice that the angle is different between the object and the image of the object. This results in two very different dimensions; in the example given the depth of field (object) is less than the depth of focus (image).

Figure 26.6 illustrates how more sensitive the central to marginal ray change is on the high power objective in relation to the lower power (illustrated with a fixed df). With the low power objective the arbitrary value of 0.03, as opposed to 0.10 for the high power objective, shows the reason why activating the microscope fine focusing mechanism has little effect on the specimen focus. The same movement with a high power objective could result in an out-of-focus image. Also note how the angle (cone) of light at the image plane varies between the two examples, the high power having a much greater 'depth of focus' in relation to its 'depth of field'. This accounts for the reason why with high power photomicrography it is so easy to achieve sharp well-focused micrographs.

Although the depth of focus with low power objectives is small it is further exacerbated because the depth of focus of the human eye becomes larger as the objective NA reduces, making focusing difficult. The depth of focus

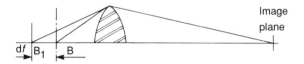

Figure 26.5 Object depth of field

is therefore a combination between the optical diffracted angle and the limit of the human eye taken as 2 minutes of arc. With low optical diffracted angles the eye has a greater depth of focus than the depth of field, which explains why supplementary focusing lenses are necessary in low power photomicrography.

Depth of focus (df) is inversely related to the NA and is affected by the wavelength of light. Since the NA is related to the optical path medium (air, water, oil, glycoral, etc.) this must also have an effect on the df. If a numerical aperture of 0.9 is taken then

$$df^* = \begin{cases} 0.27\,\mu m \text{ (refractive index 1.0) in air} \\ 0.61\,\mu m \text{ (refractive index 1.333) in water} \\ 0.93\,\mu m \text{ (refractive index 1.515) in oil} \end{cases}$$

Therefore to increase df,

(i) reduce the aperture of lens,
(ii) increase the wavelength of light, and
(iii) immerse the objective.

Remember: what affects the object df has a different effect on the image df; the two factors must be combined when considering photomicrography.

26.5 Tube Length Correction

Manufacturers design objectives to function at specified tube lengths; variations to the desired tube length must be avoided where possible. When supplementary accessories such as reflected light illuminators or analyser units are fitted to a microscope then the tube length will be increased. This will result in extremely fuzzy images, loss of par-focality and many other aberrations unless the accessories are fitted within parallel rays.

With finite tube length optics this can be corrected by the use of correction lenses within each accessory (not to be recommended with

---

*See B. O. Payne, *Microscope Design and Construction*.

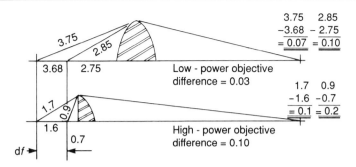

Figure 26.6 Depth of field comparison

photomicrography) or to have the objectives specially corrected for this new tube length (many manufacturers do this). Alternatively the optics could be corrected for infinite tube length, in which case these problems would not arise. Some objectives are insensitive to cover glass and tube length correction. These objectives are from 0.25 numerical aperture downwards. It follows therefore that objectives of ×10 magnification and below can be used for 160 mm in transmitted light or 210 mm in reflected light without showing any loss of resolution; however it still affects par-focality. Similarly these objectives can be used with metal specimens without a cover glass or biological specimens protected with a cover glass, and also smears on glass (haematology/cytology) without loss of image quality.

When mixing these objectives care must be taken to ensure objectives above 0.25 NA are corrected specifically for their particular applications.

This is further complicated in that objective shoulder distances, the distance from the RMS thread abutment face to the object plane, are not constant. Since some objectives are operating in air (reflected light with a metal sample) and others through a 0.17 mm thickness of glass and air it effectively changes this overall distance by 0.06 mm.

The objective par-focalizing distance is

(a) 45 mm in air
(b) 45.06 mm in air and (0.17 mm) cover glass.

The tube length is also affected by the increase or decrease that takes place using non-correcting binocular heads at differing interocular separations.

To summarize, we have medium and high power objectives whose performance will be affected by a change in cover glass or a tube length correction and on the other hand low power objectives insensitive to such changes in terms of performance but which require a very different focus with a change in tube length.

In general, take advice before mixing objectives.

26.6 Intensity of Image

This is proportional to the square of the objective numerical aperture and is also inversely proportional to the square of magnification.

Situations often occur where it is desirable to increase the intensity of light; for example, with:

(a) photomicrography to reduce time,
(b) dark ground where all plane surfaces reflect the light outside the objective field of view,
(c) imaging onto a projection screen,
(d) high power oil immersion objectives,
(e) polarized light (between crossed polars), etc.

A knowledge of how to improve intensity can often resolve the problem.

Brightness is proportional to $NA^2/magnification^2$ (transmitted light):

(a) increase NA, increase intensity;
(b) increase magnification, decrease intensity;
(c) double magnification, double NA—the result is unchanged.

Intensity is inversely proportional to the

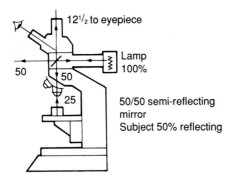

Figure 26.7 Reflected light microscope

square of the distance, i.e. decrease distance, increase intensity.

With reflected light the solution is compounded by the introduction of a semi-reflecting mirror built into every incident light illuminator. From Figure 26.7 it can be seen that a 50% reflecting mirror with a 50% reflecting specimen only utilizes 12½% of the original 100% light from the lamp. This figure does not take into account light inevitably lost through internal scatter, glass to air components, other reflecting surfaces, such as prisms, etc.

In the past it was not possible to have high efficiency binocular heads, totally reflecting surfaces, high transmission coatings on lenses, etc.; therefore all metallographs were required to be fitted with high intensity lamps such as xenon, mercury vapour or carbon arc. This loss of light in the materials microscope has to a certain extent been improved with better light transmission and shorter optical paths. However, it still presents many problems with reciprocity failure, in particular with slow films.

27

Measurements

27.1 Measuring Systems

The microscope is called upon to make a variety of different types of measurements. They can be angular as in geology or transverse (X–Y) or axial (Z) as in metrology. The measurement can be via the object stage, eyepiece or projection as in photography, CCTV or a projection screen. The chosen method in carrying out such measurements dictates the sensitivity or accuracy of the results. All measurements are ultimately traceable to the calibration standard used in graduating the measuring divisions. What is called a 'stage micrometer' is used to calibrate measurements within the eyepiece field of view (FOV); measurements greater than the FOV are calibrated to slip gauges and other metrology standards.

27.1.1 Measurements beyond the eyepiece FOV

Vernier measuring microscopes, sometimes called travelling microscopes, are manufactured specially for this purpose. The extra sensitivity comes from increased magnification facilitating a more accurate object position location and reading facility.

Other measuring microscopes employ a micrometer drum stage used for very accurate X–Y coordinates. This stage could have rotational facilities and could be used to check azimuth angles. The micrometer drums can read to 0.005 mm and models are available with digital read-out.

27.1.2 Measurements within the eyepiece FOV

All measurements within the eyepiece FOV become critical to the sensitivity of the technique and are also more operator subjective. As with the measurement of micro-indents so the need for operator calibration is also necessary in this application (see Chapter 19 for method).

27.1.3 Screen projection

If the image can be projected onto a screen then measurement within the linear part of that screen can be readily accomplished by making your own ruler or scale. For example, project a known graduated scale onto the screen and trace the graduation onto your own ruler; this ruler can be used to measure other subjects focused on the screen *at the same magnification*. (A different ruler is required for each magnification.)

27.1.4 Drawing attachment (camera lucida)

These devices permit the image of the drawing paper to be superimposed in the eyepiece FOV. A graduated scale could also be used for subject measurements.

27.2 Measurement X–Y

27.2.1 Eyepiece micrometer (graticule)

The word micrometer is sometimes confusing to the engineer who looks upon a micrometer as graduations on a threaded barrel. In this context it just means an eyepiece fitted with a graticule. (The word graticule is often substituted by the word reticle or reticule, which means a small net or grid pattern. This expression therefore should not be used to describe a graduated line.) The importance of the graticule is the accuracy of the divided lines and should be considered as accurate arbitrary divisions. In use these divisions require calibration by comparison with a preferably traceable stage micrometer; these are usually graduated in hundreds and tenths of one millimetre. The graduations in the eyepiece have to be independently calibrated for each objective and can vary between microscopes of similar manufacture. The magnification engraved on objective bodies is a nominal value and must not, for the case of calibration, be taken as a true value. The mechanical tube length, binocular separation, aperture diaphragm setting and eyepiece tube focus position can all result in magnification variations. Therefore careful notes should be made where applicable and each objective must be independently calibrated.

Calibration is carried out by superimposing the equally spaced graduations in the eyepiece with those of the stage micrometer. Although it is wise to consider the eyepiece graduations as arbitrary divisions they are nevertheless the same units as the stage micrometer; therefore with a one times objective the lines should be accurately superimposed. From this we would expect a ×10 objective to display ten divisions in the eyepiece for one division of the stage micrometer. Therefore,

$$\text{One division in the} \times 10 \text{ eyepiece} = \frac{0.01}{10}$$

$$= 0.001 \text{ mm}$$

Unfortunately in practice these divisions are not always 10, but are sometimes, say, 9.81 or 10.23, etc. Therefore,

One division in eyepiece would be

$$\frac{0.01}{9.81 \text{ or } 10.23}$$

It must be remembered that the eyepiece magnification has no effect on these graduations since the graticule is fitted into the primary image plane. The accuracy of measurement is improved if the eyepiece is a focusing type; it is however operator subjective when estimating between graduations.

27.2.2 Filar micrometer eyepiece

These measurements are made by a moving line or curtain, activated by a drum micrometer, and are calibrated in the manner given above with a traceable stage micrometer. Accuracy can be improved by calibrating to the largest dimension possible across the field of view (with flat field objectives). This method, although less operator subjective, still depends upon judging the moving line position. Greater accuracy has been proved when using a double line, with the subject not on or outside but within the two lines.

With the ×10 objective this system can read to 0.001 mm (10 μm), which is much more sensitive than the simple graticule.

27.2.3 Microscaler system

Involving the generation of moving lines on a television monitor is easier to use and interpret by virtue of increased size. These

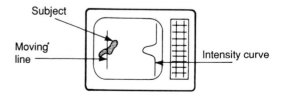

Figure 27.1 Measurement using intensity curves

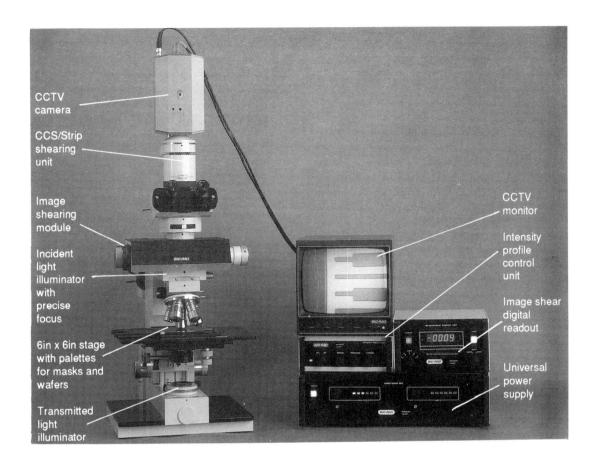

Figure 27.2 Vickers image shearing system (Photo: Vickers)

systems are very often supplemented by intensity curves which go a long way towards taking the subjective element out of positioning the lines (see Figure 27.1).

27.2.4 Image splitting or shearing

This system splits the image by means of a micrometer drum and can give the two images different colours. When observing the subject orthoscopically, readings are taken when the two images contact and when they are identically superimposed. It is a very convenient method of measuring random objects anywhere in the FOV. Some systems shear only a small band across the FOV which is excellent for shearing lines found on, for example, printed circuit boards.

27.2.5 Digital read-out

Any of the above systems utilizing a micrometer drum can be offered as digital read-out systems. The linear change from the micrometer drum, via some kind of potentiometer, feeds a zero setting digital voltmeter. These systems can be calibrated to give exact readings and also to have memory settings for different objectives.

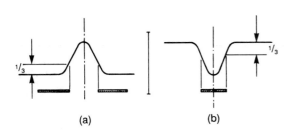

Figure 27.3 Faithful optical measurements. (a) Intensity curve hole and (b) intensity curve spot

27.3 Depth Measurements

Fine focusing controls on microscopes are usually quite accurate over small runs and can therefore be used for depth measurements. Remember to always use the highest NA objective since this reduces error from the objective depth of field.

Accurate depth measurements can be introduced by having reference targets to eliminate the objective depth of focus errors.

27.3.1 Precise focus

When it is necessary to have a close tolerance on the measured image, be it a micro indentation, depth Z axis measurement or any X–Y dimension, then it is important to focus exactly in the object plane. Failure to do this results in high or low readings. Different manufacturers offer facilities in achieving exact repeatable focus positions such as index grids that are aligned only when the fixed reference plane and image plane coincide. The sensitivity of such systems allows untrained operators to focus consistently to one-tenth of the objective depth of focus. The benefits derived from precise focus mechanisms are:

(a) more consistent sizing,
(b) increased accuracy of Z axis dimensions,
(c) low power photomicrography focusing problems eliminated (microscope manufacturers to date (1991) have not taken advantage of this last feature).

The microscope shown in Figure 27.2 adequately displays the precise optical measurement system with its strip shearing system for PCB or strip measurement. The image shearing module for fibre or random field measurement, the precise focus mechanism and the intensity profile unit along with a digital read-out system are shown.

27.3.2 Interpretation of small measurements

With optical measurements becoming more and more accurate it is inevitable that the theory of image creation and what is a true and faithful measurement becomes more important. If we consider the situation of light passing through a small hole, as in the star slide test, light when passing through that hole is manifest with the so-called 'Aery' disc due to the diffraction of light. The size of this diffracted image from the intensity curve shows how holes appear larger than an identical sized spot.

The intensity curve in Figure 27.3 shows how light intensity must be interpreted correctly; i.e. holes measure large, spots measure small.

28

Illumination Systems

28.1 Illumination Types

Illumination plays a very important role in the quality of the observed image. It must be evenly illuminated within the microscope object plane and fill the field of view of the objective back focal plane. The intensity of the illumination must be adequate and the spectral emission compatible with the specimen. A good illustration of this last point occurs when trying to read very small print: how much easier it becomes when subjected to bright sunlight.

Illumination systems in general use are based upon the Köhler method. The alternative to Köhler illumination is critical illumination or something in between, often called semi-Köhler or poor man's illumination.

Oblique illumination can also offer advantages with both the compound microscope as well as the stereo and macro microscopes. Illumination with, for example, macro photography (photomacrography) has more effect on the end result than any other controllable feature. Microscopy skills tend to take lesser prominence in the training of material scientists than is the case with life science students. This is in part because in reflected light the illuminating axis and objective axis are identical and variations in objective par-centricity are automatically accommodated. In transmitted light, adjust-

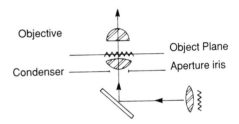

Figure 28.1 Critical illumination

ment is necessary in the substage condenser to align and focus the illumination, if necessary with every change of objective. This results in the minimum of adjustments necessary in reflected light and in many cases a total neglect still gives average to good images. However the effort taken to convert average results to optimum is very small and all users of microscopes should endeavour to ensure correct alignment and focus of the microscope illumination.

28.2 Critical Illumination (see Figure 28.1)

This is often called 'source focused' since it refers to the illuminating source being focused in the object plane. There are limitations to this system when using filament-type sources (as opposed to point-a-light-bulbs) in that the filament image exhibits colour bands and

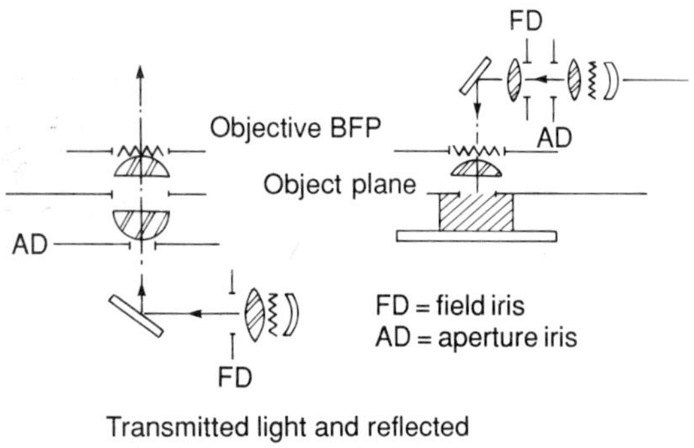

Figure 28.2 Köhler illumination

striations. If this image should not be diffused then it requires defocusing to give a non-granular structure. Critical illumination can be extremely advantageous in fluorescence and dark ground (see Chapter 32).

28.3 Köhler Illumination (see Figure 28.2)

Setting up the microscope for Köhler illumination takes a little practice and there are simple rules to help. These the life science student has to learn before achieving acceptable results from the microscope. The materialographer, on the other hand, very rarely (if ever) carries out these setting adjustments since they are generally factory adjusted. A recent survey (1987) of reflected light microscopes in use showed that a vast majority had an incorrectly focused filament and/or a diffuser inserted that destroyed the illumination quality.

To satisfy Köhler illumination the following must be observed:

(i) The illuminating source must be imaged in the objective back focal plane.
(ii) The lamp field diaphragm must be sharply focused in the objective field of view.
(iii) The aperture diaphragm must be focused in the objective back focal plane.

There are occasions with low power objectives when the filament, because it is focused in the objective back focal plane and also because of its large depth of focus, appears to the camera to be focused in the object plane. On such occasions it is best to diffuse the source, the guide being that if all else fails use the diffuser.

Some instruments incorporate a mirror behind the lamp source. This is used to even the illumination and increase the intensity. The question is often raised as to whether the reflected image should be superimposed on the source image or placed to one side. This is best decided by looking at an image in the objective back focal plane. If the source image does not fill the field then place the reflected image and source image side by side. (To look at an image of the back focal plane take out the microscope eyepiece and observe. Alternatively use a supplementary telescope in place of the eyepiece or introduce the Bertrand lens if the microscope is fitted with one.)

In reflected light the objective acts as the illumination system in that it condenses the illumination source, centring and focusing the field iris on to the object plane. Transmitted light is a very different situation, having two separate optical trains. The objective and the illuminating rays have to be carefully aligned. To do this the substage condenser must be focused up and down until a sharp image of the *field diaphragm* appears in the FOV; this diaphragm can now be centred via the substage condenser centring screws. Having aligned the system for one particular objective then the whole procedure has to be performed again with every objective change. This is necessary since objective par-centricity is governed by the cutting of the objective thread and abutment shoulder. This is a mechanical function carried out on a lathe after the lenses have been assembled. Every error is magnified by the objective primary magnification.

A simple test when looking for quality optics is to look at this objective thread (known as the RMS thread). If it is chrome plated then this last essential step in objective manufacture has been omitted.

To overcome the necessity for constant adjustments the microscope could have individual centring on each nosepiece objective station or the system be aligned to one of the higher power objectives, ignoring par-centricity errors with lower power objectives. In addition to the centring and focusing of the field diaphragm the condenser substage focusing is also to allow:

(i) different substage condensers to be used,

(ii) specimen slide thickness variations to be accommodated,
(iii) defocusing to take place, increasing the size of the illuminated image for low power observations (when a high numerical aperture condenser is in use).

The lamp condenser occasionally has a focusing device. This satisfies the requirements of true Köhler illumination which says an image of the source must be focused in the objective back focal plane. This position often varies with different objectives. To achieve the optimum from your microscope the lamp filament must be sharply focused in the objective BFP for both transmitted and reflected light use. Microscopes *without* built-in adjustment can readily have the filament refocused by simply sliding the lamp housing backwards and forwards.

Köhler illumination is sensitive to out of alignment, in that the image will sway when, for example, the bulb filament is not central. This swaying effect (as the subject is focused in and out) can be overcome by centring the bulb. It can also be obscured by introducing a diffuser into the optical path—hence the reason why a poorly aligned diffused source can go unnoticed. (Many microscopes have built-in diffusers, which fastidious users will remove.)

When working in transmitted light the illuminated field of view is dictated by the numerical aperture of the condenser. Because of physical size restrictions, high NA condensers have a smaller field of view than their lower NA counterparts. Unfortunately if high NA illumination (say 1.3) is required with low NA illumination (say 0.10) the following options are available:

(i) Change condensers (most impracticable).
(ii) Lower substage (destroys Köhler).
(iii) Remove condenser top lens. This destroys Köhler illumination but is better than option (ii) above.
(iv) Introduce supplementary lens into the illumination path (this destroys Köhler but is very convenient).

28.4 Oblique Illumination

Some 50 years ago most compound microscopes, transmitted and reflected, had the facility for fitting the off-axis illumination in order to highlight surface topography in reflected light or cell boundaries in transmitted light. It has since gone out of favour. This technique still has attractions for the materials technologist in that it can be a quick and simple way to indicate surface topography in particular with curved indulations which are not visible in true axial illumination. Examples of this would be painted or plastic surfaces and all mechanically polished multiphase surfaces.

The majority of uses for oblique illumination will be with stereoscopic or macroscopic microscopes. A good example of the use of oblique illumination is to watch a professional studio photographer at work, who will highlight subject detail by the use of many different types of lamps. Reflecting surfaces will be used below or around the subject. The light could be reflected back onto the subject from a secondary reflecting surface or directly onto the subject. The colour of the reflecting surface will be changed and the colour of the background will also be carefully matched to contrast or complement the object being photographed. Specular reflections in the subject can be exploited or extinguished using, say, a liquid layer or a light tint made from translucent material. All these techniques make the subject more pleasing to observe and in particular to photograph, but above all they can be vital in the extraction of subject information.

28.5 Diaphragm Functions

The last thing any aperture diaphragm on a microscope should be used for is stopping down light intensity. Both the field diaphragm and aperture diaphragm have an important function.

28.5.1 Field diaphragm

This is as the word implies intended to restrict the field of illumination and must match the FOV of the objective. Since the objective FOV varies with each objective it should ideally be adjusted with the introduction of each objective. This is only necessary in transmitted light since in reflected light the objective acts as its own condenser. The field iris in reflected light is somewhat redundant when set to the optimum position.

When light passes through an object it diffracts with various orders of diffraction, as will be illustrated later. It is the collection of diffracted rays that create the image. Should the illuminated field in the object plane be greater than the objective field, unwanted diffracted rays will enter the optical system, contributing nothing and introducing internal reflections and glare.

Taking this analogy further, the field diaphragm being sharp and central should be stopped down to coincide with either the eyepiece FOV (this being less than the potential objective FOV) or the camera FOV (once again being less than the eyepiece FOV).

28.5.2 Aperture diaphragm

Many, if not all, textbooks on microscopy suggest the aperture iris is used to introduce contrast to the observed image and that it should be set, for optimum conditions, to say seven-tenths open. What this does is to reduce glare caused by scattered light internally reflected from the outer surfaces of many lenses. With small curved lenses the outer ring has an obtuse angle to the incident ray which reflects in part rather than 100% transmission. The aperture iris under this consideration is used to control glare, which explains why there are numerous research programmes into lens coatings for maximum light transmission. The aperture iris has another function in that as it stops down it reduces the lens numerical aperture, giving an increased depth of focus (and reduces the effective resolution). This can be used to good effect when the subject under investigation has a greater depth than the objective in use.

As modern techniques can prepare very flat metal specimens and very thin sections for slides, these traditional methods for achieving optimum results must be questioned:

(i) Can glare be reduced by alternative methods?
(ii) If the subject is so flat should we stop down the aperture diaphragm?
(iii) Can we tolerate this loss of resolution?

One could support the use of maximum aperture with its benefits of maximum resolution if the depth of field was greater than the subject depth and provided glare could be reduced or eliminated. This subject will be addressed in the photomicrography section in Chapter 36.

28.6 Procedure for Setting up the Microscope for Köhler Illumination

28.6.1 Transmitted light applications

(i) With a specimen on the microscope stage, the field and aperture diaphragm opened wide, *focus the illuminated subject* using a ×10 objective. The quality of this image in the eyepiece is a good indication of how well the optical system is aligned. Gross misalignment will be obvious by the uneven illumination; minor misalignment can be detected by 'sway' as an image point or spot is moved in and out of focus. If the illumination withstands this test, i.e. the illumination is even and a point in the image does not sway, it signifies either a well-aligned system or that a diffuser is present in perhaps a misaligned optical train. The microscope is in effect acting like an optical bench with the objective and eyepiece creating the field axis from which all others should be aligned. Centring is therefore best carried out in the order of proximity to either of these two fixed points; i.e. centre the substage first followed by other accessories such as the lamp and then the bulb.

(ii) Close down the field diaphragm and *focus an image of the field diaphragm in the object field of view*. This is done by raising or lowering the substage condenser. If this condenser is of the flip-top design where the condenser top lens can be tripped-out to fill the field of low power objectives, it is essential the lens is *not* tripped-out. When the condenser is so positioned that it sharply focuses the field diaphragm in the object plane it automatically will have carried out two other important functions: (a) the condenser will now be positioned to operate at its highest numerical aperture and (b) an image of the aperture diaphragm will be focused in the objective back focal plane. Most manufacturers have a built-in adjustable stop arranged so that this position can be fixed and therefore the condenser substage would always be focused to this upper stop position. Fastidious microscopists prefer this stop position to be arranged 'just past' the optimum, enabling fine tuning as required. Open the field diaphragm to the extremity of the objective FOV.

(iii) *Centre the field diaphragm in the object FOV.* This is carried out using the substage condenser centring screws. The act of centring the field diaphragm automatically centres the aperture diaphragm in the objective BFP. If an image of the aperture diaphragm is not centred to the objective BFP aperture when observing, it is indicative of factory misalignment and should be corrected by an engineer independently of the microscope adjustments.

(iv) Attention can now be given to the lamp housing, the object being to focus the bulb and centre within the objective BFP. *Observe the lamp filament conoscopically and focus.* The filament is focused by adjusting the lamp condensing lens backwards and forwards or, if this adjustment is not available, sliding the whole lamp housing in or out (usually out). The observer could focus the lamp filament in the objective BFP by refocusing the substage condenser. *This must not* be used since it will destroy all previous adjustments and would require going back to step (ii).

(v) With the lamp filament or illumination source focused the object is to *centre the source*. For the illumination to be even the objective BFP must be evenly distributed throughout its aperture and not be a single-point illumination. This situation arises when arc lamps are used; to overcome this problem the lamp housing usually incorporates a reflecting concave mirror which allows a reflection of the source also to be imaged in the objective BFP. It is therefore important to have the lamp image and the reflected image position side by side in the objective BFP and *not* superimposed.

The microscope is now aligned and corrected for Köhler illumination for use with the ×10 objective. The aperture diaphragm, if now set to the seven-tenths open position, will provide a well-contrasted evenly illuminated image.

(vi) *When changing objectives* it is necessary to carry out the following supplementary adjustments:

(a) Open or close the field diaphragm to the new field extremity (wider for lower powers, more closed for higher powers). This step would not be necessary in reflected light.

(b) Centre the field diaphragm if necessary. This becomes more apparent as the objective par-centricity deviates; some instruments have individual centring objective nosepiece positions to overcome this problem. In general this step would not be necessary for routine microscope use but would be absolutely essential when a rotating stage, for example, is employed. If a rotating stage is employed, to overcome par-centricity errors the field centring can be carried out using either the objective centring or having a centring rotating stage.

(c) With low power objectives the condenser numerical aperture can be too high, therefore restricting the field of view. On such occasions it is necessary to trip-out the condenser top lens or lower the substage. This has the effect of changing the imaging position of diaphragms and is therefore not the theoretical solution, though it is quite practical. If the field of view becomes extremely large with, for example, a ×1 lens, then the condenser could be removed completely if necessary. When taking photomicrographs the field of view is restricted and could therefore accommodate a condenser normally not suitable for full field orthoscopic observations.

(d) With objectives of high numerical aperture the field of view is so small as to be less than is achievable with normal iris diaphragms. When very critical work is to be carried out it could be necessary to use a special field stop or a more precision field iris.

28.6.2 Reflected light applications

(i) Centring of the illumination substage condensing system to overcome par-

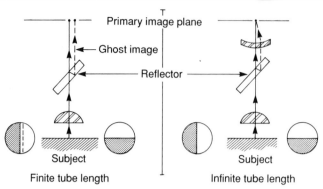

Figure 28.3 Ghost images

centricity errors is not necessary in reflected light since the objective acts as its own light condenser. The field diaphragm when set for the field extremity of the ×10 objective will not require changing with objective changes. The only adjustments therefore required are to the filament focusing and centring.

Once the specimen is in focus:
(ii) *Observing conoscopically, focus an image of the aperture diaphragm in the objective BFP*. If difficulty is experienced in seeing any filament structure it could imply that there is a diffuser present in the optical train—search and destroy! Focusing can be via a lamp condenser focusing control but it is more likely to be necessary to slide the whole lamp housing backwards and forwards until a sharp image is achieved.
(iii) *Centre the lamp filament in the objective BFP*. Most instruments have some centring facility on the lamp housing allowing this adjustment to be made. If the bulb is a precentred design, the lamp housing is devoid of any centring facility. Confirm that the filament is centred. If it is not centred adjustment by an engineer is required within the lamp housing. The microscope is now corrected for Köhler illumination for use with a ×10 objective and the aperture diaphragm, if now set to seven-tenths open, will give a well-contrasted evenly illuminated image.
(iv) *When changing objectives* no further adjustments are necessary beyond those of contrast using the aperture diaphragm.

Please note: the optimum aperture diaphragm position for transmitted and reflected light observation is not necessarily seven-tenths open; it could be nine-tenths or six-tenths, etc. The best position is the position giving the best compromise between glare (contrast) and resolution when observed orthoscopically, and should be very carefully practised. A full appreciation of looking *into the subject* and not *at the subject* can be experienced with the use of this adjustment.

28.7 Reflected Light Ghost Images

The question often arises, 'Is infinite tube length better than a finite tube length optical system for microscopy?' There are points in favour for both versions; the author favours finite systems provided there are no extra tube length correction lenses fitted. The question of a good optical system depends upon many factors, making the question of tube length type somewhat academic. There is, however, one aspect favouring infinite tube length when the optical train includes an internal reflecting/transmitting mirror.

Figure 28.3 shows how with a finite tube length the image, when reflected into the primary image plane, is accompanied by a ghost image resulting from the semi-reflecting mirror. It also shows that this ghost image is confined to the east–west direction and if the specimen is rotated by 90° the ghost image would not occur. This effect can only be seen when the subject is very black against an extremely white background. Nevertheless it is an artefact that can be overcome by the use of the infinite tube length optical system incorporating the converging lens, as shown. One could say, 'Why not have a front surface reflector and avoid the ghost?' Unfortunately front surface reflectors are less efficient, resulting in more light loss, and they also oxidize and can be damaged.

29

Eyepieces and Condensers

29.1 Eyepieces

The day when every microscope was supplied with a great host of different magnification eyepieces has gone. Today the eyepiece has to perform a specific function and magnification is kept to the useful range ($\times 10$) or conforming to a published standard such as ASTM. The field of view down the microscope is dictated by the eyepiece aperture and magnification; the eyepiece field number of 18 for a $\times 10$ normal tube eyepiece is reduced to 14 when a $\times 15$ eyepiece is used.

29.1.1 Field of view

The field of view is, as the word implies, the actual field being viewed. This decreases as the objective magnification increases. In practice the field of view is a limit of the eyepiece stop and should be quoted as an FOV number with every eyepiece supplied. These numbers vary from about 16 to 25 with a $\times 10$ eyepiece (the larger the number the larger the field of view). It is important to relate this number to the actual field size since manufacturers supply eyepieces that because of the viewing angle appear large but in fact are not. This can best be compared with observing any object at, say, 10 inches distance and then looking at the same object some 5 inches distance. It looks bigger at 5 in than at 10 in, but we know it is not. It is the apparent increase due to the increased included viewing angle.

To achieve the object FOV, the FOV number supplied with each eyepiece is divided by the objective magnification (and any other magnification factor excluding the eyepiece), giving a dimension in millimetres. For example,

Eyepiece FOV 18 with $\times 10$ objective:

$$FOV = \frac{18}{10} = 1.8 \text{ mm diameter}$$

Eyepiece FOV 18 with $\times 10$ objective and $\times 2$ magnification changer:

$$FOV = \frac{18}{10 \times 2} = 0.9 \text{ mm diameter}$$

Ultrawide eyepieces are available requiring specially large prisms in the eyepiece tube to accommodate the extra large field. The field number of 26.5 with a $\times 10$ eyepiece is reduced to 17.5 with the $\times 15$ eyepiece.

29.1.2 Eyepiece characteristics

Eyepieces with a long eyepoint are available, intended for spectacle users. Eye shields are also standard with many eyepieces, intended to cut out extraneous light.

In the past, eyepieces when used in pairs (binocular head) were selected for identical FOV magnification and centration and then engraved with an identifying number. Quality control has rendered this practice redundant. It is unwise, however, to mix different manufacturers' eyepieces in the same binocular head.

Classification of eyepieces for graticule insertion can best be described as positive or negative, although this can be confusing, as is explained later: positive when the conjugate plane (plane of the primary image) lies outside the eyepiece body, negative when within. Contrary to general belief reticules or graticules can be utilized in any eyepiece providing it coincides with the primary image plane.

Chapter 25 discussed chromatic difference of magnification and how the eyepiece was used to correct this aberration. If this correction takes place after the graticule image then it will introduce colour fringes to the graticule lines; this is one reason for correcting at the objective.

When taking photographs with the microscope a very simple method is to 'hook up' a standard single-lens reflex camera (less camera lens). On such occasions the normal microscope eyepiece must be used. There are disadvantages with this system, to be discussed in Chapter 36, necessitating the long tube method where the objective to eyepiece conjugates are reproduced in the camera plane. Special eyepieces with finite focusing give sharper results and eliminate the much to be avoided correction lenses. Camera eyepieces are generally of low power from, say, ×2 to ×5 magnification.

Another candidate for special attention is the CCTV eyepiece (or relay lens). This gives parallel light with the appropriate corrections compatible with the television camera, giving far superior results when compared with standard microscope eyepieces. The following is a list of eyepiece types that could be encountered:

(i) *Huygenian*. This is a simple inexpensive design with a field limiting diaphragm between the upper and lower lens. Non-compensating, it will therefore show chromatic difference of magnification when using high NA objectives.

(ii) *Compensating*. This is a more complex design to compensate for chromatic differences of magnification, often introducing this aberration to any fitted graticule.

(iii) *Focusing*: This incorporates a mechanism allowing a graticule in the primary image plane to be focused. When graticules are used in the normal compensating eyepiece, the edge of each line or scale could be strongly coloured, i.e. the eyepiece removes chromatic differences of magnification from the specimen and in so doing introduces the same in subjects positioned in the primary image plane. The focusing eyepiece overcomes this aberration by correcting for chromatic differences of magnification in the eyepiece front lens prior to reaching the graticule. The focusing device is to facilitate sharp focus of the graticule but is also used to accommodate user eye defects, therefore maintaining a fixed microscope tube length. This last point is essential in photomicrography when using a fixed camera plane.

(iv) *Widefield*. This, as the name implies, means an eyepiece with a large FOV number. The word widefield is often misused since there seems to be no standard. In the absence of any standard, the normal field should be up to 19 mm, the widefield from 19 mm and the ultra-widefield from 23 mm diameter.

(v) *Ramsden*. This is similar to the Huygenian; it is simple, non-compensating but with the conjugate plane outside the eyepiece lens system (the Huygenian is within).

(vi) *Kellner*. This eyepiece is a well-corrected Ramsden (improved chromatic).

(vii) *Filar micrometer*. This eyepiece makes use of a movable line or curtain, activated via a screw or micrometer. The design is of the Kellner type with the conjugate plane outside the eyepiece system. All users of such equipment as the Vickers hardness tester will be familiar with this design of eyepiece. Compensation for chromatic differences of magnification has to take place after the primary image plane. Measurements with this type of eyepiece are more accurate because of the micrometer and the accurate line positioning—more so than with the fixed graticule eyepiece.

(viii) *Reducing (negative) projection*. Because the reducing eyepiece, i.e. one with a diverging lens acting as a projection lens, is called a negative eyepiece it is wise to avoid the words positive or negative in any description of an eyepiece. This eyepiece would then become a projection or reducing eyepiece.

(ix) *Flat field*. These eyepieces are designed to reduce the field curvature created in the primary image. They are to be used with caution in photomicrography since the eye is more accommodating than the film plane.

(x) *High eyepoint*. These eyepieces have an extended 'exit pupil' distance (long eyepoint) and are designed for spectacle users. Having been involved in the microscope industry when these eyepieces were first designed, it always came as a shock to find, even with long eyepoint eyepieces, that spectacle users still removed their glasses prior to using the microscope. The author, now at the age of needing glasses for reading, removes his before looking down the microscope; there is no substitute for direct vision. The majority of eyepieces used in the middle and top quality microscopes are long eyepoint designs and have proved to be more comfortable than the traditional distance. They do, however, require the familiar rubber eye cups to avoid extraneous light entering the optics.

Other eyepieces the user may come in contact with are:

(a) Pointer eyepiece
(b) Low power projection
(c) CCTV relay eyepiece
(d) Comparison eyepiece
(e) Image splitting eyepiece
(f) Graticule, goniometer or reticule eyepiece

29.2 Substage Condensers

29.2.1 Requirements

As indicated in Chapter 28, the condenser acts as an objective in reverse. The requirements of the condenser to fill the objective FOV with the correct cone of light (NA) and also image the field diaphragm often result in a compromise. For example, to fill the field of low power objectives necessitates a low NA/large field condenser. When used with high power objectives, this condenser, because of its low NA, is inadequate, resulting in a loss of resolution.

To overcome the problem of the low NA, condensers with a small field/high NA are available. The problem arises when large field, say ×3 objective, and high NA objective (0.7 upwards) are both required. This is where the compromise starts if changing condensers is to be avoided:

(i) *Lower the substage*. This increases the FOV, reducing the NA which is required for low power objectives. Unfortunately it then images the field and aperture diaphragm along with the filament image in the wrong plane.

(ii) *Remove the condenser front lens*. This can be done by having a trip-out or flip-top design or, on the high NA achromatic condensers, unscrewing the top lens cell. This again has the disadvantage of destroying the features of Köhler illumination, as in step (i) above.

If it is necessary to take a photomicrograph under circumstances (i) or (ii) above, attention must be paid to even out the filament structure. It could be necessary to introduce a diffuser into the optical path. *Note*: do not execute (i) or (ii) above if the FOV is greater than the camera format FOV.

When using condensers having a numerical aperture less than the objective NA, the resultant numerical aperture is the summation of both divided by two. The main requirements of the substage condenser are:

(a) Filling the back aperture of the objective with light.
(b) Giving a full FOV.
(c) Having an adequate cone of light to match the objective NA.
(d) Not destroying corrections the objective has been selected to overcome, such as chromatic and spherical aberrations.

If the condenser is to satisfy the NA or cone of light requirements of the ×100 with 1.30 oil objective then the condenser itself must have an equivalent NA. High NA condensers therefore require oiling to the specimen slide in order to achieve its full potential. Unlike oil objectives which are not suitable for use when dry, high NA condensers can be used in air, the resultant NA, however, being say 0.95.

Because the oiling of condensers is 'messy' it tends to be overlooked for routine work, which is understandable. If, however, improved sharpness and/or resolution is required and in particular when taking photomicrographs the condenser must be oiled.

29.2.2 Types

Condensers are usually one of three types:

(i) *Abbe*. This is of a simple two lens design, inexpensive and adequate for student use. It exhibits spherical aberration and has no chromatic corrections and requires diffusing or lowering the substage to fill the FOV of objectives $< \times 4$.

(ii) *Aplanatic*. This condenser has been corrected for spherical aberration (green) but still has little chromatic correction. It has, however, been corrected for coma.

(iii) *Achromatic*. This is corrected both for spherical aberration (green) and also for chromatic aberration (green only). This condenser gives good images of, for example, the field iris but its major advantage is in allowing the objective to resolve fine detail without fringes associated with uncorrected illumination. Because of this last point it tends to be the high NA design and in consequence will not fill the FOV below the $\times 10$ objective. Its maximum NA is achieved by oiling.

The choice of condenser (see Table 29.1) is dictated by the requirements of the specimen; as with objectives it is not necessarily the most expensive that is the best. To have specially corrected fluorite or apochromatic lenses would require the better corrected condenser (achromatic). This assumes that the benefit of higher corrected objectives was dictated by the specimen. If the colours and contrast in the specimen are to be created optically as in the polarizing microscope then the benefits of condenser corrections could be of secondary importance and the Abbe or aplanatic condensers would be preferred.

29.2.3 Substage condenser accessories

The condenser usually has some centring facility; on some of the older microscopes it has a tilting mechanism for oblique illumination. Simple filter holders are to be found as an integral part of the substage or attached to the condenser. When attached to the condenser they can also be utilized for low power dark ground stops (sometimes called dark field). Also stationed at the substage will be the polarizer when required. This glass polarizing disc can be fitted into simple filter holders or can be of the built-in graduated type.

Specialized techniques utilizing different types of condensers are to be found in later chapters.

Table 29.1 Choosing a condenser

Substage condensers	Corrections		Advantages
	Chromatic	Spherical	
Abbe	Nil	Nil	Inexpensive
Aplanatic	Little	Green	General purpose
Achromatic	Green	Green	Better corrections and high NA

30

Introduction to Interference

30.1 Diffraction Theory

When an object such as a diffraction grating or a diatom (*Pleurosigma angulatum*) which has a large number of symmetrically disposed and evenly spaced markings is examined by transmitted light, it will be seen, on removing the eyepiece and looking at the objective back focal plane, that the light consists of a central beam of white light surrounded by a number of symmetrically disposed spectra.

The number and arrangement of the spectra depends upon the shape and size of the markings. Their distance from the central beam (zero order of diffraction) is greater as the marking (gratings) become finer. We therefore have a central beam called the zero-order diffraction beam and subsequent first, second and third orders of diffraction.

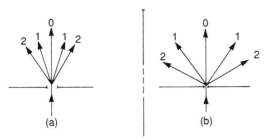

Figure 30.1 Angles of diffraction. (a) Coarse gratings and (b) fine gratings

Figure 30.1 shows the orders of diffraction for two different size gratings. Note how the included angle or cone of diffraction is smaller for the larger of the two holes. When looking at a plan view of the spectra, as imaged in the objective back focal plane, the number and pattern of each order of diffraction will change with the shape of the object hole.

According to the theory of microscope vision a true image will be achieved only if all the spectra are collected by the objective and recombined with the central beam (zero order) in the primary image plane.

From this we can conclude a two-part function:

(i) creating the image and
(ii) the quality of the image.

30.2 Creating the Image

Figure 30.2 illustrates how the image is created. Consider a ray of light (X) travelling upwards past the condenser. On reaching the object the zero-order ray transmits without deviation (Z), the diffracted ray (Y) taking another route; both rays combine in the primary image plane.

The quality of the created image depends upon the ability of the objective lens to collect all the diffracted rays. Since light diffracts at an increasing oblique angle as the gratings

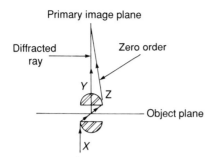

Figure 30.2 Creating the image

become finer it follows that progressively higher numerical aperture objectives are required to resolve finer detail. (*Note*: magnification does not resolve finer detail, it only magnifies the created image.) Because of the physical size restrictions in the optical train it is necessary with the higher numerical aperture objectives to reduce the working clearance in order to collect the diffracted rays.

Figure 30.3(a) illustrates the collection of zero- and first-order rays only; this would not be adequate to create a good well-structured and resolved image. The image could be resolved by increasing the diameter of the lens; unfortunately this would result in very expensive and impracticable large fields. Figure 30.3(b) shows a much improved situation where all the diffracted rays are collected in the objective front lens, the working clearance (wc) being reduced. There is therefore a relationship between the objective NA and the physical working clearance—as the NA increases the working clearance decreases. The diffracted rays, once collected into the objective front lens, must not be restricted as they pass through the optical train if a loss of optical resolution is to be avoided. Peripheral diffracted rays are restricted, for example, if the aperture iris is employed or the fitting of funnel stops for darkground applications are present.

It is possible to illustrate the loss of optical quality occurring when middle-order diffracted rays are obstructed by fitting a stop into the objective back focal plane.

30.2.1 The microscope image

The previous consideration was for light passing between a grating. Light still diffracts when passing through a specimen, only this time the diffracted rays are retarded in relation to the direct ray, relative to the refractive index of the subject, i.e. refraction.

Thus an image as we understand it in a microscope is due to the interaction of secondary light waves originating from light absorbing or retarding features of the specimen with those that have passed uninfluenced through the specimen. Should the secondary waves be less than a quarter lambda out of phase, they will, when recombined, fail to be visible to the eye and no apparent contrast change will occur.

This happens often in life sciences where the cell refractive index differences are too small to show any contrast—hence the reason for staining. It also happens to the geologist with minerals that are closely indexed and to the materials scientist where the surface light or angle differences are too small. All of these situations will create an image but the eye is not sensitive enough to see them. There are fortunately optical techniques available that enable the created image to be visible.

30.3 Phase Contrast

The phase contrast microscope is used to enhance small phase shifts, thereby depicting height differences when used in the reflected light mode and refractive index differences in the transmitted light mode. The phase contrast microscope shifts the relative phase of the secondary waves by a quarter lambda (green light), giving a total phase difference (optical path difference) in the region of a half lambda and thus interfering destructively (see Figure 30.4).

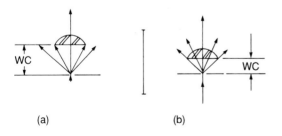

Figure 30.3 Collecting diffractions. (a) Incorrect and (b) correct

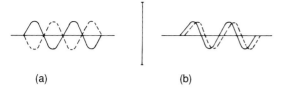

Figure 30.4 Summation of amplitudes. (a) Destructive and (b) complementary

Though used extensively in transmitted light, phase contrast is sadly missed as a reflected light technique. In detecting the optical path differences between surfaces there is much information that could be derived from abrasion rate differences occurring in the mechanical preparation of specimens. This would avoid having to resort to differential interference contrast.

Optical path differences (OPD) of less than a quarter lambda are not able to be detected by the eye, but the eye can detect changes in amplitude (light intensity) and wavelength pitch (colour). The phase contrast microscope makes use of the eye's sensitivity range of grey light levels.

When considering two rays incident upon a surface (Figure 30.5) the resulting light will be the summation of amplitudes; e.g. two waves in phase upon a surface will increase the amplitude and two waves a half lambda out of phase will extinguish the amplitude (nil amplitude). Both amplitude and phase shifting are adjusted in the phase contrast microscope.

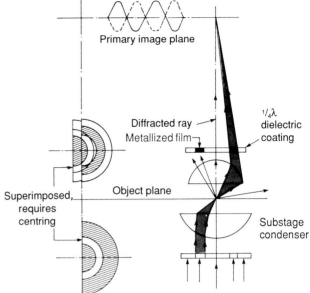

Figure 30.6 Positive phase contrast microscopes

30.3.1 Transmitted light phase contrast

Consider the direct ray (Figure 30.6) as it passes through the substage condenser to be an annular cone of light unhindered as it passes through the specimen, the objective and the metallized film. This metallized film reduces the amplitude without affecting phase retardation. This is necessary in order to balance the amplitude of the zero-order (direct)

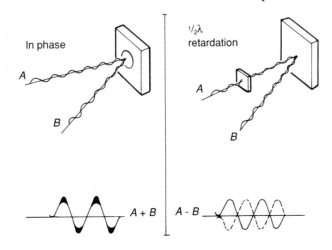

Figure 30.5 Phase retardation. (Resulting light equals the sum of amplitudes)

ray with the secondary rays. Secondary rays are always reduced in intensity.

Now consider diffracted secondary rays occurring from the object, passing through the objective and through a dielectric coating which retards the already retarded wave by another quarter lambda, thereby giving destructive interference when recombining with the direct ray at the primary image plane. The diffracted ray, now having an amplitude similar to the direct ray but retarded by a half lambda, is extinguished when recombined.

The metallized film is intended to balance the amplitude of the primary ray with that of the diffracted secondary rays, which themselves depend upon the specimen. Some specimens such as asbestos having a high absorption require special metallized coating to balance the amplitude.

Figure 30.6 shows the two annular rings, the bright ring of light from the condenser and the dark ring from the objective metallized film. These two rings imaged in the objective BFP *must* be concentric and be an exact match. It is necessary to change the condenser ring with every change of objective.

It is also possible to have what is called negative phase contrast, where the diffracted ray is retarded by three-quarters lambda, reversing dark subjects to light. This is based on the identification advantages when observing a bright subject on a dark background.

Phase contrast finds application in the field of pollution analysis, particle size and counting, occupational health as well as a host of uses in life science where specimen staining is not possible. Although good for revealing structural detail phase contrast exhibits halos within the image and must therefore be used with caution when used for particle sizing.

30.3.2 Reflected light phase contrast

The foregoing has been based on transmitted light; the explanation, however, holds good for solid specimens. Think of the rays as being deflected rather than diffracted. The microscope construction is different in that the phase shift is done outside the objective. Remember that in reflected light phase contrast detects different height levels. This technique has to a large extent been replaced by surface interference, to be investigated later.

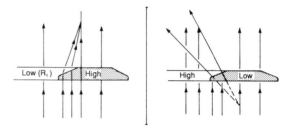

Figure 30.7 The Becke line

30.4 The Becke Line

Microscopy very often involves the identification of a subject as well as object morphology. Identification in transmitted light hinges very much on the refractive index.

Identification in reflected light can be its reflectivity or optical characteristics but more often is based, as well as structure morphology, on its hardness or resistance to abrasion. Hardness is adequately accomplished via the pyramid or knoop hardness test. Abrasion characteristics can often be related to the mechanically prepared top surface (hence the reason for wishing that phase contrast was still available). The Becke line test can recognize angle surfaces and hence height differences in reflected light and refractive index differences in transmitted light.

Figure 30.7 illustrates a very quick and easy test in identifying which of two subjects has the greater refractive index. The edge of a small transparent object acts as a prism or lens, deviating an incident parallel beam towards the subject having the greater numerical aperture. The most favourable conditions are achieved with parallel light; therefore one should close down the aperture iris, using a relatively high numerical aperture objective (NA 0.65). For ease of observation and increased sensitivity use a green monochromatic filter. This technique is used when looking at unstained cells, fibres, oils, etc.

Immersing the transparent object in different liquids of known refractive index until the Becke line no longer exists fingerprints the object refractive index.

30.5 Dispersion Staining

When light is refracted the angle of refraction varies with wavelength; this is called dispersion.

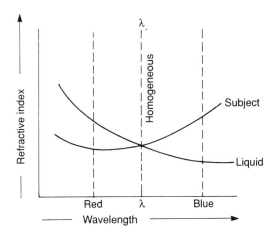

Figure 30.8 Dispersion characteristics

Unlike the Becke line which is generally a qualitative test, dispersion staining is quantitative—a means of tracing the refractive index. The refractive index changes with the wavelength of light used and it is this dispersion characteristic that we make use of. The term dispersion staining is used to describe the colour effects produced when a transparent, colourless subject, immersed in a liquid having a refractive index near to that of the subject, is viewed orthoscopically in transmitted light.

At the intersect of two different refractive indices, light is defracted. If we now immerse the subject into a liquid whose refractive index lambda (see Figure 30.8) is identical there will be refraction at all wavelengths, excepting the homogeneous wavelength lambda. If, for example, we were able to place a 'stop' into the objective which extinguishes refracted rays, the subject would at the liquid/subject interface exhibit only the colour of the wavelength lambda.

To produce these colours the subject and immersion liquid must therefore have different dispersion curves which intersect at one point to give optically a homogeneous condition. As a result, light of compatible wavelength will pass through the subject interface without deviation; light of other wavelengths will be deviated in relation to their refractive index difference.

From the layout drawing (Figure 30.9a) notice a *small circular hole* in the objective back focal plane (BFP); its purpose is to allow the homogeneous unrefracted wavelength to pass through to the primary image plane, the refracted reds and blues being restricted. As a consequence of this the subject is perceived as having a white image with a coloured edge matching the homogeneous wavelength. The general background colour would also be white.

Imagine now the above situation, only this time a *central stop* (Figure 30.9b) is used instead of the small circular hole in the objective back focal plane. The homogeneous, unrefracted rays are now stopped, the image being created by the refracted reds and blues. In this case the subject would appear dark with purple borders seen against a dark background.

With polarized light and an anisotropic subject (having just two vibration directions) different dispersion colours will be obtained dependent upon orientation of the specimen. This technique is used in the identification of fibres and is developed further in Chapter 34. It can also be seen how identification can be made by means of a known refractive index liquid, with reference to boundary colours.

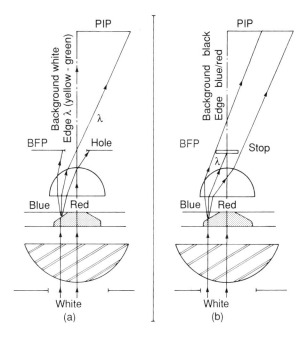

Figure 30.9 Dispersion staining microscope

31

Surface Finish Interference

Interferometry is a technique for observing minute unevenness of specimen surface by changing it to interference colours or fringes.

Applications for surface interference would be in the measurements of engineering surfaces where the peak-to-valley roughness would be of primary interest. Bearing materials often have a limit on the tolerated surface roughness or the size of rogue damage. Semiconductor technology is another application where small layer thicknesses can be accurately and non-destructively measured. Tensile slip bands are often visible in soft materials. Relief grinding or polishing can be measured. Surface coatings such as those used in phase contrast (previous chapter) where quarter lambda is necessary can be readily checked using interferometry.

31.1 Quantitative Z Measurements

Materiologists often require quantitative information relating to the surface topography in the Z direction (depth). The choice of technique very much depends upon the accuracy required and the depth of measurement. The simplest way to make depth measurements is to use the fine focusing on the actual microscope stand. The restrictions using this technique are:

(i) Errors occur because of the objective depth of field ($\times 10$ objective, $8.7\,\mu m$).
(ii) Microscope focusing is not always arithmetical.
(iii) It is sometimes difficult to focus on reflecting surfaces.

To a degree much can be overcome by using higher numerical aperture objectives ($\times 100$ objective has a depth of field of $0.5\,\mu m$). The microscope focusing error is often acceptable within a specified range which the manufacturer should be able to supply. Point (iii) above can be overcome by having a target acting as a secondary telescope set at exactly the same optical distance as the objective. The accuracy of this method would be $0.5\,\mu m$ at best.

To improve on the $0.5\,\mu m$ accuracy it is necessary to go to surface interference techniques, which in order of accuracy are:

(i) Multiple beam (after Tolansky)
(ii) Double beam (after Mirau)
(iii) Double beam (after Michelson)

These three methods are to be investigated after a résumé of interference principles.

31.2 Interference Principles

The superimposition of two or more waves (interaction of primary with secondary) is

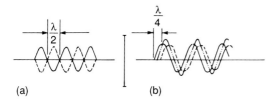

Figure 31.1 Sine waves. (a) Extinguished and (b) amplitude increased

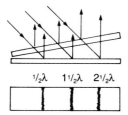

Figure 31.2 Interference patterns

termed interference. If the two wave crests coincide, their amplitudes add up. Conversely, the amplitudes are subtracted from each other when wave crest and trough oppose, causing extinction when the OPD is half lambda. Partial amplification or extinction results from all intermediate positions. Light interference only occurs when the interfering wave train originates from the same source.

Figure 31.1 illustrates pictorially how two sine waves with half lambda out of phase extinguish each other and quarter lambda out of phase complement with an amplitude increase.

A simple form of interference has been used in the metrology laboratory for many years. It is based on the superimposition of glass blocks or optical flats over the sample under test. This test can be used in the materials laboratory to show the flatness of preparation.

By simply using two glass slides in monochromatic light (Figure 31.2) an interference pattern can be achieved. Light partly reflects from the first glass slide and partly transmits to the lower slide. Light from the lower slide reflects back (in part) and when recombined with the upper reflected beam will interfere constructively or destructively.

Because this is a double reflecting system, half lambda out of phase is actually quarter lambda out of phase. When observing each extinction the actual 'out of phase' will be half lambda and not lambda.

In practice it is sometimes an advantage to use white light. This gives a total spectral effect around the orders of extinction and helps to determine first order, in particular with confusing regular images.

It can be seen how this technique is used as a general metrology tool; it is this same basic principle that is used with microscopes to give very sensitive and accurate results. The principle is to create a reference beam and to have this beam combine with a secondary beam that has traversed the specimen. Having done this, if the two beams have the same OPD, i.e. the two beams are identical in length, the image will be devoid of fringes. In order to create fringes it is necessary to tilt one of these surfaces (the specimen or the reference). The degree to which one of these surfaces is tilted determines the number of fringes in the field of view. The degree of surface change is then measured by the fringe pattern.

The example shown in Figure 31.3 is of a stepped subject followed by a V grooved surface. When the mercury green filter (or sodium light) is used the first-order extinction and all other positions (½–1½–2½) are visible as black lines, the first order being the darkest. When the OPD exceeds one wavelength it is not always obvious which are the first-order rays. This can be remedied by taking out the filter and using white light; the filter should be repositioned, however, for accurate measurements. The two examples given show an OPD of half lambda; in the Z axis this is quarter lambda. Had one of the observed surfaces been a sphere, as in a ball bearing, the interference rings would be concentric.

31.2.1 Equipment requirement

Interference techniques are generally supplied as accessories to the routine compound microscope and require few if any special requirements beyond:

(i) Interference filter
(ii) Special objectives
(iii) Measuring eyepiece

The Mirau system requires the addition of a tilting stage.

280 Surface Preparation and Microscopy of Materials

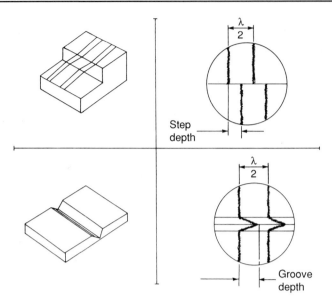

Figure 31.3 Applications of interference

31.3 Equipment Types

31.3.1 Michelson method (double beam)

This method is useful for low power (×10 objective) observations. It requires no special microscope fittings since the reference surface is outside the objective and can be used as the tilting surface. Applications are in the 2 to 10 μm range.

Figure 31.4 shows how the incident ray from the illumination after passing the objective meets a beam-splitter, sending one beam to the reference mirror and the other via the specimen. These two beams recombine at the primary image plane, the distance travelled by both beams being identical. In order to create interference patterns the reference mirror is inclined, the manifest interference lines and spacings being related to this angle of inclination. The reflectivity of the reference mirror must take into account the subject reflectivity in order to balance the two resultant amplitudes. Since the beam-splitter is required below the objective the system is limited by the objective working clearance—hence its application to low magnification objectives only.

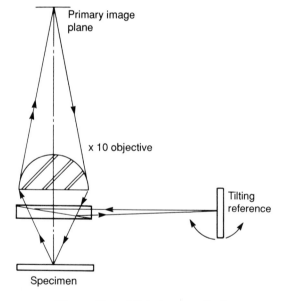

Figure 31.4 Michelson method

31.3.2 Mirau method (double beam)

This method, illustrated in Figure 31.5, is used with the higher numerical aperture objective and is a more sensitive design. The reference surface is built into the objective. This

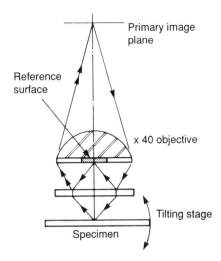

Figure 31.5 Mirau method

necessitates the use of a special microscope tilting stage in order to create the interference patterns. The intensity of the direct and reference ray now becomes more important and in order to balance these two amplitudes, different light absorbing coatings are offered for the reference surface. This system has applications in the 0.5 to 2 μm range.

31.3.3 Multiple beam (Tolansky)

This is a much more sensitive system than the double beam methods and is used in the detection range of 0.01 to 1.0 μm. A series of reference mirrors from 4 to 90% reflectivity are used to balance the reflectivity of the specimen to the reference amplitude. These reference mirrors are actually birefringent prisms, the orientation of the fringes being dependent upon the mirror rotational position. The distance between fringes is governed by the level of this reference mirror. The inclination of the reference mirror eliminates the need for special inclination microscope stages.

The accuracy of all fringe forming systems can be improved by the use of a filar micrometer eyepiece.

Operation of this technique requires specimen contact of the selected reference mirror as illustrated in Figure 31.6. Care must therefore be taken to scan the specimen with the reference mirror holder in the upper position; this should only be lowered and contacting the specimen when the target area is stationary. The intervals and directions of the fringes, now visible, can be changed by means of the adjusting screws on the reference mirror holder. Raise the reference mirror before scanning other areas of the specimen. Since this technique is extremely sensitive to very small differences in optical path difference it is essential to isolate the equipment from any vibrations.

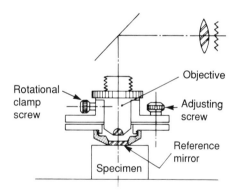

Figure 31.6 Multiple beam (after Tolansky)

32

Contrast Interference

32.1 Differential Interference Contrast (Nomarski)

This can be described as phase contrast in colour. It converts the grey levels caused by optical path differences into chromatic differences. In reflected light it converts not differences in height but angles into colour differences. In transmitted light it is the refractive index difference that introduces the colours.

32.1.1 Principle (reflected light)

Differential interference (after Nomarski) makes use of a birefringent beam-splitter (see Figure 32.1(a)), where the incident beam is split into two: the ordinary ray and the extraordinary ray. The shear or separation between these two rays is less than the limit of resolution for the objective being used; therefore no double image should be orthoscopically observed. (In practice the degree of shear varies with the beam splitting wedge position and when set for first order grey extinction a double image is visible, it is therefore *essential* to remove the wedge from the optic axis when not observing in DIC.) This shear, however, is directional and specimens that are not uniform should be rotated to find the best position. These two rays are polarized,

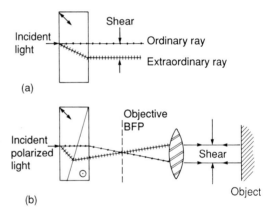

Figure 32.1 Creating the OPD. (a) Single beam-splitter and (b) modified beam-splitter (↘ and ⊙ indicate the direction of the optic axis)

the planes of vibration being perpendicular to each other. Figure 32.1(b) shows the modified beam-splitter consisting of two wedge-shaped uniaxial prisms, whose optic axes are at right angles to each other. The incident polarized beam is still split into two as in Figure 32.1(a) but this time the second prism causes the two rays to intersect at the objective back focal plane. The ordinary and extraordinary rays will when recombined have travelled different optical path distances; these distances will also change with different wedge positions. In order to equalize the intensities of the two rays, the wedge is placed at 45° to the incident

Plate 13

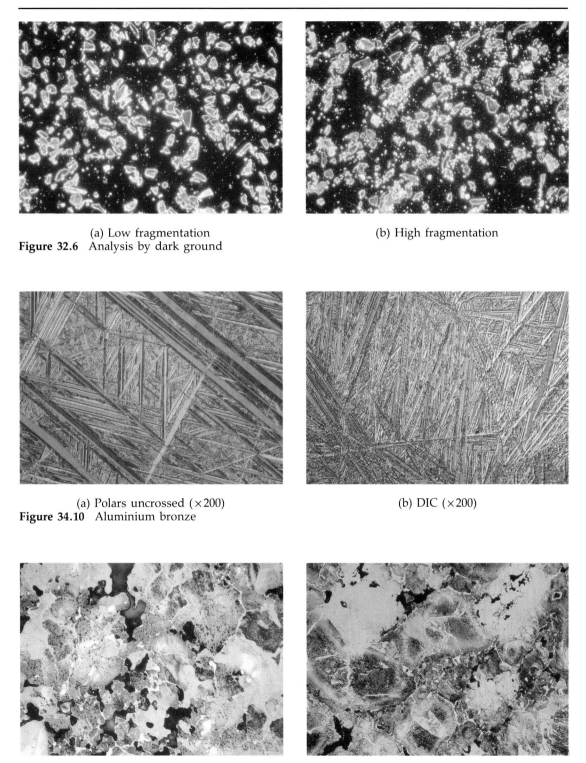

(a) Low fragmentation　　　　　　　　　　(b) High fragmentation
Figure 32.6　Analysis by dark ground

(a) Polars uncrossed (×200)　　　　　　　　(b) DIC (×200)
Figure 34.10　Aluminium bronze

(a) Incorrect orientation (×400)　　　　　　(b) Correct orientation (×400)
Figure 34.11　Sintered iron

Plate 14

Figure 34.12 Roman glass: uncrossed polars (×200)

Figure 34.13 Titanium: crossed polars with sensitive tint (×200)

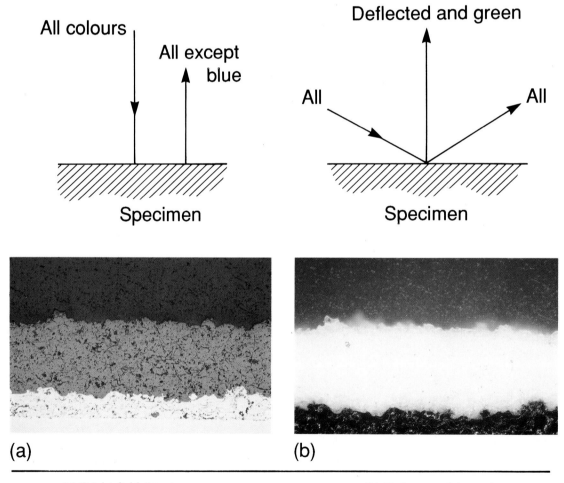

(a) Bright field (grey) (b) Dark ground (green)
Figure 35.4 Dark ground excitation of blue dye

Plate 15

(a) Full aperture, uncrossed polars (×400)

(b) Reduced aperture without polars (×400)

Figure 36.4 Optically induced glare

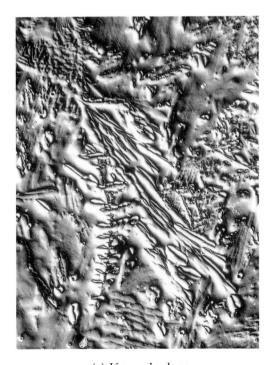

(a) Upwards slope

(b) Downward slope

Figure 36.7 First-order grey in two positions

Plate 16

Figure 37.2 Micrographs showing growth of sodium silicate crystal (×50)

Plate 17

Figure 38.5 Iron oxide/quartz matrix, polarized (×200)

Figure 38.6 Iron oxide/quartz matrix, dark ground (×200)

Figure 38.7 Carbon–carbon, polarized (×200)

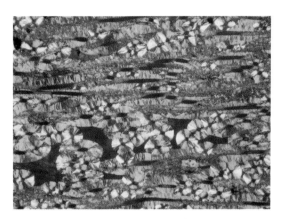

Figure 38.8 Carbon–carbon, polarized (×100)

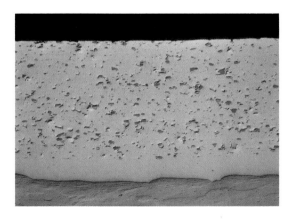

Figure 38.9 Bearing material, DIC (×400)

Figure 38.10 Glass matrix composite, DIC (×200)

Plate 18

Figure 38.11 Metal matrix composite, DIC (×200)

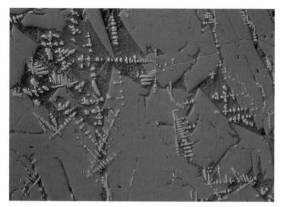

Figure 38.12 Roman slag, DIC (×200)

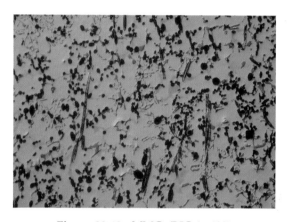

Figure 38.13 MMC, DIC (×400)

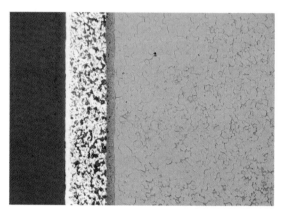

Figure 38.14 Coated ceramic, BF (×400)

Figure 38.15 Coated ceramic, DG (×400)

Figure 38.16 Coated ceramic, DIC (×400)

Plate 19

Figure 38.17 The Christmas Tree: silica crystals (×50)

Figure 38.18 Cleavage: silica crystals (×100)

Plate 20

Figure 38.19 The Cavern: covellite (×50)

Figure 38.20 The Flying Moth: broken crystals (×400)
(Photo: Trevor Bousfield)

Plate 21

Figure 38.23 The Fourth Horseman of the Apocalypse: Roman glass, uncrossed polars (×200)

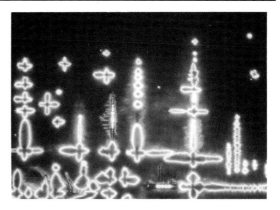

Figure 38.25 Blackpool Illuminations: iron oxide, DG (×1000)

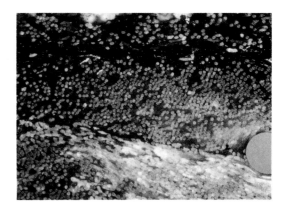

Figure 38.26 The Sky at Night: carbon fibre in resin (×100)

Figure 38.27 Attack on Planet Earth: combined subject exposure (×100/1000)

Plate 22

Figure 38.28 The Mountain Range

Figure 38.29 The Isle of Skye

Figure 38.30 The Sea Bed

Figure 38.31 Going for Colour

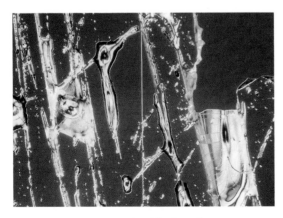

Figure 38.32 Double Interference
(Photo: Trevor Bousfield)

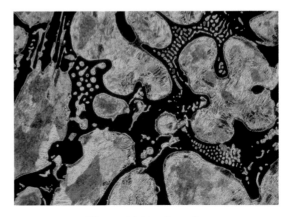

Figure 38.33 Amoeba

Plate 23

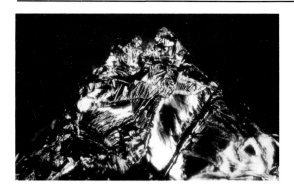

Figure 38.34 Diamonds Are Forever

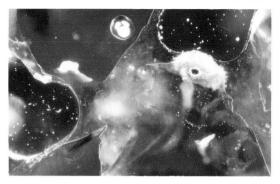

Figure 38.35 The Easter Chick (discovered in vitrified alumina in Figure 14.13 (Plate 9)

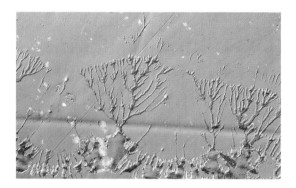

Figure 38.36 The Stormy Plain (×50)

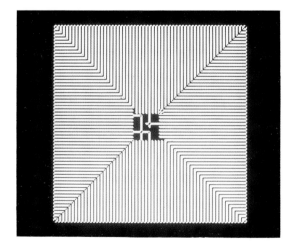

Figure 38.37 Wafer Gratings, DG (×200). (Photo: Trevor Bousfield)

Figure 38.38 Wafer Gratings, crossed polars (×200). (Photo: Trevor Bousfield)

Plate 24

Figure 38.39 Poster display

Figure 38.40 Poster display

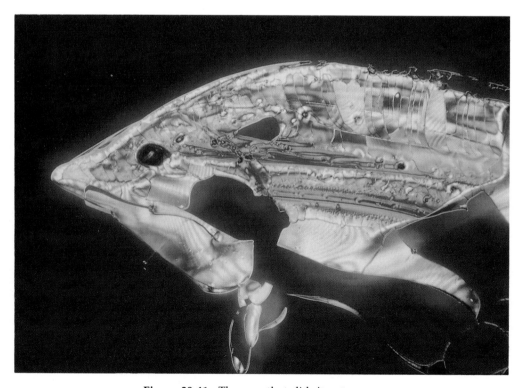

Figure 38.41 The one that didn't get away

polarized light direction. These two rays when recombined in the primary image plane, although of different OPD, will not interfere because their directions of vibration are perpendicular to each other. In order to observe interference colours the analyser must be introduced.

When this technique is utilized in the reflected light microscope, the polars are crossed and the wedge is at 45°. The background colour will no longer be extinguished but exhibits a colour relative to the OPD of the birefringent wedge, the background colour changing with every lateral movement of the wedge. In order to have one single colour across the whole field of view, the plane of fringe localization must coincide with the objective back focal plane.

Both transmitted and reflected light systems are therefore based on two requirements:

(i) The specimen is observed between crossed polars.
(ii) A birefringent wedge (beam-splitter) is placed in the illuminated beam before and after reaching the specimen. In reflected light this requirement is satisfied with just the single wedge fitted above the back focal plane of each objective (see Figure 32.2).

Simply speaking, the wedge splits the light into two parallel waves vibrating at 90° to each other. Having passed through or been reflected from the specimen these waves recombine to give a general background colour. This background colour should be even across the whole field of view. This last point is important since some instruments offer 'up and down' adjustment to the wedge.

To change the colour of the background the wedge must be very gently traversed across the optic axis. Applications in transmitted light detect refractive index differences and in reflected light detect differences in surface slope.

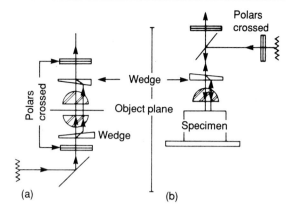

Figure 32.2 Nomarski microscope. (a) Transmitted light and (b) reflected light

32.1.2 Reflected Light

Observation in reflected light shows the background colour as indicative of the wedge position. Changes to this colour occur with every optical path difference. This being so, the materialographer will be able to detect slope or angle differences *not*, as is often portrayed, step differences. If *both* beams hit the same surface level their optical path difference is the same as *both* beams hitting another stepped level.

Fortunately most surfaces being investigated by the materialographer constitute a series of curves (a curve being an infinite variety of slopes). These curves are inevitable when mechanically preparing a two- or multiphase material, due to preferential polishing.

This qualitative study renders a sharply defined relief-like image with an excellent variable range contrast of zero-order to first-order interference colours. The interference relief-like images exhibit different colours with the degree of relief and will all change with a change in the background colour.

Orthoscopic observations, although very impressive, can leave the observer with a

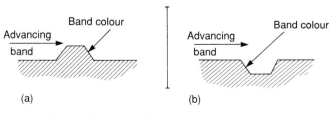

Figure 32.3 (a) Hill and (b) hole identification

dilemma; is it a hill or a hole (see Figure 32.3). This is the kind of problem the American astronauts faced when landing on the moon since we recognize hills and valleys by the reflections caused by the sun. When clarification of such matters is necessary, it is a hill when the side farthest away from the advancing band changes to that colour ahead of the advancing band.

When highlighting surface topography it is more pronounced to have the wedge just short of the first-order extinction position. This grey level position is the natural position for black and white photomicrography, giving the best three-dimensional image.

If the reader will rotate this book 180° then look again at Figure 32.4 the matrix valley will be manifest as hills.

32.1.3 Setting up the microscope

Some equipment manufacturers suggest that the change in interference colours should be achieved by rotating the polarizer or analyser. This can lead to less than optimum results; therefore the following is intended to overcome this possibility:

(i) Having focused the subject in bright field, open the aperture and field diaphragm to the maximum objective aperture.
(ii) Introduce and set the polars (crossed) to extinction.
(iii) Very carefully introduce the wedge until the first-order grey has been achieved. This gives the best position for contrast and three dimensions in grey levels. For black and white photography introduce the green filter.

The best position for first-order grey is when the raised subject within the specimen has a small white or bright upper surface with a small shadow on the lower surface. It is possible to have an enlarged bright upper surface with an equally enlarged lower shadow; this undesirable condition is typical of gross specimen relief or an incorrect wedge position. The technique is to firstly set the wedge to give total background extinction; this will give a *flat* image manifesting two small white lines around the subject periphery (Figure 32.4(a))—very similar to dark ground. As the wedge is advanced the upper surface becomes bright at the expense of the lower surface, which dulls. It is the subtle positioning of the wedge from the total extinction position that gives the most rewarding three-dimensional realistic image (Figure 32.4(b)). Remember also that the shading effect giving the three-dimensional image is not actually in a north–south position, but at 45° in accordance with the prism direction.
(iv) To introduce colours move the wedge into the most complementary position. To take advantage of the best three-dimensional position the wedge should be left in the first-order grey position and the sensitive tint plate introduced.
(v) Finally adjust the aperture diaphragm as necessary.

Examples of reflected light DIC can be found in Section 38.4.

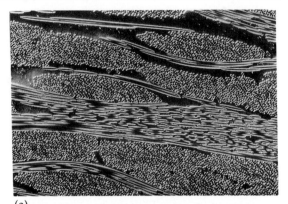

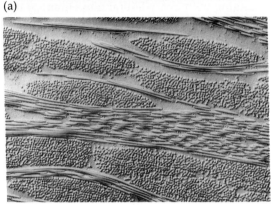

Figure 32.4 Wedge positioning. (a) Background extinction (×100 DIC) and (b) grey extinction (×100 DIC)

32.1.4 Transmitted light

When observing the ray path (Figure 32.2) we

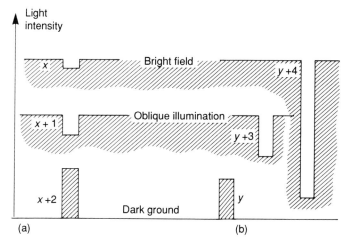

Figure 32.5 Light intensity comparison. (a) Scratched and (b) etched

see that the primary wedge (under the substage condenser) must be focused into a position ensuring it is imaged in the objective back focal plane. The secondary wedge, in the objective back focal plane, can be traversed to accommodate the chosen background colour. Imagine the light as it passes through the specimen; when these two rays pass through objects having different refractive indices their optical path difference will be different, giving that particular area a different colour to the background. This new colour and the background colour change as the wedge position changes and is indicative of a refractive index difference. The advantage of this system over phase contrast is its ability to show very small refractive index differences as colour differences that the eye can detect rather than undetected grey level differences. Unlike phase contrast, DIC does not exhibit specimen 'haloes'; nor does it restrict objective numerical aperture. This results in better clarity, image brightness and resolution.

32.2 Dark Ground

Often called dark field, dark ground is a technique where the subject stands out bright against a dark background. This is done for ease of identification with, say, small bacteria in transmitted light or fine scratch-type surface defects in reflected light. Tiny particles, precipitates or inclusions, too small for clear vision in bright field, can often be rendered visible in dark ground. It is also used, for example, in the electronics industry as an aid to measuring the width of plated layers.

It is wise to remember the use of critical illumination when using dark ground techniques. It can make a big difference to image brightness. When introducing the dark ground stop to a bright field image, one *must also fully open both field and aperture diaphragms*. Fine tuning for photomicrography will necessitate aperture diaphragm adjustment for maximum background contrast, the field diaphragm remaining open.

Figure 32.5 is representative of light intensity emanating from a scratched surface. The white line intensity against a black background, as seen in dark ground, gives the greatest contrast $(x+2)$. It is, however, the lowest in intensity. If dark ground was to be used on an etched subject, a grain boundary, for example, then the boundary could exhibit the greatest intensity. It would, however, manifest the least contrast, i.e. use the technique that gives the greatest contrast.

Analysis by dark ground is illustrated in Figure 32.6 (see Plate 13). The material under investigation is a composite with silicon carbide particles in an aluminium alloy matrix. An example of a well-prepared sample and procedure is given in Chapter 13 (Example 13.15). The problem when preparing this composite is in impressing the silicon carbide chip as the surface is ground. In bright field or DIC these fragmented particles look very much like the silicon carbide particulate and can give a totally erroneous Vf value. As

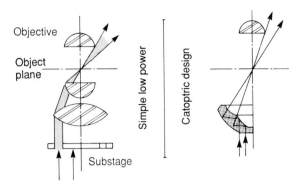

Figure 32.7 Dark ground condensers

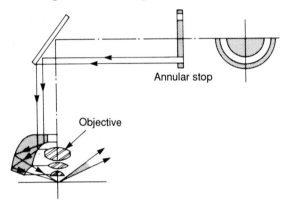

Figure 32.8 Reflected light dark field

shown in the two micrographs (Figure 32.6(a) and (b)), dark ground leaves the observer in no doubt as to which particle is a reimpressed fragment and which is the original silicon carbide particulate. This method of analysis was used to illustrate the preparation faults in a proposed method which, until that time, had been considered to be a good and faithful preparation.

The principle of dark ground ensures that no direct light from the illumination system enters the objective. Only light that has been diffracted as in transmitted light or deflected as in reflected light enters the objective. *It is most important* when setting up the microscope for dark ground applications to open *fully* the microscope field and aperture diaphragms; they can be adjusted if necessary when the subject has been imaged.

32.2.1 Transmitted light

This technique uses standard objectives but a special substage condenser. The dark ground condenser provides a hollow cone of light having an inner angle that corresponds to a numerical aperture exceeding that of the objective being used.

When following the ray path from the substage (Figure 32.7), if no object is present the orthoscopic image should remain black since no direct light enters the objective. It is only when an object diffracts or disperses the light that it can enter the objective and be observed. A patch stop can be fitted to standard condensers but this is usually limited to 0.6 NA and below. Because of internal reflections within the standard condenser some unwanted light enters the objective, particularly at medium powers. Therefore the design of the catoptric condenser is ideal for objectives up to 0.8 NA.

For high power dark ground (NA 0.5 to 0.95) a special dark ground condenser is used which must be oiled to the specimen slide for maximum NA and amplitude transmission. With these condensers it is essential, with higher than 0.95 NA objectives, to use a 'funnel stop' or built-in iris diaphragm within the objective. When using high NA condensers and objectives, extreme care is necessary to accurately align both to one single axis; failure to do so can completely nullify the specimen image.

32.2.2 Reflected light

In this application the use of special objectives incorporating a catoptric lens is necessary as well as the need for an annular ring in the illumination system. These dark ground objectives are suitable for both bright field and dark ground. They occasionally have a restricted NA in bright field.

Figure 32.8 shows the optical path. The subject, if devoid of any surface irregularities, will appear black. It is surprising, however, how what appears to be scratch free in bright field looks like the plan view of a main line railway station when observed in dark ground. The objective and annular stop adjustments so essential in transmitted light are not so necessary in reflected light. Because dark ground in reflected light is activated via one simple control, the necessity to open out all diaphragms can be overlooked, resulting in no image at all.

Examples of dark ground can be found in Section 13.5 and Chapter 38.

33

Video Imaging and Archiving

33.1 Closed Circuit Television (CCTV)

Closed circuit television was first introduced in microscopy some twenty years ago as a teaching aid in universities. Both life science and the traditionally named metallurgical facilities made good use of the large screen. Sometimes as many as four monitors were placed around the lecture room. When these systems were fitted with mechanical pointers within the microscope they offered a major step forward in visual aids. In those days the system was restricted to large groups due to the expense; they were also extremely critical to colour shifts with changes in light transmission and colour temperature. Having installed such a system in a northern university and balanced the camera optics and illumination system, a plate was fixed onto the monitor indicating the controls had been set and must not be touched. On a subsequent visit to this university department the picture on the screen was so poor as to prompt the remark 'who is the fool who has been messing about with this?' I am told that the 'Prof' who was within earshot quickly retreated. Thankfully the cost and the sensitivity have been much reduced. Today CCTV has many applications including video recording, measurement, multiviewing, archiving, etc.

33.1.1 Black and white

In general, most routine microscopes can accommodate black and white television systems providing the microscope has the mechanical structure to carry the camera. Ideally the microscope should be set for Köhler illumination and should have some facility for focusing the primary image at infinity into the camera. The light intensity from the microscope can be varied and is usually set for normal orthoscopic observation. With some very light sensitive systems this can be too bright and requires a reduction in intensity to achieve a satisfactory picture on the screen. The field diaphragm and aperture control are employed as for Köhler, the field iris therefore sharply focusing on the monitor when the specimen is in focus. The monitor field of view is often restricted, requiring the field iris to be adjusted less than the orthoscopic position. It is normal practice to enhance the image on the monitor by adjusting the contrast controls, often giving contrast beyond that of the normal orthoscopic image.

33.1.2 Colour

Colour CCTV is different to black and white in that it is essential to make all the necessary chromatic corrections to balance the illuminating source with that of the camera response (just as is necessary in colour photomicrography).

When the television crew interview a local resident the pictures are often taken outside the house rather than inside because the camera is balanced for daylight.

Firstly the microscope must be a well-corrected system offering Köhler illumination with chromatically compatible optical components (condenser, objective and eyepiece). The colour temperature of the lamp must be set and *unaltered*. This method, although ideal for photomicrography since the high lamp output gives the desirable short exposures, is often too bright for CCTV applications. To overcome this, the lamp *must not* be turned down; select the appropriate neutral density filter or purchase special CCTV objectives. These CCTV objectives balance the intensity such that changing objectives offers lower and identical amplitudes to the camera (light output would normally be less for a ×100 objective compared with a ×20 objective).

33.1.3 Quality of picture

The quality of both black and white and colour pictures can be improved by increasing the screen lines from the conventional 625. The chromatic quality for colour systems, however, is dependent upon the type of camera used, which varies from a trielectrode single tube camera, used with single three-colour correction monitors, to the three-tube camera where an independent three-system three-colour monitor is employed. The choice of camera and monitor will depend not only on cost but also on how sensitive the material being analysed is to chromatic errors.

The geologist, for example, whose method of analysis is based on discrete colour changes would have a different priority to the materialographer whose use of colour could be cosmetic. On some cameras, the colour balance can be achieved automatically by activating the 'auto-white' station, provided that white is present in the observed subject.

It is possible to focus the primary image of the object directly into the camera by replacing the microscope eyepiece and the camera lens with a special TV eyepiece. This is the recommended route but alternative methods such as collecting the primary image with the camera lens (the eyepiece removed) or collecting the primary image using a standard eyepiece (the camera lens removed) are also sometimes necessary.

Pictures can be enhanced electronically both vertically and horizontally. This renders pictures more distinct by analysing for instance the vertical portion of the video image between scanning lines and adding this to the video signal.

33.1.4 Magnification

Monitor magnification is often taken too seriously, since to double the magnification you could double the size of the screen while giving exactly the same information (still only having 625 lines). To have any real meaning magnification is best when based on the objective multiplied by the eyelens factor.

The reason for the last statement is because occasions arise where it is specified that the subject must be observed under a given magnification. This is an expression that usually relates to a particular objective numerical aperture to achieve a level of distinct vision. When it is expressed as a magnification and incorrectly interpreted for CCTV use, the subject could be unresolved.

33.1.5 Measurement

Measuring with the screen is a well-practised procedure and can be achieved by having a superimposed graticule or by calibrating your own scale (to be used as a ruler). The latter is done from the screen superimposition of a stage micrometer or alternative traceable reference. Remember that the actual dimensions of either the graticule or the stage micrometer will alter with every change of objective. It is possible to introduce a control unit to the camera signal in order to generate clear reference lines in both the vertical and horizontal modes. These lines can be calibrated to each objective with a memory system allowing direct reading to take place. A more sophisticated version shows an intensity curve as the line traverses, enabling accurate edge positions to be found, measurement no longer being operator subjective.

Video pointers are also available as a visual aid to the teaching requirement. The information can also be coupled to a commercial video cassette.

33.2 Methods of Archiving

Graphic archiving is the term used to describe the retention of the observed image and can be used for the recording of photographs or digital images from video printers. Current technology is allowing another form of archiving, that of optical disk storage. These disks can store black and white or coloured images taken directly from the television video camera. They can also be text, time and date generated, and can currently store 1000 colour or 2000 black and white images.

A quality control department requirement often includes recording test information related to materials. This information will include the image of the microstructure but this is a small part of the often complex test procedure (physical and chemical). It is possible therefore to have a software generated laboratory management plan where every test requirement will be displayed on a television screen, the progress of the test procedures and results being recorded along with the microstructural image. This total archiving is the way laboratories will be managed in the future and will therefore have an influence on the manner and type of micrographs taken. The following is a résumé of today's available options with price comparisons based on 1990 prices.

33.2.1 Polaroid

This is quick and simple to use, but requires care to overcome the gross colour shift when working in the area of reciprocity. Quality is very good, but perhaps not quite good enough for the most discerning, unless the positive/negative film is used. A disadvantage to the use of Polaroid is the current lack of inputting text to the photographed image.

    Cost: Black and white £ 0.62 each (type 667)
           Colour £ 0.96 each (type 669)
  Capital outlay (say) £1000.00–£2000.00

33.2.2 Kodak Ektachrome 100 plus (5×4)

This is quite an expensive procedure. It requires care to correct for colour shift but the superb quality, colour saturation and excellent tonal range make it (along with other types) the choice of the professionals.

    Cost: Film £ 1.02 per shot
  Capital outlay (say) £1000.00–2000.00

The cost could be reduced considerably by using a smaller format 35 mm; added to this cost must be the price of printing.

Although it is not instant its quality ensures a future as a technique for producing micrographs. When it is necessary to generate text on the photographed image a 35 mm system must be adopted.

(Since writing, this facility has become available on all format cameras.)

33.2.3 Video printers

These are available in colour or black and white prints from a variety of video sources. The resolution of the black and white picture can be as good as 1280 dots (horizontal) by 1160 lines (vertical), each dot capable of 64 grey levels. The colour print can be as good as 1280 dots by 576 lines with 64 density levels in each primary colour. Creating pictures from video signals overcomes the problems of exposure time relative to reciprocity. It is necessary to use the correct illuminating colour temperature prior to picture conversion, but to a certain extent this can be corrected within the copier processor.

Video printer sales in the United Kingdom amount to millions of pounds. They are obviously finding many applications in recording images from the microscope. Black and white prints are inexpensive and the quality could be considered as adequate. When compared to a black and white Polaroid print they are much inferior. Colour video prints, on the other hand, are half the price of a Polaroid picture with a similar quality ratio. Before all the video printer manufacturers make a personal attack on the author, let me say that this is a developing technology. If total laboratory management via computerized software is ever implemented then video printers will be the only option.

As with all video created signals the picture can be enhanced through brightness, contrast and sharpness.

    Cost: Black and white £0.05 each
           Colour £0.49 each (640 dots, 576 lines)

From the above it can be seen how competitive the black and white prints are.

Capital cost for this system is very high if you include the TV camera and monitor into the calculation. Assuming they already exist then the capital costs are as below:

Black and white £1000–£2000 for the large format high resolution system
Colour £2000–£4500

33.2.4 Video archiving

The quality of the picture and the optical disk image retention has little, if any, picture quality loss. Its function is extremely simple: just press the button; you have stored exactly what you see. This must have many attractions, and will, once the laboratory management software systems are implemented, be the favoured image storing method. The optical disk has a capacity for storing 1000 colour or 2000 black and white images; these images when recalled show no loss of quality but if you wish to reproduce the image via a video copier then the quality is influenced by the resolution of the processor.

Cost: Black and white £0.085 each
Colour £0.17 each
Capital outlay £6000 (not including video printer and video camera)

33.2.5 Summary

(i) Polaroid is expensive but instant and quality is good.
(ii) Conventional photomicrography is still the best for quality.
(iii) Video printers are an excellent economic buy for black and white.
(iv) Optical disks are inexpensive yet retain good picture quality. Unfortunately the capital outlay is high. Once the capital outlay includes the software package for laboratory management, this will become the favoured technique. Hopefully by that time the quality of video printers will have improved, avoiding the need for conventional photography.

34

Polarizing Light Microscopy

34.1 The Equipment (see Figure 34.1)

The first-order polarizing microscope is different to the conventional compound microscope in the following ways:

Polarizer. This is to be found below the substage and is graduated, 90° indexed and can be swung out of the optical axis. It is also necessary to have a polarizer in the incident or reflected light illuminator; this should also be rotatable and removable. The polarizer is normally set to the north–south vibration direction. The observer looking into the microscope is looking north.

Analyser. The analyser, situated between the objective and the eyepiece will be rotatable and graduated. The analyser vibrates east–west for extinction. Compensation slots are usually housed in the analyser block at 45° to the analyser vibration direction, situated below the analyser and orientated south–west.

Rotating stage. The rotating graduated stage which is used for specimen orientation will have centring screws in the absence of any objective centring. It is undesirable to have both objective and stage centring since one must be a fixed reference. This is sometimes overcome by having just one fixed station in the objective nosepiece which would be used for ×40 high NA objective. The $X-Y$ traverse of the specimen is achieved through an attachable mechanical stage fitted to the top surface of the rotating stage.

Objectives. Objectives for polarized light are said to be 'strain free'. This simply means that they exhibit less strain than those used for normal conventional use; strain free is often quoted as to within 1/500 wavelength. It is possible to use standard objectives and not experience any problems providing they are not fluorite glass combinations. One simple check is to observe extinction having first removed the specimen. The illumination is set at maximum voltage; the orthoscopic image should be black. Manufacturers look for extinction ratios between the polarizer and analyser of 20 000 to 1.

Bertrand lens. Minerals exhibit what are called interference (conoscopic) figures in the back focal plane (BFP) of the objective. The Bertrand lens is an intermediate lens between the objective and the eyepiece, which when 'tripped in' allows the observer to focus on the objective BFP instead of the object plane.

These lenses are often focusable, since the BFP varies with different objectives. They can be built into the microscope stand or be integral within the monocular head. In the

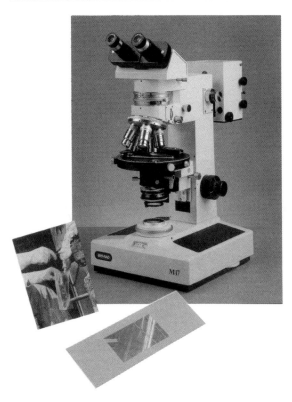

Figure 34.1 First-order polarizing microscope (Biorad Ltd)

absence of a Bertrand lens it is possible to replace the microscope eyepiece with an auxiliary telescope, sometimes called a phase telescope. This is preferable to a non-focusing Bertrand lens. When using the microscope with the Bertrand lens in position one is said to observe the object 'conoscopically'.

Substage condenser. It is important to keep all optical paths free from strain. This also applies to the substage condenser. Strain-free condensers are available but once again this does not imply that a non-strain-free condenser necessarily exhibits more strain.

Viewing heads. These are available as monocular, binocular or trinocular heads. Care must be taken to ensure that the polarizing effect down each tube of a binocular head is equal. This is not always the case and accounts for the apparent colour or intensity difference between eyepieces.

34.2 Practical Objectives

Microscope manufacturers offer polarizing accessories to fit routine orthoscopic instruments as well as supplying dedicated instruments called first-order polarizing microscopes. Polarizing microscopes have become an everyday tool for the geologist and petrologist where subject birefringence of the anisotropic specimen allows mineral identification when observed in transmitted light. Transmitted polarized light is also used in the identification of difficult types of mineral fibres such as asbestos. In fact the use of transmitted polarized light is used in a host of applications. Some of the most colourful and dramatic effects are achieved using chemical crystals. The recrystallization of ascorbic acid from solution when viewed between crossed polars can often be seen winning micrographic competitions. Polarized light when used in life science is also well documented, as in fact are most applications in transmitted light. Reflected light applications are not so well documented, which could account for their lack of use in microstructural analysis of solid subjects. This is a great pity since a completely new world is awaiting the microscopist with the introduction of the polarizer, analyser, tint plate and birefringent prism. The author uses these accessories every day and the following will illustrate some of the advantages achieved with simple accessories. Identifying:

(i) Non-metal inclusions in a metal
(ii) Non-mineral inclusion in a mineral
(iii) The integrity of free graphite in metals
(iv) Contrasting phases in a non-etched material
(v) Highlighting by colour grain orientation after anodizing
(vi) Highlighting birefringent subjects separately or within an isotropic material
(vii) Using the sensitive tint plate to add colour to grey level images
(viii) Combining with a birefringent prism to give three dimensional information
(ix) Separating dark holes (porosity) from dark phases
(x) Observing optical activity in fibres
(xi) Impressed abrasives from secondary phases or precipitates
(xii) Striation—bones and teeth

Other advantages are:

(i) Improving subject contrast by reducing glare
(ii) Extinguishing directional scratches

The geologist will identify minerals by morphology and colour relative to thickness and optical path difference between the two propagation directions. Polarizing can also be used in testing the strength of fibres. The chemist will notice colour changes with changes in substances or detect the absence of crystalline properties. The rheumatologist will be looking for specific optical responses to a crystalline joint.

The general objective is to make use of the specimen birefringence or bireflectance for the purpose of identifying, analysing and enhancing the observed image.

34.2.1 Polarizing principles

In order to fully utilize the polarizing microscope an understanding of specimen and instrument principles is necessary (see Figure 34.2). Consider the polarizer in the transmitted light mode as being set in the north–south direction. The polarizer will only transmit light vibrating in one direction, the remainder being absorbed. When this plane polarized light transmits along the optic axis and reaches the analyser, set to transmit only light vibrating in the east–west direction, all light will be absorbed; i.e. when the polars are crossed the light is extinguished.

To understand what happens to the polarized light as it transmits through a specimen we need to understand the two different types of specimens.

Isotropic

When light passes through an isotropic material it continues to vibrate in the same direction and axis when leaving as when entering. Light normally vibrates in a multitude of directions normal to the propagating ray and when transmitting through an isotropic material will continue on its same axis and vibration directions.

When light that has been plane polarized passes through an isotropic material it

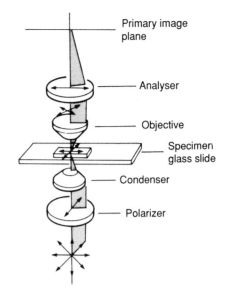

Figure 34.2 Crossed polars

continues on its same axis, vibrating in the created propagated direction without any change of orientation or increase in the number of propagations.

Anisotropic

When light passes through an anisotropic subject the situation is very different; it propagates into two vibration directions only. Therefore light propagated in a multitude of directions will, on passing through an anisotropic material, only vibrate in two directions. Light that has been plane polarized will also, on passing through an anisotropic material, only vibrate in two directions. Anisotropic materials are not always uniaxial; i.e. the single propagating axis may be resolved into two or three axes.

If we look in detail at the two vibrating directions we find they have a difficult wave front. Light passing along one vibration direction does so quicker than the other vibration direction. These two vibration directions are orientated at 90° to each other. Double refraction such as occurs in an anisotropic material is said to exhibit birefringence.

Back to our microscope. If we now place an isotropic material between cross-polars the object will be extinguished. If, however, we use an anisotropic material as the specimen between cross-polars the birefringent part of the specimen will split the light into two beams that are out of phase, perpendicular and vibrating at 90° to each other. Some of the light will now pass through the analyser where it recombines in the primary image plane to be imaged in the eyepiece. Therefore we now have two rays both resolved into the same plane which when recombined interfere to give polarization colours.

The factors governing these polarization colours are:

(a) thickness of the materials section and hence the reason why all geological thin sections are made to one thickness (30 μm or grey extinction for quartz) and
(b) specimen material birefringence (difference in refractive index) and dispersion.

It follows that if the thickness is known and is standard, the interference colour must be a result of a particular optical path difference (OPD), which in turn is the result of birefringence. Minerals can be identified by their manifest colour or refractive index difference associated with the OPD of the two vibration directions.

Summary

If an anisotropic material is viewed between crossed polars it will be seen when rotating the microscope stage to have two extinction points at 90° to each other. At 45° to these extinction points the subject will be at its brightest. The retardation (R) is relative to the birefringence and thickness and is expressed as $R=(N^2-N^1)T$ where T = thickness and N = refractive index. When the polarizing microscope is used in a qualitative mode the integrity of the colours can be sacrificed. This can be achieved by rotating the polarizer or the analyser. Rotation by a few degrees will lighten the dark shades and slowly change the colours; to rotate by 90° will allow all previously extinguished light to pass through. These colours can be enhanced by the insertion at 45° of a sensitive tint plate positioned anywhere between the polarizer and the analyser. A particular advantage of the first-order red sensitive tint plate is the conversion of first-order grey levels to a blue (extinction of black will change to magenta).

When light passes through a birefringent material, it does as stated above—change the vibration direction into the two propagation directions relative to the orientation of the positioned mineral. If the lenses in the objective were to be made from, say, fluorite which is birefringent, then the two propagated rays would be reorientated by the fluorite glass, making the geologist's task of identifying by specific colour impossible. Asbestos identification also by colour would equally lead to an erroneous analysis. This is the reason why only achromatic lenses are used on the first-order polarizing microscope (fluorite free).

Pleochroism

When rotating an anisotropic subject in plane polarized light (polarized only) look for any change of colour occurring within a specific grain.

It has been said that anisotropic minerals have two directions of propagation. It is these two directions that are being observed when looking for different colour tints. This effect is similar to a dichroic material where some wavelengths are reflected and others transmitted; this phenomenon is called pleochroism.

34.3 Techniques

34.3.1 Anisotropic thin sections

As has been mentioned, anisotropy has been exploited in chemistry, fibre identification, crystallography and geology. Microscope examination can be grouped into one of two main headings: orthoscopic observation for object plane analysis and conoscopic observation for objective back focal plane interpretation. The observed subject will be single crystals or combined crystals as in rocks and minerals. Rocks can themselves be subdivided into igneous or sedimentary classifications, unless they are metamorphic, which is a combination of both.

Igneous

This is the name given to rocks formed in the earth by heat and pressure, i.e. volcanic rock. The type and size of the crystals formed in this rock depend on heat and pressure which is indicative of the distance away from the centre of the volcano. Crystals formed nearer the centre have larger sized grains than those farther away. Figure 34.3 gives an indication of how, for example, grain size is an indication of just where the rocks originate. We can group by the amount of SiO_2 and pleochroism is another factor; together with quartz grey extinction at 30 µm all form part of the geologist's diagnostic tools. (Grey extinction is sometimes quoted as <30 µm; this figure would not take into account the thickness of the specimen to glass slide bonding agent.)

Sedimentary

As the name implies, this type of rock originates from the formation of sediments. Erosion by weather or water washing over igneous rocks will over a period of time cause small crystals to be washed away. These small particles, when combined with others, will eventually create another new rock, i.e. sedimentary.

Metamorphic

This type of rock includes those that were both igneous and sedimentary in origin but their characteristics have changed through being subjected to heat and pressure.

34.3.2 Compensation

We have mentioned the phenomenon of an anisotropic material. Its response to incident light is to transmit the incident ray in two distinct vibration directions at 90° to each other (see Figure 34.4). These two vibrations are designated as fast and slow directions, which is another way of describing an optical path difference due to a different refractive index between the two directions. This optical path difference (OPD) is a major key to unlocking the identity of the specimen. If, for example, we could introduce a retardation plate to the advanced direction that was equal to the OPD of the extinguished slow direction then the advanced rays would be extinguished. This is in fact what happens when we slide an optical wedge called a compensator along the fast vibration direction. The birefringent material when rotated between crossed polars exhibits bright and extinguished positions every 90°. The brightest position occurs at 45° from extinction; it is this position where the fast and slow propagation direction is in line with the

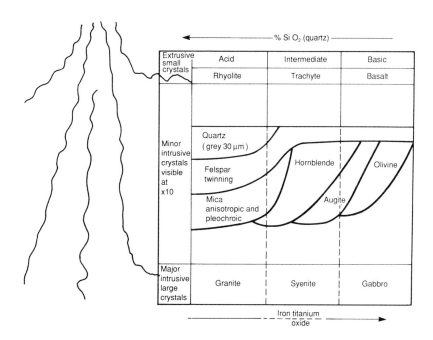

Figure 34.3 Igneous rocks

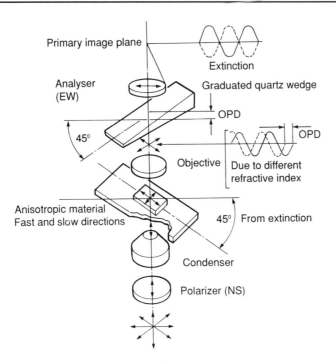

Figure 34.4 OPD compensation

compensator. In this position it is possible by sliding the uniaxial quartz wedge to extinguish the slow direction. The position of the quartz wedge in the optic axis is indicative of the OPD between the two vibration directions. From this information mineral or crystal identification can be made. When sliding the wedge into the optic axis at 45° to the crossed polars to extinguish the slow direction, a series of interference colours relating to the retardation of the two beams occurs.

The order (first, second, third, fourth) of these colours will indicate if the wedge is compensating or complementing the two beams. When the fast vibration direction of the wedge is arranged with the slow direction of the specimen the order of colours will be lowered and hence can be extinguished. If both fast vibration directions are present then the advancing colours (blue, yellow, orange, red, purple, etc.) will not extinguish; the specimen will have to be rotated 90° to allow compensation.

34.3.3 Sensitive tint plate

This is one of the most used accessories yet one of the least understood by the materialographer. It consists of a parallel plate of quartz (not a wedge as is the compensator), the thickness of which is equal to a path difference of one wavelength for green light. When the tint plate is positioned at 45° to the crossed polars the extinguished black light becomes blue/red— hence the reason for this plate being called a first-order red compensator. Its introduction enriches the subject colours, often making normal tints into vivid shades; its use by the geologist though is in detecting birefringence.

When light passes through a material of given refractive index it has a specific dispersion characteristic. It is the degree of dispersion that dictates the strength of colour when observing orthoscopically. When these colour differences are very weak it is necessary to introduce the sensitive tint to reorientate the extinguished ray into the first-order red position with subsequent shifts to all other colours.

34.3.4 Quarter wave plate

Crystals that themselves give high colours are less suitable for compensation with the sensitive tint. The quarter wave plate, giving a grey interference tint, can be used to determine the optical sign by raising or lowering the interference colour. The quarter wave plate is also used to produce circular polarized light when positioned at 45° to the plane polarized vibration direction—useful in outdoor photography.

Many snapshots taken by amateur photographers are, quite naturally, shot whilst on holiday. If holidays are by the sea, on sunny days (unknown to the photographer) the polarized reflections coming from the water will destroy the collected image and result in a foggy picture. Rather than blame the camera, it is better to avoid taking shots facing water, or, better still, fit a circular polarizer in front of the camera lens—the results will be unbelievably better.

Another extremely useful application of the quarter wave plate is in reducing glare due to back reflection from objectives. This is generally a reflected light phenomenon and is particularly important when the specimen reflections are low, i.e. coals, coke, ceramics, etc. Back reflections from objectives are inevitable in relected light and can cause the optical designer many problems. Some lower power objectives, for example, though suitable for transmitted light, are impossible to use in reflected light for no other reason than glare.

If, for example, 3% of the incoming illumination was scattered within the objective and 60% of the light hitting the specimen was reflected back into the primary image plane, the glare would have less prominence. If the specimen was to reflect back only 5% of the incident light then glare is a problem. This can be overcome by having a quarter wave plate between the specimen and the objective and set at 45° to the crossed polars. The incident beam will be circularly polarized as it transmits through the quarter wave plate and converted into plane polarized light vibrating at 90° to the original direction on its return. This beam is now transmitted by the analyser which extinguished the non-orientated glare.

Some manufacturers have introduced this technique with much success to stereoscopic microscopes when using the so-called coaxial illuminators.

34.4 Crystallography

Crystallography is a subject on its own and many books have been dedicated to just this area of science. It is not intended therefore to give anything but the briefest of introductions.

Most particles with the exception of glasses and some polymers are crystalline. Mineral crystals are inorganic chemical compounds and as a result of nature they tend to be anisotropic (but not necessarily). Crystals can be separate or in aggregates, their sizes varying from, say, 1 μm to many feet. Crystals exhibit many different properties and it is these the crystallographer must define for optical analysis.

Crystals are characterized by repetition of their constituent atoms in a three-dimensional array, quite unlike randomly placed atoms in a liquid or other non-crystalline material, as shown in Figure 34.5. The atomic arrangement in solid materials is obviously very important in identification. There are six systems, differing in atomic arrangement, three illustrated in Figure 34.6.

The spacing of atoms along crystallographic axes determines not only the crystal shape and angles but its optical properties. Each crystal in the cubic system has identical spacing of atoms along its three crystallographic axes. Therefore the optical properties of all cubic crystals are identical in all three axial directions. It follows that when the condition varies (as in the case with non-cubic crystals) so do the optical properties. The crystal shapes and properties vary enormously and it is sufficient to say it is these variations that the crystallographer is detecting in identifying a given crystal.

Although corresponding faces of different crystals of the same substance frequently exhibit considerable variations in shape and size their angular relationships do not vary. The determination of crystal form, habit and cleavage does not require the use of polarized light; therefore the analyser is removed.

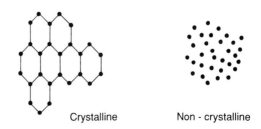

Figure 34.5 Constituent atoms

34.4.1 Conoscopic interference figures

Interference figures or patterns occur in the back focal plane of the objective when true crystals are positioned in the optic axes. The appearance of these figures depends upon (a) how many axes the crystal has, i.e. uniaxial one/biaxial two, etc., (b) the inclination of the axis relative to the microscope optic axis and (c) the sample thickness and degree of birefringence.

If the direction of the section is exactly perpendicular to the optic axis the interference figure remains stationary while the stage is rotated.

Uniaxial interference figures

(a) Section normal to optic axis; thick section or crystal of high birefringence
(b) Section normal to optic axis; thin section or crystal of low birefringence
(c) and (d) Movement of isogyres of uncentred figure without change of orientation on rotation of microscope stage (axis off centre)

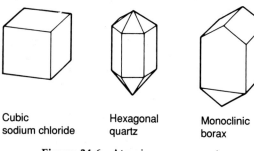

Cubic sodium chloride   Hexagonal quartz   Monoclinic borax

Figure 34.6 Atomic arrangement

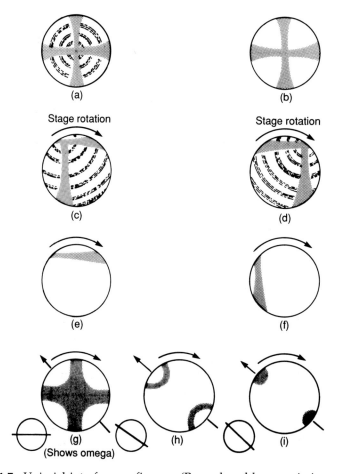

Figure 34.7 Uniaxial interference figures. (Reproduced by permission of Biorad)

(e) and (f) Movement of isogyres without change of orientation of figure outside field of view (axis not normal)

(g), (h) and (i) 'Flash' figure of sections cut parallel to optic axis; slight rotation causes rapid separation of isogyres

Biaxial interference figures

The rings of an interference figure are indicative of birefringence and thickness (or a combination of both); they also orientate about the crystal axis. It is not surprising therefore to find two sets of rings when observing a biaxial crystal, as shown in Figure 34.8.

(a) and (b) Section to the acute bisectrix; behaviour of figures on rotation of stage

(c), (d) and (e) Disposition of the isogyres at (c, e) extinction and (d) 45° positions

34.4.2 Determination of optic sign

This is the determination of the fast and slow directions caused by birefringence. Figure 34.9 shows how the quartz wedge, quarter wave plate and sensitive tint can be used.

34.5 Fibre Identification

Inhalation of noxious dust particles, in particular asbestos, can result in fatal respiratory disorders. The polarizing microscope is used in such fibre identification. There are several different types of asbestos (amosite, chrysotile, crocidolite); 'blue' asbestos (crocidolite) is considered to be particularly hazardous. Fortunately, the optical properties differ from other types, allowing initial analysis by polarized light supported by dispersion staining. All forms of asbestos are birefringent, the colours observed at the 45° position to extinction being a characteristic of particular types. Crocidolite exhibits pleochroism and sometimes gives a blue extinction whereas other types give total extinction. Crocidolite fibres are unlike any other type of asbestos in that they are *in most cases* length fast, i.e. the refractive index along the *length* of the fibre is lower than that observed across the *width*. This information could be derived by compensation as explained earlier. There is, however, a much quicker method and that is by observing interference colours using a sensitive tint plate, the polars being crossed. Crocidolite will be tinted yellow-orange when the fibre *length* is aligned with the slow direction of the tint plate and a bluish tint when the fibre *width* is aligned with the slow direction of the tint plate.

The reverse effect will manifest itself for other types of asbestos. Unfortunately some

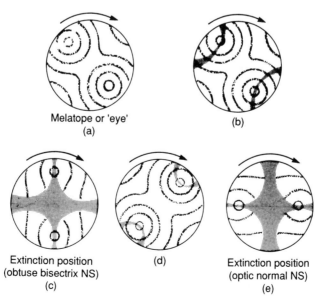

Figure 34.8 Biaxial interference figures. (Reproduced by permission of Biorad)

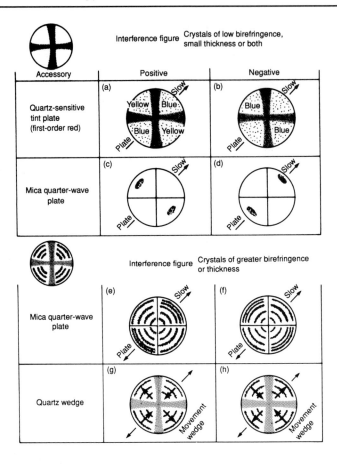

Figure 34.9 Determination of optic sign. (Reproduced by permission of Biorad)

crocidolite fibres are found to be length slow. Dispersion staining enables a more positive identification of all length slow and fast fibres.

34.6 Conditioning Samples for Polarized Light Applications

From what has gone before we could conclude that only samples exhibiting birefringence or bireflectance would manifest interference colours when observed in polarized light. It is possible to create this condition by electrolytic etching or chemical etching. Isotropic metals can be grain anisotropically orientated by anodic coating which is from quarter lambda thick. Aluminium alloys are a good example.

34.6.1 Anodizing

This has been used for many years to study the grain structure of aluminium. This electrolytic technique produces an interference film on the metal surface resulting in clearly defined coloured grain structures. This anodic oxidation is not anisotropic but it does cause light orientation at the metal/oxide interface, the degree of orientation varying between grains. The polarizing microscope can of course capitalize on this, the colours being readily changed by varying the uncrossed condition and also by introducing the sensitive tint plate. The anodized layer on some occasions exhibits dichroism (transmits some wavelengths, reflects others). This is used to advantage on stereoscopic microscopes where two oblique illuminators are used at 180° to each other— one lamp using, say, a red filter, the other a green.

Anodizing can also be successfully employed with other isotropic metals such as copper and titanium. It can also enhance the birefringence of some anisotropic materials.

34.6.2 Etchants

Although cubic isotropic crystal structures do not respond to polarized light they will on some occasions respond to tint etching where grain contrast can be observed with normal bright field illumination. A number of chemical etchants have been found to render many metals active to polarized light. An everyday example of this is etched pearlite which emits different interference colours relative to the uncrossed polar condition. Once again it is not the etchant that has created an anisotropic condition; certain wavelengths of light are extinguished or absorbed at specific polarizing directions. Sintered iron is used as one of the examples to follow, illustrating the resultant glare effect when pearlite is excited at specific angles. Pearlite under polarized light is also illustrated in the photomicrography chapter (Chapter 36) to demonstrate subject resolution relative to glare.

Copper, magnesium and titanium are all materials suitable for polarized light receptive etchants.

34.7 Practical Examples

34.7.1 Aluminium bronze

The micrograph of aluminium bronze was photographed using ×20/0.4 NA objective. This subject when observed in bright field illumination shows only grain boundaries, the overall background colours being uniform. A very different situation occurs with the introduction of the uncrossed polarizer and analyser, as illustrated in Figure 34.10 (see Plate 13). This is achieved by carefully orientating the polarizer to create an interesting well-resolved needle-like structure. Figure 34.10(b) is included since it illustrates what not to do. The micrograph is arguably very interesting as a picture but does little to illustrate the integrity of the subject. The object behind all optical techniques is to reveal more useful structural information, with the aesthetics taking second place. What DIC has revealed is the unwanted information relating to the mode of sample preparation.

34.7.2 Sintered iron

Most materialographers will at some time in their careers have looked at an etched steel (2% nital) similar to Figure 34.11 (Plate 13). In this particular case the samples were prepared to faithfully reveal the level of porosity. The purpose behind etching the subject was to closely investigate the integrity of the preparation in close proximity of each hole. Whenever a ductile material exhibiting holes (pores) is mechanically prepared the result can be lidding, resulting in a low incorrect porosity level, or alternatively edge rounding caused by nap polishing, resulting in high incorrect porosity results. By etching it was possible, via the pearlite laminations, to confirm that neither of the two conditions above had taken place.

Careful orientation of the polarizer will give a variable colour range. Note how the general structure within the pearlite has been lost (Figure 34.11(a)), with unresolved pearlite flares in the blue region and some areas of porosity losing their contrast. Figure 34.11(b) reveals the maximum information yet is still aesthetically pleasing.

34.7.3 Roman glass

This material was prepared with a view to observing the ageing effect. The subject was extremely brittle and friable causing undesirable scratches which could be confused with the striations. The manganese ingress being investigated could also be confused with induced porosity. With bright field illumination the dark phase is visible on a featureless background (Figure 34.12; Plate 14). Stopping down the aperture iris revealed lamellar striations in the glass and also reduced the glare but did little to suppress the surface scratches. Glare was reduced by introducing the polarizer, the analyser being set just short of extinction to contrast the lamellar striations. (The subject would be totally extinguished between cross-polars.) In this condition unidirectional scratches running 90° to the striations had to be reduced in prominence by rotating the specimen in relation to the polarizer vibration direction.

34.7.4 Titanium

Pure titanium is soft and can, by using classical material removal techniques, be deformed at the worked surface. This is overcome by using an electrolytic material removal procedure or an etch polishing technique. Figure 34.13 (Plate 14) shows titanium that has been electrolytically polished and etched. The grain structures can be displayed in a variety of colours by rotating the uncrossed polars, in particular when the sensitive tint plate is in position. Had this subject not been electrolytically polished, the massive twinning could have been the result of induced residual stress.

Other examples of the use of polarized light can be found in Chapters 37 and 38.

35

Fluorescence, Reflectance and Con-focal Microscopy

35.1 Fluorescence Microscopy

Fluorescence microscopes are used to detect the specimen reaction to a specific wavelength of light. It has been explained how light can be retarded when passing through the specimen, how the specimen can absorb different wavelengths and how they can be refracted, but on this occasion we are looking for an actual change in wavelength as the light passes through the subject; viz. light entering the specimen is blue yet is reemitted as green. This phenomenon is known as fluorescence and the light entering the subject is known as the excitation wavelength. This excitation wavelength must be matched to the particular receptive excitation of the specimen in order to achieve any response. Some materials, for example, can only be excited if radiated with ultraviolet light, others with violet, blue or red. Some can be excited with a broad band of blue (425 to 445 nm) while others are best suited to a specific 435 nm. What can be said is that the exciter wavelength is always shorter than its response. If the specimen continues to glow when the exciter beam is extinguished, which is most unlikely, the phenomenon is said to be phosphorescence.

35.1.1 Two basic types

Fluorescence can be one of two types, either primary or secondary. *Primary* is when the specimen itself fluoresces when appropriately excited; it will not have been induced by chemical additions such as dyes. Some naturally occurring minerals come into this category and can be excited in the ultraviolet (365 nm) region. This type of response is sometimes referred to as autofluorescence. *Secondary* fluorescence is when the specimen has been induced by the introduction of a dye or other such chemical means so that its existence can be detected when excited with the appropriate wavelength. Secondary fluorescence has been the domain of life sciences but is now being increasingly used in the materials field. The range of possible dyes goes from 365 to 510 nm, so it is essential before employing such techniques to establish (a) what wavelength is appropriate for excitation and (b) what type of bulb is rich in the required excitation wavelength. Once this information is established the task is quite

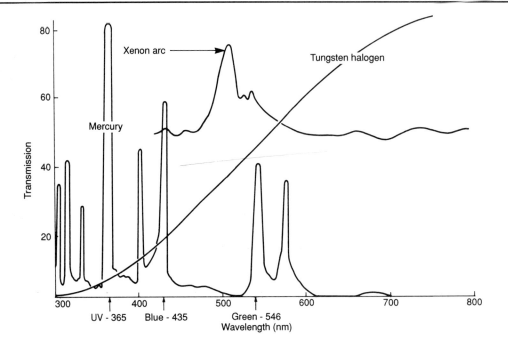

Figure 35.1 Transmission curves for different lamps

simple: (a) use an excitation filter in the illumination that transmits the excitation wavelength and (b) cancel this excitation wavelength by using a barrier filter situated in the optical train before the eyepiece. When operating in reflected light the necessary built-in semi-reflecting mirror will often be of the dichroic type. What this does is to reflect the excitation wavelength and transmit all others, i.e. only the specific excitation wavelength is incident on the specimen. When light reflects back from the specimen the dichroic mirror allows all non-excitation wavelengths to pass through, the excitation wavelength being reflected out of the optical path. In theory this would ensure that none of the harmful short wavelengths of light enter the eyepiece; in practice this is not so and hence the reason for still incorporating a barrier filter.

When a specific excitation is required from a broad band excitation filter then suppressor filters can be used. A parallel to this takes place in coloured photography where for different reasons one would wish to suppress a particular colour such as, say, red when one would use a red filter.

35.1.2 Microscope types

Fluorescence microscopes comprise two types: episcopic (reflected) and diascopic (transmitted). It has also proved advantageous to combine both episcopic fluorescence excitation with diascopic phase or differential interference contrast. In the materials field oblique light fluorescence is also utilized.

35.1.3 Applications in reflected light (industrial)

Although primary fluorescence has been used in mineralogy for many years there is now a growing tendency for its use in the secondary fluorescence mode in detecting porosity of ceramics, concrete, plasma coatings, etc. In spite of the author's views (see Chapter 33) this technique is enjoying popular use.

It is said to be used to assist in the identification of the true structure by adding a fluorescent chemical to the vacuum impregnation resins. When excited with short wavelength light the reemitted light will be in the longer visible region and will be confined

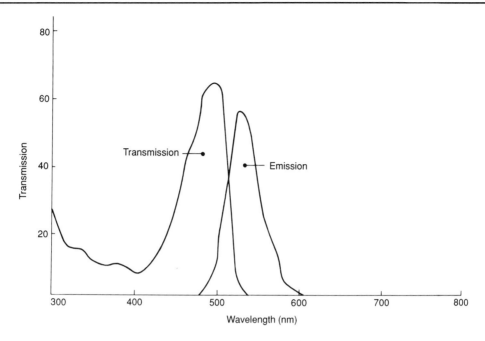

Figure 35.2 Transmission curve for specific dye

to the areas of resin. Before any filter combination can be selected it is essential to note the excitation peak for the dye and its emission peak.

Some dyes require excitation at a very specific wavelength, others are broad band excitable. With the specific dyes, sometimes called conjugates, it is necessary to ensure that the lamp spectral response is suitable. Figure 35.1 gives the spectral response from a mercury vapour lamp; these lamps are favoured for excitation in the specific regions of 365 (ultraviolet), 405 (violet) and 435 (blue). Peaks in between these values would require a different type of lamp. When broad band excitation is required the mercury vapour lamp, although suitable, is not necessarily the best.

Figure 35.1 also shows the transmission curves for the tungsten halogen bulb and the xenon arc: the tungsten lamp has little ultraviolet and only one-quarter of the mercury transmission in the 435 nm blue and the xenon is totally without an ultraviolet response, yet nearly equal to the mercury vapour lamp at 435 nm. To illustrate the importance of matching the lamp to the excitation response, the reader is asked to identify a suitable light source to satisfy a dye for material science applications where the excitation wavelength is 500 to 520 nm. From Figure 35.1 it can be seen that the tungsten halogen would be perfectly satisfactory, the xenon arc is much better and the mercury vapour lamp is of no use.

Some dyes have excitation peaks and emission peaks quite close to each other and these are usually in the longer wavelength region. One such dye has an excitation peak of 515 nm with an emission peak of 535 nm. With such a long excitation wavelength it is hardly necessary to excite with any of the short wavelength light usually necessary in fluorescence. The choice of lamps would also *exclude* the high cost, dangerous high pressure lamps such as the mercury vapour lamp. The quartz halogen or tungsten filament is rich in this region. Since the harmful short wavelengths of light do not require isolation the need for a barrier filter is not so great, thus allowing the background non-fluorescing subject to take the colour of the excitation region. This practice is to be avoided if there is any chance of isolated short wavelengths entering the eyepiece. A recent publication (1989) for such a dye says 'when excited by ultraviolet light',

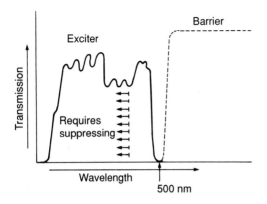

Figure 35.3 Transmission curves for filters

and then goes on to describe the excitation and emission peaks. This statement is extremely misleading and is the traditional way of describing the light necessary for fluorescence. The danger of misunderstanding such statements is when you use ultraviolet light (MV lamp). Not only is it not necessary but it could be extremely harmful if not adequately suppressed or extinguished in the barrier filter.

With most fluorescent dyes it is possible to obtain a transmission curve specific to that dye. One such example is shown in Figure 35.2 with its excitation and emission peaks. Given this information the reader can select from Figure 35.1 the most appropriate lamp. The next question is, 'What does the exciter filter transmission look like?' One such curve is shown in Figure 35.3.

Notice from Figure 35.3 how the exciter filter excites with the required dye excitation; it also isolates shorter wavelengths (ultraviolet and violet). The unwanted shorter wavelengths are suppressed prior to reaching the specimen by filtration, as shown in the diagram. Having isolated the specific excitation wavelength, this must be extinguished before leaving the eyepiece. The barrier filter, as shown, allows transmission of all wavelengths longer than the excitation wavelength and extinguishes all light of shorter wavelengths.

Excitation characteristics can be exploited in the standard reflected light microscope without resorting to special exciter, dichroic and barrier filters. One of the disadvantages of a fluorescing material is its tendency to 'glow' which can destroy the sharp features of a carefully resolved subject. It is this glowing effect, where light is not only changed in wavelength but also diffracted, that makes dark ground fluorescence microscopy an interesting technique. From the sketch and micrograph in Figure 35.4(a) (Plate 14) the reflected light from the specimen includes all wavelengths except the reorientated blue.

This image, when observed orthoscopically, is usually a grey colour and looks no different to a non-dyed resin. From Figure 35.4(b) (Plate 14) it can be seen how the non-dyed part of the specimen would be reflected out of the orthoscopic image. The only light entering the objective would be the green fluorescing reorientated rays from the resin and white light deflected from the sample.

35.1.4 Applications in reflected light (life science)

Although this book is dedicated to industrial applications the following is included as an illustration of the diverse applications for such an important optical technique.

A most important function of the fluorescent microscope is in the detection of what are known as antigen–antibody reactions. When a person or animal is infected with a disease, antibodies to that disease are produced and found in the blood. The serum taken from such an animal is called an immune serum or antiserum (immune since it is these that cause immunity to the disease—hence the term immunofluorescence). The antibodies produced are entirely specific for the disease which caused them to be produced. When this antiserum is mixed with its specific antigen (the disease-causing agent) a reaction will take place and the antibodies will become attached to the antigen. In practice some infected tissue from the patient is mixed with a known antiserum. The tissue will react only if the antibodies in the serum are specific to the disease from which the patient is suffering. The problem is, how can this reaction be observed?

Serum antibodies can be 'labelled' by chemically combining them with fluorescence dyes. This process is called conjugation. The fluorescence microscope can therefore detect the presence of any tissue reaction and hence a specific diagnosis.

Fluorescence, Reflectance and Con-focal Microscopy

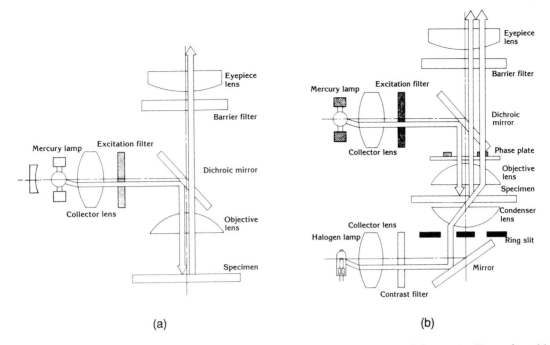

Figure 35.5 Fluorescence of optical train. (a) Episcopic and (b) episcopic and diascopic. (Reproduced by permission of Nikon)

35.1.5 The equipment

Figure 35.5 shows the optical train with the appropriate accessories for epi-fluorescence and the same situation with the addition of the diascopic mode. Note how the mercury vapour lamp is shown for all epi-applications. This type of lamp is more applicable when working with specific dyes. The majority of materials applications would be in the blue to green broad band excitation region and therefore would be *better* suited to the quartz halogen lamp.

Microscopes designed for fluorescence microscopy usually incorporate modules where the appropriate excitation, dichroic and barrier filter are built in, ensuring the correct combination and ensuring all harmful rays are extinguished before leaving the eyepiece. *Beware:* if you are adding fluorescence excitation to an existing microscope please ensure that all short wave rays are extinguished. Filters suitable for a quartz halogen lamp are not necessarily suitable with, say, mercury vapour (MV). (The MV lamp is rich in ultraviolet and at 435 nm it is a factor of about 10 times brighter.)

35.1.6 Objectives

Some manufacturers produce special fluorescence objectives which have an improved transmission ratio for ultraviolet and short wavelengths. The materialographer will not require these if exciting in the longer (blue) region. If the excited wavelength is to be extinguished by the barrier filter, which is necessary when exciting with short wavelengths, then image intensity will be much reduced. Since objective intensity is related to $NA^2/magnification$ it is wise to select the highest possible NA relative to the magnification. If image intensity is still low the total magnification can be reduced by using a lower magnification eyepiece (intensity is inversely related to the square of the magnification).

When taking photographs the exposure time will inevitably be long. To overcome this and also reduce reciprocity it is wise to avoid the use of large format cameras in favour of 35 mm.

35.2 Reflectance Microscopy

Many different materials have different

reflecting powers; some are bireflecting and will on rotation in polarized light show a difference of intensity or colour. Reflectance is the ratio of the intensity of the *reflected* wave to that of the *incident* wave and is called the reflecting power. Very often the reflecting power is given relative to a specific incident wavelength and varies with different colours. Measurement of reflectance is carried out in both air and oil, the latter being preferred for consistent results; R is less in oil than in air. Reflectance can therefore be used in the identification of minerals and is often associated with other mineral characteristics such as a micro indentation value. One everyday use for reflectance is in the grading of coking materials used in the manufacture of steel.

35.2.1 Specimen preparation

This plays a very important part in the integrity of the results. Minerals only reach their full reflecting power when the surface is completely plane and polished. Well-polished materials exhibiting polishing relief will result in erroneous readings. Hard minerals take longer to polish than softer minerals. Softer minerals could develop lamellar; some could tarnish with the polishing media or atmosphere. It is therefore important to have correct and consistent results; the need for traceability in all our sample preparation is paramount in reflectance. The practised method is to have a known certified standard of reflectance as near as possible to the reflectance of the prepared materials standard. The material standard must be prepared and its reflectivity checked against the certified standard. This method of auditing is to confirm the validity of the preparation.

35.2.2 The equipment

Reflectance microscopes look very much like a conventional compound microscope with the addition of a highly sensitive photomultiplier and a pin-hole centrable stop unit fitted in place of the camera unit. Pin-hole diaphragms vary in size from 50 μm upwards. The object area is centred within the eyepiece cross-wire and the appropriate diaphragm detected and positioned. The illumination must be stabilized. It is common practice to have the detector unit hooked up to some data processing system. When this technique is used in a quality control environment the equipment should have some automatic focusing arrangement. This is not just to make the system operator friendly, it is because reflectivity varies with different positions of focus.

35.3 Con-focal Scanning Optical Microscope (SOM)

Con-focal microscopy has until now been confined to life science applications. It could, however, be a valuable tool in materials science and is currently being investigated for use in quantifying surface topography of opaque materials.

Microstructural analysis and surface topography play a major role in the investigation of materials. These surfaces to be examined have usually been mechanically worked as in specimen preparation or modified as in hardness testing, wear testing, fatigue and stress testing. One of the themes of microscopy is to look within the subject (and not at it); this is what SOM offers. It allows a specific depth of field to be optically sliced, imaged and then recombined with as many other slices as the specimen will accommodate. The principle of SOM involves building up an image made from small pin-hole illuminated areas. The system incorporates a chopping device or spinning disc with thousands of pin-holes rotating at, say, 1000 rev/min. As the disc rotates these pin-holes move across the whole field of view, providing a full real-time image. The depth of the optical slice is determined by the working numerical aperture of the objective lens, the size of the illuminated path and the illumination wavelength and can be as small as 0.2 μm. The slices are combined to give an extended depth of field. What then are its advantages over, say, first-order grey DIC, surface interference and stylus profilometers? The first point is that the depth of field is not restricted to the objective depth of focus in the image plane. Surface interference is excellent when the surface topography variations are simple, such as a scratch on a smooth surface, but if the surface is a complicated irregular pattern the image would be confused.

First-order grey DIC would cope with this irregular surface providing the scattered light was not too great. Therefore subjects such as ceramics exhibiting this surface condition could best be accommodated using SOM.

Since the normal orthoscopic image is the result of combining rays emanating from the whole objective depth of field it is difficult to extinguish selective rays vibrating in the same direction. This would not prove to be a problem with slicing. Subjects which could benefit from this technique would be the low reflecting cokes and coals.

Many brittle subjects also fall into this low reflecting category. When these subjects are worked, material is removed by cracking. It is the length and type of crack that is proving to be technically very interesting if only it could be quantified. Using slicing methods SOM can often make the crack more visible by eliminating unwanted information resulting from other slices.

Since one of the principal advantages of SOM is its extended field of view it is well suited to the analysis of surfaces that have large changes in height beyond that of the normal objective depth of field. It is in this area that surface interference and differential interference are unable to cope and stylus profilometers will only give single line information. One good example would be a hardness indentation.

36

Photomicrography

36.1 Introduction

'Every picture tells a story.' This is very true in photomicrography, but it is often the wrong kind of story:

(i) Poor sample preparation
(ii) Microscope alignment errors
(iii) Poor film resolution
(iv) Filament structure imaged in picture
(v) Exposure incorrect
(vi) Dirt on the optics showing as ghosts in the picture
(vii) Colour shift due to reciprocity

The list is endless. Unfortunately when the microscopist takes a photomicrograph it is going to be judged by professional people—not like the family snapshot that is only seen by granny.

The quality of a technical report is very often judged by the impact of the photomicrographs. They must be good so as to enhance the written text. With a little care, guidance and awareness of the possible pitfalls perhaps we too can become professional photomicrographers.

The displayed micrograph should be accompanied by a measured bar scale. This bar scale can be superimposed onto the micrograph or directly underneath and should be expressed in micrometres. Since the scale of the micrograph is only relating to size it is important to denote the lens resolving power. This can be done by indicating the microscope objective lens numerical aperture. Magnification has traditionally been quoted with published micrographs. Although this is not necessarily a favoured approach, it should, when used, be quoted as the unity magnification, i.e. objective magnification $\times 10$. Unity magnification allows the resolution to be, at worst, approximated. Some microscopes permit the imaging of a graticule into the camera plane, in which case it will be necessary to calibrate the graduations against a known standard. When the micrograph is without superimposed graduations the bar scale can be inserted from a similar magnified micrograph of a stage micrometer. The author has a set of micrographs taken at different magnifications of a stage micrometer from which a bar scale can be easily transferred to any future micrograph.

36.2 Technical Considerations

Figure 36.1 shows the optical layout within a compound microscope which is typical of a photomicrographic system. The microscopist has a choice when selecting a photomicrographic system: he can opt for the independent camera viewfinder as in Figure 36.1 or focus

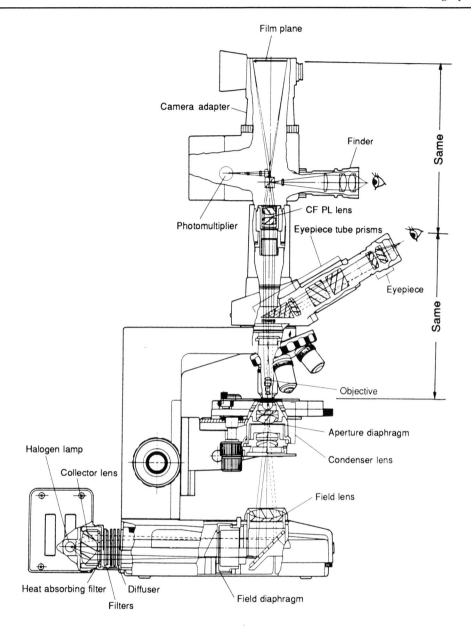

Figure 36.1 Photomicrographic system. (Reproduced by permission of Nikon)

through the binocular tube when the camera is a built-in integral part of the instrument construction. Both systems require careful diopter adjustment which will be covered later. If the user is a complete novice it is wise to use the independent camera viewfinder. When observing the vertical optical path it will be noticed that the primary image is collected by the photo-eyepiece; it then passes a prism system before being focused onto the film plane. This prism allows the object to be viewed in the camera viewfinder and is tripped-out of the optical view when exposing the film; this is an important feature. The obvious benefit is less glass to air elements resulting in sharper pictures. Above this prism system is the photomultiplier, detecting not just the level of illumination to the film plane but also the chromatic quality. This is the correct place to make such measurements since it will no longer be influenced by glass surfaces. The optical layout shown is for a 35 mm camera system. Large format cameras would require a diverging lens in order to fill the larger field. This lens obviously remains in the optical field during exposure and is the major cause of dirt in the optics, imaged as ghosts in the developed picture. This problem does not arise with horizontal optical trains.

When focusing through the camera viewfinder the specimen area of interest is often found to be positioned at an undesirable azimuth angle, i.e. not in line with *XY* coordinates. Unless the microscope is fitted with a rotating stage the whole camera assembly must be rotatable. The system illustrated is a simple direct optical train; other designs are more complicated such as Figures 23.1 and 23.2 where the optical train takes many different routes. This is done to make the instruments more modular and robust. It does, however, involve more glass to air surfaces not all tripped out during exposure and involves a greater optical path distance. For unity magnification the camera is positioned at a distance equal to the orthoscopic distance, as shown in Figure 36.1. When it is necessary to increase this distance it must repeat the full distance in order to achieve the same conditions; this again necessitates more glass to air surfaces and a loss of light intensity.

36.2.1 Vibrations

The materialographer is often involved in long (one minute plus) exposures when taking a photomicrograph; therefore the microscope should be sited in a vibration-free environment or placed on an antivibration platform.

36.2.2 Camera eyepiece

The compound microscope produces a virtual image with parallel rays emerging from the eyepiece. The photomicrographic camera requires the image of the object to be focused in the camera plane. This can be achieved by refocusing the orthoscopic image to convert the virtual to a real image. Unfortunately, this refocusing when using high NA objectives is usually accompanied by considerable spherical aberration. When using a normal eyepiece for photomicrography it will therefore be necessary to introduce a supplementary system of lenses to focus a real image into the camera plane, preventing the introduction of spherical aberration. This perhaps explains why the results achieved, using a single lens reflex camera on top of a standard microscope eyepiece, can be somewhat short of perfection. The most favoured approach is to use a camera eyepiece that projects a real image focused in the camera plane.

36.2.3 Measurement of exposure

As indicated above, take these measurements in an area as close as possible to the film plane.

36.2.4 Focusing

The human eye is not very sensitive when trying to focus images with a higher depth of focus than its own. In addition the high depth of field exhibited by low power objectives is manifest as much less in the film plane. We must therefore always use the largest magnification and objective NA that is tolerable. If this still necessitates the use of low power objectives additional accessories should be available enabling the focus to be accurately adjusted.

36.2.5 Diopter adjustments

When using the camera viewfinder the format graticule must be sharply focused independent of the subject image. In the centre of this graticule will be noticed a double cross-line. To focus this cross-line turn the diopter adjustment ring until the double cross-lines can be *clearly* recognized as two distinct lines (when out of focus they appear as one).

What is happening when carrying out this exercise is that the viewfinder front lens is moving backwards and forwards until it represents an identical optical distance to that of the film plane, while taking into account the variations in different observers' focusing abilities. Therefore the viewfinder position will vary from user to user but the instrument fine focusing control will always remain constant.

When all diopter adjustments have been made, focus an image of the subject. When this has been carried out try the *parallax test*, i.e. move the eye side by side and up and down. If the subject appears to sway or a single point moves then the object is not correctly focused and requires correction.

Remember, what the eye sees is not always what the camera sees. Take, for example, photomicrography in differential interference contrast. When this technique is adjusted to emphasize its three-dimensional capabilities (first-order grey) there is a tendency for the microscopist to focus on the background or the top of the feature of interest. Although the object depth of focus shows everything else to be in focus the image depth of focus at the film plane may not. On such occasions it is wise to take a microscope fine focusing reading for both 'hill and hole' and photograph at the average position. This knowledge can be put to good use when wishing to dampen or obscure the background. To do this focus sharply on the point of interest and rotate the fine focusing control one-quarter of the depth of field in the opposite direction to the background.

36.3 Microscope Photographic Systems

All microscopes are suitable for accepting cameras since they are able to project a *real* image onto a screen. The camera film is substituted for the screen; all that is required is a lightproof enclosure.

The simplest form is to adapt a single lens reflex (SLR) camera via a camera adaptor to an existing microscope eyepiece tube. The camera lens is not used since the microscope eyepiece is all that is required to project an image into the camera. Focusing is done through the camera viewfinder (hence the reason for a reflex camera). Exposure is through the camera focal plane shutter which on many SLR cameras is automatic.

We now have an automatic camera costing about one-tenth of that which the microscope manufacturer would charge. So what is wrong? It might be adequate, but it would help to take into consideration the following:

(i) The microscope eyepiece is not a finite secondary imaging system and therefore the image could lack clarity.
(ii) A restricted field of view could occur.
(iii) Magnification would be arbitrary.
(iv) Only at unity magnification (conjugates of the primary image) can we be sure of not incurring the optical designer's aberrations. The optical designer when computing any optical system is carefully balancing many aspects. The designer could, for example, make additional chromatic corrections at the expense of other aberrations. Increasing the working clearance or increasing the numerical aperture could incur other undesirable features. The designer can, rather than design-out an aberration, reimage into a non-conjugate plane. (This reimaged plane could coincide with the SLR.)
(v) The exposure is integrated on light intensity rather than specific features which could be a disadvantage.
(vi) Precise focusing is not easy through the viewfinder.
(vii) Focal plane shutters operating as an integral part of the camera system *can* cause vibration at the time of exposure.

The above should give some indication of why special microscope camera systems are necessary for optimum results.

There are additional accessories for use with SLR cameras which overcome some of the shortcomings given above. A special long tube camera adaptor can be used which repeats the microscope conjugate distance giving unity magnification. Additional improvements are achieved by using a finitely corrected camera eyepiece.

Microscopes designed to take cameras usually have separate ports where the camera chosen can be permanently fitted. These cameras can be 35 mm, Polaroid or cut film holders having the traditional (MPP) 5×4 format. Exposure units can be built in or free standing, automatic or manual, integrated or spot. The light from the microscope to the camera can be from 100% to nil. The microscope could have a light-reducing slide stopping extraneous light entering the eyepiece during long exposures.

A finite camera eyepiece focusing system can give superior results and is therefore recommended. The magnification in the camera plane takes on a much greater importance than is necessary; i.e. the magnification should be such as to represent the near whole size as imaged in the orthoscopic field using a ×10 eyepiece. This usually requires the use of low power camera eyepieces (×2). For reasons such as objective field curvature it may be necessary to increase the camera eyepiece magnification reducing the FOV. When the camera eyepiece magnification is increased, the field of view is reduced from the achieved orthoscopic using a ×10 eyepiece and empty magnification is incurred.

36.3.1 Selection of film

Quality micrographs usually result from using the slowest ASA on the largest format film. The ISO/ASA is related to the light sensitivity of the film; 3000 ISO/ASA is more sensitive than, say, 100 ISO/ASA, the latter requiring the longer exposure. It usually follows that the lower the ISO/ASA the higher the film resolution. If selection was as simple as the suggestion above then there would be a very limited choice of film and they would all be low ASA. Unfortunately many factors influence the selection, not least of these being the available light and its quality. If light levels are low then higher ASA films become necessary to avoid prolonged exposure times.

Film characteristics are illustrated in Figure 36.2 where the range of correct exposure is within the density slopes α and α_2. Figure 36.2 is also used to illustrate the sensitivity to film density. D-max (1) represents the highest density that a film can exhibit. In a positive film this represents the darkest area of the subject (in a negative it represents the brightest). D-min (5) represents the lowest density that a film can exhibit corresponding to the brightest highlights of the subject. Figure 36.2 shows the D-min values to be similar, the D-max being greater with the low ASA film. Imagine a subject to have subtle grey level changes that require recording, say, positions (1), (2) and (3). The low ASA film with the greatest slope difference will give the best contrast. At this point the reader could be saying 'why don't we use the low ASA films exclusively if this gives a more contrasting result?' The answer is, they are more critical to exact exposure conditions and require longer exposures. Table 36.1 illustrates some interesting features relative to high and low ASA films.

Black and white films

These are usually of the negative type and should be extremely fine grained. Different types are available with varying contrast. Some are very high contrast and require exact exposures. Black and white films are available with sensitivity to specific wavelength, e.g. blue and ultraviolet sensitive films. Others such as the panchromatic are sensitive to a wide range of wavelengths but are more sensitive to green. The recommendation for filters is (a) *always* use green unless the film has a specific

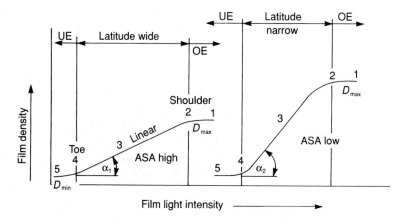

Figure 36.2 Film characteristics (UE = underexposure; OE = overexposure)

Table 36.1 Film features

Feature	High ASA	Low ASA
Light	Sensitive to low level light	High level necessary
Exposure latitude	Wide Tolerant to exposure error	Narrow Intolerant to exposure error
Contrast	Weak	Strong
Film resolution (grains)	Poor (large grains) Low film resolution	Good (fine grains) High film resolution
Uses	Poor light level such as dark ground, polarized light, DIC, etc., where there is a chance of vibration or movement	Whenever possible

non-green sensitivity and (b) *never* use blue because the film is generally insensitive to this wavelength. This last point is important since the image to the eye always looks better when a blue filter is present. There is an exception to the non-use of blue filters, however, and that is when it is required to emphasize a red subject (blue complements red).

Film Choice

Name	Manufacturer	ISO/ASA	Type
Panatomic-x	Kodak	32	High resolution
Neopan SS	Fuji	100	General use
Plus-XPan	Kodak	125	General use

Contact local film suppliers for a more comprehensive list.

Colour film

Colour films are much more sensitive to the correct wavelength balance of light. This is an inherent characteristic related to the response of the film to the emitted spectrum and can vary between different film suppliers. The chromatic response of the film is influenced by the colour temperature of the illuminated light which changes as the lamp voltage changes. It is essential therefore to know what lamp voltage setting more closely matches the kelvin scale or colour temperature of the selected film. The choice of film type is based on one of two film colour temperatures: (a) 4500 to 5400 kelvin for daylight and (b) 3200 kelvin for tungsten. Both microscope illumination and film type have to be matched in colour temperature and this is done by filters to be explained later.

Colour films are also supplied as a negative or reversal film. Colour reversal films are used to obtain colour slides and are generally recommended for excellent colour reproduction, resolution and colour contrast. These films are usually ASA 100 or less. When developing prints from other colour reversal film the colour shift can be quite pronounced and it is essential if this is to be avoided to supply the printing laboratory with a comparison colour match.

Negative films, on the other hand, are favoured for reproduction since they can be more readily adjusted in their colour balance during the printing process. The negative films also tend to be less sensitive to incorrect colour temperature balance. Whether the film is a negative or reversal it is wise to remember that the process laboratory, if not given a comparison print or colour slide, have no idea what the true colour should be.

Film choice

Kodachrome	25/64	Pro
Ektachrome	50	(Tungsten)
Ektachrome	100	Plus
Fujichrome	50	Velvia

Note: this is only for guidance; films are constantly being updated.

Polaroid film

Polaroid offer many instant films in various formats with an equal variety of contrasts. They can be positive or positive/negative. The image of Polaroid must *not* be judged by the snapshot quality produced by their instant cameras. Photomicrographs taken in black and white can challenge its rivals. The D-max/D-min ratios are wide enough to satisfy the most fastidious materialographer. The quality of colour prints is very often a great disappointment to many—this is in most cases the result of reciprocity failure. The correct colour correction (CC) filters will very quickly restore the user's faith in Polaroid colour.

36.4 Filters

Good photographs (micro or macro) require the observance of three major disciplines: good simple optics, fully corrected illumination and correct light filtering. Light filtering can be required for many different reasons and these will be discussed in detail. It could be said that light filtering is only necessary when conditions short of optimum are prevailing. We should therefore always attempt to improve those conditions prior to compensating. One example would be to use the correct colour temperature film rather than compensating for the illuminating source and vice versa. Another would be to increase the light gathering quality of the lens before compensating for reciprocity failure. These and many other examples could be used but we will still have to accommodate conditions where correction is necessary.

Those material scientists who have taken Polaroid *colour* pictures only to be disappointed by false colour will swell the ranks; this includes the writer. The reflected light microscope can at best only return about 12% of the originating light intensity. The need for filter corrections for reciprocity is therefore paramount. The writer contacted (September 1990) most major manufacturers/suppliers of microscopes and the major suppliers of film material for specific guidance and references to correct for reciprocity colour shifts. The following are some of the replies:

(i) Quote: 'With our automatic exposure meter we compensate for reciprocity.' This is half true in that the exposure time will be increased, but with this increase in time comes the need to rebalance (by filtration) the illuminating source which the automatic exposure meter *will not* do.
(ii) Quote: 'We know about these filters, have you tried your local camera shop? They stock "Wratten filters".' This again is true but a normal camera shop is unlikely to be able to offer guidance in reflected light microscopy and will not have any filters small enough to fit into the illuminating filter holders—that is assuming they have any at all.
(iii) Quote: 'I think I know what you mean, I will put some technical information in the post for you.' True to his word, one day a mini library arrived (with a note saying would I return it after use). Unfortunately as good as the information was, it made no reference to filters for reciprocity.

The above is one of the reasons for making this special section on filters.

The final example, if in fact this is necessary, is to quote a real life story the writer experienced some decade(s) ago as product manager of a British microscope manufacturer. A camera and film manufacturer had developed a new camera and wanted to illustrate its potential for photomicrography. We had in the past experienced great difficulty, in particular with fluorescence micrography, in reproducing these instant pictures with a true colour rendition. This new camera was to overcome all this!! The end result was that two hours later all we had was a collection of miserable pictures. Yes, the camera was very good but without reciprocity compensating filters the results were not. The demonstrator suggested the problems could exist in the microscope and that he had not experienced similar problems in the past. His portfolio of excellent examples were taken under high light levels, avoiding such problems.

36.4.1 Reasons and types of filters

The information supplied with a filter can include what is called a filter factor. This factor is used to multiply the normal exposure time when using the filter.

Colour temperature correction

The filters used to carry out this function are often called colour conversion filters and are designed to correct the colour balance of the illumination so that it matches the colour balance of the film. The filters are designed to:

(a) Increase the blue when converting tungsten illumination (3200 K) to daylight film and

(b) Increase the red-yellow when converting a daylight illumination (fluorescent tubes) to tungsten film.

Filters increasing blue to raise the colour temperature of illumination	Filters increasing red-yellow to lower the colour temperature of illumination	
80A 80B 80C 80D	85A 85B 85C	Filter code
4.0 3.4 2.0 1.3	1.3 1.7 1.7	Filter factor

The filters usually used in microscopy are:

80A to raise the temperature and
85B to lower the temperature.

If a daylight film is to be used with a tungsten filament lamp it will be necessary to (a) set the lamp voltage to the 3200 K position and (b) insert filter 80A into the illumination path. If, for example, the colour temperature should be lower than 3200 K it will be necessary to insert additional filters.

Fine tuning of the colour temperature is achieved by the use of 'light balancing filters'— the 81 series to lower and the 82 to raise the effective colour temperature of the light source. If, for example, we wished to raise the effective colour temperature we would supplement filter 80A with filter 82.

Colour compensating filters (designated CC)

These are used to suppress or enhance different colours in the finished photograph. This must only be carried out once the correct colour temperature of the illumination has been achieved. It could be true to say that we in the materials field should not use such filters since it would not be faithfully reproducing the true coloured image. If, however, this practice improves contrast, which it often can, then it must be acceptable. Normally any specific colour can be suppressed by using filters of a colour that is complementary to that colour (see Figure 36.3).

The filter strength increases from CC05 to CC50. To carry a full set would require 36 different filters; the 6 above are mid-range filter strength. Filter strength increases with number.

Contrasts in black and white

There is a tendency for the microscopist to set up a microscope for Köhler illumination and then adjust the aperture iris to achieve *contrast*. On occasions this is necessary but full aperture and maximum resolution should be pursued first. Note the following:

(i) Most subject detail stands out more if we render the specimen darker (reduce the exposure).

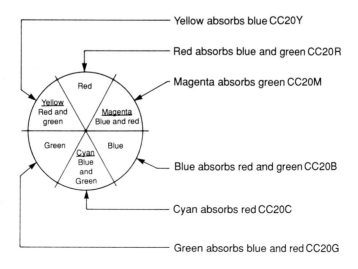

Figure 36.3 Complementary colours

(ii) To enhance contrast it is more effective to darken, by filtration, that part of the specimen containing the most important information. Complementary filters will darken the subject when photographing in black and white. Not all subjects manifest themselves in colour; it may therefore be necessary to optically stain in order to enhance the required detail.
(iii) Should the subject be pale and of one colour then a filter of the opposite (complementary) colour will enhance specimen detail.
(iv) A very dense subject can be diluted and thus reveal more detail by using a filter of the same colour.
(v) Contrast using a black and white film; in particular a black and white subject *must* encompass filters that take into account the optical corrections of the objective being used. Achromatic objectives are corrected for spherical aberration in the green *only* region. The green region is a narrow band spectrum and responds best to narrow band illumination. The green 58 broad band filter is often chosen; supplementing the 58 green with a 15 deep yellow will narrow the spectrum band transmission. Apochromatic lenses are corrected for not just the green region but also the blue; improved resolution can then be achieved by using a shorter wavelength filter. The danger with the blue is that it is not always compatible with the balance of some films. (Panchromatic is highly sensitive at the blue end of the spectrum.)

Contrast in colour

Contrast filters are not intended to change the overall colour balance of the picture as do colour compensating filters. They are used to contrast particular colours, viz.

Wratten 22 Deep orange	Blue-green absorption
30 Light magenta	Green absorption
35 Purple	Total green absorption
45 Blue-green	Ultraviolet and red absorption

Special filters

(i) Ultraviolet radiation from the illuminating source can cause a bluish haze to appear on the developed film; eliminate with a Wratten 2A filter.
(ii) Didymium filters enhance reds and blues.
(iii) Neutral density filters control the amplitude of the illumination.
(iv) Heat absorbing filters reduce infrared radiation. Along with polarizing plates they will require compensation to eliminate green; use CC5M for example.

Reciprocity failure

Having first explored all avenues to increase the level of illumination, when exposing films beyond their linear intensity/exposure characteristics we need to (a) increase the exposure time and (b) use colour correction filters to compensate for colour shifts. This is the most important consideration for the materialographer when taking, for example, 'instant' photographs. Films are usually rated for exposure of between 0.2 and 0.5 seconds; unfortunately we often have to use 10 to 30 second exposures, and occasionally as much as one hour.

Film manufacturers could be encouraged to quote a more comprehensive guide to CC filters to overcome reciprocity. Table 36.2 is one such example which is included with the Polaroid type 59, ISO 80/20° sheet film pack, intended for microscope users. For example, taking the 30 second exposure, to correct for reciprocity (a) use the modified film speed setting of ISO 16/13° and (b) insert filters CC10M+CC10R into the illumination path.

Table 36.2 is intended for use when the specimen colours are a direct result of the specimen and have not been optically stained. Since the materials colour application will be predominantly optically stained it will be

Table 36.2 Guide to CC filters

Exposure time (s)	Filters	New film speed setting
1/8	CC20C+CC30B	ISO 25/15°
1	CC30B	ISO 25/15°
5	CC20M+CC10B	ISO 20/14°
10	CC20M	ISO 16/13°
30	CC10M+CC10R	ISO 16/13°

necessary to reduce the blue at short exposures and be blue free from 5 seconds onwards. It will also be necessary to introduce the red filter at the 2 to 3 second exposure and increase in strength the red, perhaps reducing the magenta as exposure time increases.

Available filters

The recommended filters given are based on the Kodak range. The author is not aware of any microscope manufacturer offering a suitable range. Unfortunately one will have to buy sheets of 50 mm × 50 mm and cut to size to suit your own particular microscope.

To conclude this section, the writer now using CC filters has overcome a lifetime of reciprocity colour shift problems. It is hoped that the reader will equally benefit and perhaps suppliers of film will be more forthcoming with CC corrections to overcome reciprocity failure.

Optical techniques—film and filter guide-lines

To offer instant solutions to illumination chromatic corrections is not possible. The following may, along with the previous documented information, guide the user in the right direction:

Technique	Guide-lines
DIC	*Black and white film*: use green filter. Kodak Technical Pan Film/ASA 100 for extra contrast.
	Colour film: use CC5M to compensate for green from heat absorbing filters or polarizers. Use a blue filter if voltage is low.
Fluorescent/ dark ground	Use high sensitive films ASA 400. If the area to be exposed is small in comparison to the background the total integrated exposure will be too long; therefore reduce by one stop.
Reduce format	If the chosen format is 5×4, change to 35 mm. This will increase light intensity by a factor of as much as 25/1. This can totally eliminate reciprocity problems.

36.4.2 Data imprinting

Printing information on every micrograph via a data-back facility is an extremely beneficial accessory to all cameras. This practice has previously been restricted to the 35 mm camera but the author has now used a large format data-back system, currently available (June 1991), making it possible to data all micrographs. Unfortunately, this camera back was not available when the micrographs for this book were produced. Had it been available such information as the following could have been printed on the edge of each film:

(i) Random alpha numerical data (eight digits)
(ii) Shutter speed 0.01 seconds to 16 min 39 seconds
(iii) ISO film speed
(iv) Exposure compensation
(v) Frame number/film number
(vi) Bar scale value in micrometres
(vii) Photographic magnification
(viii) Photometric value

(For publications the bar scale would be chosen.)

36.5 Photomicrographs— Check-list for Success

Will your photographs be worth 1000 words? Remember, when you take photomicrographs at work you become a professional photographer. Before embarking on future ventures give thought to the following metallographer's house rules.

36.5.1 The equipment

Do we need expensive equipment to achieve good results? No, anything beyond a well-corrected objective, a finite focusing eyepiece and the camera back is superfluous. Sophisticated metallographs are very convenient but for every extra glass to air surface there corresponds an extra loss of image quality.

To complement the basic microscope kit one must have some viewing eyepiece system to view the primary image. This will inevitably involve the use of an axial prism which must be removed from the optic axis during exposure time. Shutter activation can be manual or electronic and is best positioned as far away from the focal plane as possible. To check that all prisms, mirrors, etc., have been removed from the camera objective axis, the camera focusing eyepiece image should go black during exposure. The camera back must be at unity magnification, which is achieved when the camera is positioned in a duplicate conjugate plane to the object primary image distance. The camera field can be varied by the eyepiece magnification, the eyepiece being specially designed to focus on the secondary image plane (normal microscope eyepieces are not designed to do this). Determining the film exposure should be made as near to the film plane as possible. The system should be free from vibration.

36.5.2 Cleaning

A microscope service engineer cannot be called upon every time a photograph is taken. Therefore, lens cleaning needs attention, but please only use a fine haired brush and a lens blower. Lens tissue with a little alcohol can be used by the more experienced, but only if all else fails. Dust in the final print is obviously imaged. Cleaning should therefore only take place at or near conjugate planes. These are: (a) objective front lens; (b) eyepiece top surface; (c) converging or diverging lens surface nearest the camera; (d) filter/polarizing slides that are situated near the field diaphragm.

Procedure

Using a well-illuminated source room look for glass surface reflections allowing dust to become visible. This in itself is an extremely sensitive test and the user must practise with different reflecting angles before acquiring the expertise of detecting small dust particles. The light source for this test could be daylight and fluorescent tube or any simple light bulb. Well-corrected illuminators are not to be recommended; the light intensity obscures the dust. If left open to the atmosphere, a clean lens surface could be dusty within minutes. Using a fine haired brush and/or blower remove any dust from the camera diverging lens usually present with large format cameras. This is usually the major source of imaged dust. The camera eyepiece top lens and the objective bottom lens must next be inspected.

If any of these surfaces are stained then careful wiping with lens tissue is necessary *but* not until all dust and *grit* have first been brushed off. When wiping any lens it is often wise to use a circular motion starting at the centre and moving outwards. With small diameter lenses it will be necessary to use a piece of dogwood (soft wood) as a support for the lens tissue. Do not use pencils or matchsticks as they are too hard.

Now look at all other surfaces near a conjugate plane, i.e. filters, polarizer and in particular any imaged graticule. Make sure the lamp condenser is clean and also any other surface near the field diaphragm. Finally, blow any dirt away from the reflected light mirror but avoid making contact with this surface. If necessary this procedure should be carried out every day. When photographs are taken regularly have the glass surfaces professionally cleaned more frequently than the normal servicing period.

36.5.3 Microscope set for Köhler illumination

It cannot be taken for granted that most microscopists understand Köhler illumination. On a recent tour of the major cities of Britain, the majority of materials scientists admitted to having no formal training in microscopy.

The basic fault with microscopes designed for the materialographer is in the illumination; it either has a diffuser fitted and/or the filament is not imaged in the objective back focal plane. The diffuser should be removed, by force if necessary, and the filament image corrected by sliding the lamp housing, usually away from the lamp abutment face, until sharply focused. This latter adjustment is worth checking with each objective since objective back focal planes vary. The quality of the illumination plays one of the most important roles in forming good images.

36.5.4 The specimen

Before taking a picture we need to ask ourselves what information we require from the specimen. This will have dictated the method chosen for its preparation. For example, if we require to illustrate, say, a friable structural phase then the preparation techniques could result in a scratch substrate. If this is necessary in order not to destroy the friable phase then we may have to find ways of extinguishing part or all of this background when photographing (polarized light). This same subject taken in a dark field would obviously be unacceptable. If we need to illustrate structural phase changes then the preparation techniques would include a prolonged polishing step with a nap-type cloth. Optically this would best be observed in the first-order grey position with differential interference contrast. Fine surface topography changes would require a nil-nap-type cloth and be optically observed in, say, first-order red DIC etc.

36.5.5 Quantitative

There are many things the microscopist can do that actually have an effect on picture quality beyond that of aesthetics:

(i) *Lamp*. The colour film chosen will be well balanced for a particular colour temperature of illumination which must be observed to achieve true colour rendition in the photograph. The lamp must be set to this temperature. If the intensity is too high for orthoscopic observation it must be reduced with neutral density filters. When working with black and white film any colour temperature can be used but a green filter must be introduced to be compatible with the objective correction for spherical and chromatic aberrations (the film is also more sensitive to green). Spectral response from the lamp must be compatible with the coloured film material if the image is created directly from the lamp and is not a result of optical staining (interference colours such as DIC). A mercury vapour lamp with its peaked spectral output is unsuitable for colour photography. However, if the specimen colour is achieved through optical staining it is perfectly satisfactory. The colour temperature of the lamp must be capable of being set to the film temperature, i.e. daylight films (5400 K) for xenon/electric flash of fluorescent lamps and tungsten films (3200 K) for filament bulbs. Tungsten filaments require setting to 3200 K and the voltage setting should be provided. If in doubt set to maximum volts. Filters are available for converting daylight to tungsten and vice versa should it prove necessary.

Even though the illumination has been set to the correct colour temperature to match the film the results are not always a perfect colour reproduction. A simple test is if the colour temperature is too high the whites will be bluish. If the colour temperature is low the lighter colours will have a hint of red.

When the microscope creates colours such as in polarized light, Nomarski, fluorescence, etc., then this is to be considered as a daylight response irrespective of the illuminating source; i.e. colour achieved by optical staining should be photographed with a daylight film irrespective of the illuminating source.

(ii) *Objective*. Select the objective with the highest numerical aperture compatible with an adequate field of view. By so doing one is ensuring sharp photographs since the depth of focus increases in the camera plane as the depth of field in the object plane reduces. Objectives with low magnifications and hence a low numerical aperture have a large depth of field in the object plane but a smaller depth of focus in the camera.

With objectives having a numerical aperture of 0.10 and less (magnifications of $\times 5$ and less at 160 mm tube length) the depth of field is 55 μm and greater. Unfortunately this is not the case in the camera plane where it is considerably less. The $\times 100$ objective, on the other hand, having a numerical aperture of 1.25 has a depth of field of about 0.5 μm with a corresponding camera plane depth of focus much greater. It is easy to see the reason why the experienced operator achieves sharp photographs at high magnification. Low magnifications, which you would expect to be easy, come out blurred and out of focus.

If one is forced to take low magnification (NA) shots then great care is necessary in focusing exactly in the centre of the object depth of field position. Optical aids are available on some microscopes to achieve this. Without such luxuries it is necessary to take a fine focus reading in the upper and lower limit to find the centre position.

Objective chromatic corrections must also be considered. In general, use achromatic lenses unless absolute crispness of fine detail is essential when apochromatic or semi-apochromatic lenses are advisable.

The eyepiece will have to be low magnification to ensure an appropriate field of view or framing image. The eyepiece should also compensate for chromatic differences of magnification.

(iii) *Technique.* Select the appropriate technique for revealing the structural information, i.e. bright field, dark field, polarizing, DIC, etc.

(iv) *Field diaphragm.* This must be set within the orthoscopic field of view and touching the periphery of the camera format. The camera format is usually an integral part of the viewing eyepiece. The reason for stopping down the field diaphragm is to restrict unwanted scattered light, created outside the camera format, from entering the optical path.

(v) *Aperture diaphragm.* The choice is to stop down the iris to the conventional two-thirds open position, with its so-called increased contrast or to pursue maximum resolution photomicrography. The preference of the author would be to always target for the latter. The first consideration before adjusting the aperture diaphragm is, 'Does the subject require additional depth of field?' If the *thin* sectioned specimen is extremely thin (which is a tendency) or if the *thick* section is very planar then the specimen does not demand this increased df. With the aperture iris set open at 100% of the objective BFP diameter, carefully observe the orthoscopic image for glare.

To alleviate glare, if present, explore the use of the polarizer with its various azimuth positions; finally introduce the analyser (uncrossed). If glare can be reduced by this technique it allows micrographs to be taken at full aperture, giving maximum resolution.

The introduction of the polarizer can with certain specimens aggravate the glare situation. One such example is pearlite in steel (Figure 36.4; see Plate 15). The full aperture with the analyser and polarizer uncrossed is shown at (a) where the pearlite appears unresolved, yet at (b) without the polarizer and at seven-tenths aperture pearlite appears resolved.

(vi) *Film selection.* This depends very often on circumstances. If instant results are required then one must use Polaroid or a video printer. Given a choice, go for the large format (5×4) cut film material with its excellent resolution or 35 mm transparency when light levels are very low.

The 5×4 cut film is shortly to be made available in double pack form which eliminates the necessity for dark room loading and means one cut film holder is all that is necessary.

Metallographs very often incorporate a diverging lens to image the full field of a large format camera. They also involve the use of a fully reflecting mirror to project the image into the camera (hence the term projection microscope) (see Figure 36.5). If, for example, the exposure time proved to be 500 s, this can be reduced to as low as 20 s by using a 35 mm camera (with the same ASA, same field of view). A sign of reciprocity failure is when the developed colour prints have a green cast.

(vii) *Exposure time.* This must be short and where possible must not incur reciprocity failure. Exposure is relative to intensity and time ($E = I \times T$) as an arithmetical curve over a given distance beyond which corrections are necessary. It is most important to remember that corrections for reciprocity failure by increased time do not take into consideration the film's spectral response; this may also require filter correction.

When using a full field integrated exposure device, as opposed to a spot measuring system, it will be necessary to compensate for variations in imaged contrast. Integrated exposure meters are calibrated to give the correct exposure when the object is uniformly distributed in medium contrast throughout the

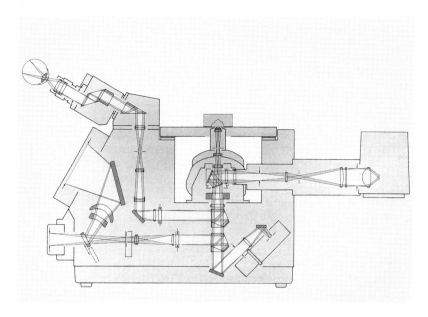

Figure 36.5 Projection microscope. (Reproduced by permission of Carl Zeiss)

whole FOV. Therefore, predominantly bright subjects will require less exposure than is calibrated and dark subjects more. In general, a very useful guide to exposure is to underexpose for positive film and overexpose for negative film. Very often, when exposing for plated layers, the area of interest has a different light level to the rest of the viewed image; the exposure should therefore be adjusted accordingly. By doing this you will achieve a picture with more contrast. When developing the print it is always possible to increase the illumination if the film has too much contrast, but it is impossible to introduce contrast if the film is 'washed out'. Exposure adjustments of half a stop are recommended as a general practice. To reduce reciprocity use one whole stop *but* notify the developing laboratory to enable a ×2 development to be used.

Figure 36.6 is an example of the more sophisticated microscope incorporating three different cameras (one not shown). The automatic exposure panel located at the front includes functions such as 1% spot metering, 30% integrated metering, automatic exposure lock and automatic compensation for film reciprocity failure. (Remember the automatic compensation for film reciprocity does not take into consideration colour shifts; these still require compensation by CC filters.)

36.5.6 Qualitative

This is the stage where the operator can create the aesthetics so necessary to produce photographs with life. Interesting photographs are important and the photographer is to be admired as long as the structural information is not compromised—this is the time to be imaginative.

(i) *Filters*. These can be used to advantage in highlighting or suppressing natural colours emanating from the specimen. Blue, for example, would suppress a copper phase yet enhance the features of, say, a grey structure.
(ii) *Polarizer*. This can be used not just in the control of glare and scratch directions but also with subjects exhibiting pleochroism. Find the most interesting position by specimen orientation.
(iii) *First-order red compensator*. This is used to add the rewarding warmth of selected colour background.

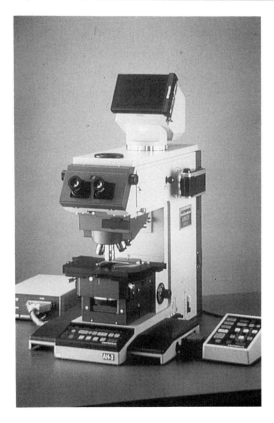

Figure 36.6 Advanced photomicrographic functions. (Reproduced by permission of Olympus)

(iv) *Analyser.* This is always interesting when combined with the polarizer but please experiment with different orientations with or without the sensitive tint.

(v) *Birefringent wedge (DIC).* This can be used on its own or with just the polarizer or analyser to give what appears to be an interesting picture, but beware of the ghost image often present when both polarizer and analyser are not present. The most interesting and rewarding structural information is achieved when the birefringent wedge is placed between crossed polars, and is then very carefully moved until the first-order grey is achieved. The first-order grey is achieved by very carefully moving the wedge from the extinguished position; this can, as shown in Figure 36.7 (see Plate 15), emphasize slopes as upwards or downwards.

A perceived hill or hole is related to how the brain recognizes shadows; i.e. when the shadow is below the subject it is recognized as a protrusion, when the shadow is above the subject it is recognized as a hole. This is also the case with a micrograph. Therefore by rotating 180° it is possible to reverse the surface perception. This is a disadvantage, however, as the micrograph then looks upside down.

The first-order red sensitive tint when introduced to this combination results in a blue (not red) background. Rotations of the polarizer will then result in interesting colours without a great loss of structural depth. Because the wedge causes interference in one axis only, it is wise, when possible, to rotate the subject into its best position.

(vi) *Increased film speed or contrast.* Having exhausted all the filtering avenues to improve contrast there are still two more to consider:

(a) increasing duration of the first development stage and
(b) change in developer type or temperature.

A photomicrographer's preference may make it desirable to increase the effective speed or contrast by underexposing the film. This is achieved, for example, by setting the ASA rating at 100 for a 50 ASA film and then increasing the duration of the first development step for a ×2 increase in nominal film speed. Although this method improves specimen subject contrast it does have a 'trade-off' with a slight reduction in background density. This is immaterial in bright field since the specimen subject is rarely reproduced at that level of blackness. First-order DIC with its extremely long exposures can also benefit—not necessarily for improved contrast but the reduction in exposure time.

(vii) *Dark adapt.* Having made all the necessary adjustments to create that special micrograph, all that is now necessary is to fine tune. To carry out this to the optimum one must turn out all the lights and dark adapt. When fine tuning in a darkened room it is possible to train the redundant eye to focus at infinity while observing the object through the viewing eyepiece. The final adjustment can now be made to :

(a) the focusing eyepiece graticule and
(b) the object. Learn to carefully look into the subject and not at it. When satisfied with the field and conditions, *shoot*.

37

Inverted Techniques

37.1 Explanation

This is a new and interesting technique for the study of thin translucent materials. Instead of using conventional transmitted light, which reveals little if any information beyond peripheral morphology, the idea is to partially reflect light back from the top surface of the glass/sample interface. Although colloidal silica solution will be used throughout the illustrations these techniques will have a much wider appeal and application. (It is the solution that is crystalline; the amorphous silica forms the visually dark central print.)

37.2 Bright Field

Figure 37.1 is the inverted bright field ray diagram showing the partial transmission and reflected rays. The two recombined waves are confined to the glass-to-air surface and air-to-crystal surface, which applies similarly with light propagating from below the slide (as shown) or from above. Figure 37.2(a) (see Plate 16) shows the existence of interference patterns from within the columnar grains.

37.3 Dark Ground

Figure 37.3 shows the ray diagram for inverted dark ground; the resultant image is displayed

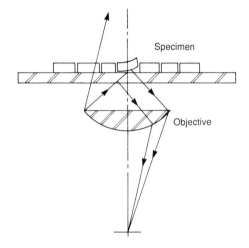

Figure 37.1 Inverted bright field—detects crystal boundaries and surface interference

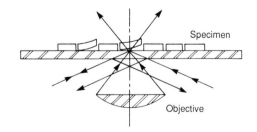

Figure 37.3 Inverted dark ground with low NA—detects crystal boundaries, dislocations and dichroism

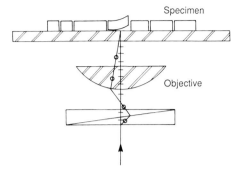

Figure 37.4 Inverted differential interference contrast—detects surface slope, surface interference and dichroism

in Figure 37.2(b) (Plate 16). It was not surprising to reveal the crystal morphology but it was very surprising to see the so-called dichroism effect. This effect could be the result of inclined crystals.

37.4 DIC

The principle is shown in Figure 37.4 with Figure 37.2(c) and (d) (Plate 16) used as examples of differential interference contrast with an extinguished background. Notice the green shades; this is usually a tell-tale sign of reciprocity failure. With exposures of 20 minutes this is hardly surprising. The 100 ISO film used is colour balanced for a maximum exposure of 1 second. This effect could have been overcome by changing the microscope/format/film or more colour compensating filters. The use of a first-order red sensitive tint plate into the extinguished background is to warm the image producing the spectacular colours as shown. No reciprocity problems have occurred even though this daylight film was used at its 100 ISO setting without compensating filters for an exposure time of 30 seconds.

The advice supplied with the film when using a tungsten light source is to fit a Wratten filter 80A and exposure at an effective ISO speed of 25. The reason for ignoring this advice is that the colours are optically stained and as such must be considered as compatible with daylight film (even though they were created from light emanating from a tungsten filament source).

It is possible to change the subject and background colour by changing the birefringent wedge position. Examples of this are shown in Figure 37.2(f) and (g) (Plate 16). By carefully adjusting the wedge position it is possible to selectively extinguish different features within the subject structure. Examples of this are illustrated in the next chapter.

38

Photomicrography in Practice

38.1 Background

I recall the reaction of my peers, many years ago, when first photographing materials that orthoscopically manifest themselves as boring grey level subjects, yet are ideal subjects for optical staining into coloured images. Optical staining has long been used as a diagnostic tool in life sciences, the aesthetics being an added bonus. This was not the case with the classical metallurgist whose training demanded planar grey level pictures with no 'frills'. Thankfully this is not quite the case today but there still remains a doubt with many microscope users in the materials field when making an analysis beyond the subjective cosmetics of colour.

Materials are not normally coloured when observed down the microscope, nor do the life science specimens appear coloured without staining, be it chemically or optically. In order to identify or enhance our knowledge of the subject we need to make full use of both of these staining techniques. Much has been documented on chemical staining; the following will therefore concentrate on optical staining.

Before we run away with the idea that it is colour or nothing let me put in a plea for good black and white images—they too can be revealing in the correct mode. The bacteriologist will, for example, use dark ground to highlight the tiny bacteria against a black background. This technique can be equally informative to the materials technologist when identifying, for example, very small particulate or precipitates (think of the visible stars against a clear night sky). Dark field techniques can also be very rewarding if the subject is partially translucent; instead of having a black background against a bright deflected area you have the addition of a refracted image. One of the most impressive examples of black and white images is to observe between crossed polar spheroidal graphite cast iron; the giroscopic effect is quite dramatic. Non-bireflecting materials can be extinguished between crosspolars; we therefore have an ideal background to enhance mineral inclusions that are bireflecting. In this last example we can reintroduce the extinguished background (if so desired) by using an additional oblique illumination.

Chapter 32 made particular reference to what is called 'first-order grey differential interference contrast'. When taking black and white micrographs, this technique enhances specimen contrast and image depth, often preferred to the flatter more colourful wedge positions. Microscopy is about looking *into* the subject and not *at* it; the first-order grey DIC technique certainly does that, as it gives a three-dimensional effect to the smallest of surface irregularities by revealing information not available without it. This 'in-depth' knowledge of the material structure is very important, but unfortunately to the uninformed observer its appearance is synonymous with a poorly prepared (gross relief) subject. This technique can make a 2 μm difference in topography look like the surface of a mountain terrain.

The writer recalls some of his best grey level DIC micrographs being suppressed in case they were to be interpreted not for the information they unearthed but for the possible incorrect interpretation others could conclude relating to the technique of preparation. This leads me quite naturally into the area of photomicrography. We must not only make the best use of our microscope techniques but also photographically record the result; Chapter 36 covers the techniques involved.

Photomicrographs can be very beautiful. The writer has a collection sized 20 in × 16 in permanently displayed in an entrance hall giving pleasure to all who pause and immerse themselves in the colours and endless shapes. When you ask the layman what he or she thinks of a particular micrograph you have replies that vary from 'a stormy night' to 'a cornfield in mid-summer'. An experiment using 10 different people observing 24 different micrographs and asking for their favourite 5 to be individually listed resulted in 2 of the micrographs being chosen by all 10 people and another one chosen by 8 of the 10. The rest were made up using nearly all the different micrographs. It was not surprising to see the spread of interest but it was surprising to see most people sharing the same views with 3 out of the 24 pictures.

Every good musician must share his talent with others and not just perform for himself. Similarly in microscopy there is no substitute for actually looking down the microscope and into the subject—let the lay person into this fascinating world and share it with your non-scientific colleagues.

The following will, I hope, contribute to a better understanding in this specialized field of microscopy and perhaps encourage the user to experiment and contribute to the many different photomicrographic competitions (Nikon, Olympus, Polaroid, Buehler/Metaserv, Kodak, International Metallographic Society, etc.). These competitions usually receive many entries in transmitted light, in particular in polarized light, but where are those wonderful optically stained shots in reflected light?

38.2 Equipment

The word photomicrography is synonymous with expensive equipment. Ask a microscope manufacturer (or supplier) to send you information on his metallographs and you will receive the most comprehensive sophisticated instrument available. Because of its sophistication, in the optical path it will have many extra mirrors and glass-to-air surfaces such as correction lenses, magnification changers, zoom lenses, reimaging lens, etc. All these 'extras' degrade the image and result in a lower quality image than the simple microscope with its direct optical path. What the metallographs offer is stability (from vibration) and versatility.

If we consider commercial pressures then to take a good photograph we need good and expensive equipment! This is *not true*; all that is required is an objective and a finite focus eyepiece imaging onto the camera film plane, with the appropriate connections.

The quality of the picture is dependent upon the care taken in preparing the sample—care ensuring Köhler illumination and the resolution of the film being used. The following micrographs have therefore been taken from well-prepared samples using an inexpensive upright microscope on a 5 in × 4 in cut film transparency (35 mm would give similar results).

38.3 Choice of Film

Irrespective of the type of camera it is always wise to give preference to the slowest film (ISO 50 to 100) because of the fine grain. With colour emulsions in photomicrography the choice is restricted mainly to transparency films; these films are balanced for light of a particular colour temperature. From Chapter 36 you will see how the colour balance is based on not only the microscope illuminating colour temperature but also the manner in which the specimen colour is created (via chemical staining or optical staining).

Most winning entries in photomicrographic competitions make use of the high colour contrast films. The writer has for this book used Kodak film Ektachrome 100 which, although not so contrasting, does give a more subtle and faithful reproduction. There is a case when taking micrographs for competitions to create the most dramatic effect, and if this means exaggerating the colours, via film or filter, then so be it.

The two most used formats are 35 mm and 4 in × 5 in; 35 mm tends to be the modern preference, perhaps because of convenience. The writer, however, always uses cut-film holders, in part because photomicrography is not a regular or large volume requirement but also because this large format has given some excellent results.

38.4 Black and White Micrographs

Do not shoot in black and white as you will not win any competition. This is *not true*. Had Figure 38.1 been presented at the last Buehler/Metaserv competition then I would have given it a prize, but then not only am I biased but also not a judge.

Figures 38.1 and 38.2 are examples of 'in-depth' studies using the first-order grey differential interference contrast position. Most photographic competitions displaying the grey level three-dimensional effect are usually the result of an SEM; although SEM pictures are extremely eye catching they are everyday happenings and should not be compared with the more difficult creations resulting from the optical microscope.

The ceramic particles in Figure 38.1 reveal a wealth of information not visible in bright field. The advice to all practising materials microscopists is to look *into* the subject and not

Figure 38.2 Plasma spray (× 200)

just at it; Figure 38.2 is an excellent example of this where an analysis has to be made relating to integrity of the sample preparation on a plasma-sprayed coating. Chapter 12 on coatings indicated the need for accurate porosity readings. These readings, of course, would be false and pointless if the preparation technique had induced or enlarged the visible dark areas. The first-order grey exaggerates any surface height difference that occurs with mechanical preparation due to varying abrasion rates. This allows the microscopist to readily detect those holes that are resin ingressed and, what is more important, those holes that are not. To concentrate on the unfilled holes allows an analysis relative to shape and composition, depicting true or specimen induced porosity. There are many alternative proposals for detecting sample integrity of coatings, but there is no real substitute for an in-depth DIC investigation.

The cross-section of an avocado pear (Figure 38.3) is in fact that of iron oxide; the effect is created by observing between cross-polars. It makes a welcome change from the traditional SG iron photomicrograph. Figure 38.4 illustrates the sensitivity to which features can be highlighted in a truly three-dimensional mode, clearly revealing every phase characteristic. The silicon oxide (white) is shown in its globular and dendritic modes, the interesting structure of the iron oxide quartz matrix being related to differing abrasion rates. This subject is nearly featureless when observed in bright field.

Figure 38.1 Ceramic MMC—DIC (× 200)

Photomicrography in Practice 331

Figure 38.3 Iron oxide—crossed polars (×100)

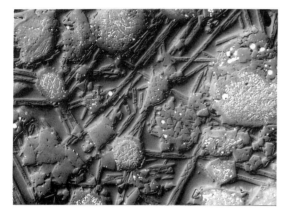

Figure 38.4 DIC (×200)

38.5 Micrographs in Colour

The geologist makes full use of colour in the identification of minerals; it is not just a pretty picture. We in the materials field must equally make use of colour and hopefully not just to make a pretty picture. The examples that follow will illustrate the 'information gains' that colour can give. If by chance they are more aesthetically pleasing, this is taken as a bonus. When photographing for a competition one could sacrifice information for aesthetics. The writer when making a micrographic collection noticed the predominance of blue; this is not surprising since blue resolves lined detail better than, say, red. In order to create a balance of colour it was therefore necessary to introduce some reds.

When preparing for a micrographic competition the technical content plays a major role. The micrographs (or macrographs) must therefore be telling a story beyond just the photograph. It could be the difficulty of the actual material to prepare or the choice of optical staining to reveal specific information; it could be the optical technique relating to improved structural analysis. Some of these points will be investigated below.

Above all, when preparing for a competition make sure the material used as a subject is capable of responding to optical staining. Some competitions put less emphasis on technical content and more on aesthetics. For this type of competition you could use those special 'once in a lifetime' shots, such as the final micrograph to this book. Alternatively, experiment with crystals, such as the colloidal silica, as in Chapter 37.

38.6 Optical Technique

The chosen technique should be undertaken with a knowledge of what one expects to reveal from the specimen and the appropriate optical technique compatible with those expectations. With the material iron oxide in a quartz matrix (Figures 38.5 and 38.6; see Plate 17) at least two important considerations need investigating:

(a) General morphology
(b) Subsurface information

(i) The dendritic structures would best be observed using normal reflected light techniques. However, since this material could respond to using crossed polars (bireflecting) this must also be investigated. In this case having established a lack of response to polarized light within different dendrites, the choice of colour is required. Select the most faithful shade and hence the choice of blue (reds could look better but dendritic morphology would not be as sharp).

(ii) Since some light will pass through the quartz matrix, dark ground is a worthy contributor since it will identify:
   (a) top surface reflections (black)
   (b) top surface deflections (white) and
   (c) refracted/reflected light.

Figures 38.5 and 38.6 (see Plate 17) are identical fields of view yet dramatically different pictures. Figure 38.5 is taken in polarized light using a first-order red sensitive tint plate (polars not crossed) and clearly shows the iron oxide dendrites in the quartz matrix. Figure 38.6 in dark ground shows only the iron oxide top surface in dark field, the edge of the dendrites highlighted with white prism-like reflections appearing within the red granular matrix.

Note: the white prisms are optical interference patterns and are not part of the specimen morphology. This micrograph is technically very important and for those students interested in microtopography is worthy of an explanation. Multiple-beam interferograms of such materials as diamond and quartz have revealed a most striking combination of trigons when observed by double beam interference techniques. The trigons within trigons illustrated in the micrograph indicate the type of slope (plane or curved). Where separate trigons are displayed and associated with what looks like progressively reduced intensity of the original trigon, a different explanation is necessary. Quartz often manifests growth centres that have striation steps. These striations are causing partially transmitted light to reflect back and recombine with the top surface reflected ray, causing a lower intensity interference image. As the partially reflected ray continues past the first striation and reflects back, interfering with the top surface reflection, this additional image is even lower in intensity and takes a new position. This action continues until the image intensity is too low to become visible. Figure 38.6 was taken using a 0.40 NA objective, the reflected illumination being 0.40 to 0.65 NA. To use a higher NA would result in increased surface reflection (at 0.9 NA it is 100%); to use a lower NA would result in too much transmission. Without the knowledge of this phenomenon it would be quite easy to make an incorrect analysis, i.e. the white prisms you see are not really there!!

38.7 Magnification Selection

All objectives have a specific working depth of field, i.e. the limit beyond which the subject structure becomes blurred. Figures 38.7 and 38.8 (see Plate 17) are examples of a well-prepared carbon/carbon composite material. This material is very birefringement and rewards the observer with an infinite variety of colours simply by rotating the polarizer in combination with the analyser and sensitive tint. Figure 38.7 could on first sight be the first choice because of its attractive colours, but a close inspection of this higher magnified shot shows a lack of clarity in structural detail. This is overcome by increasing the depth of field, i.e. reducing the objective magnification, as in Figure 38.8.

If the colour in Figure 38.7 is preferred, there is no reason why this should not be used in Figure 38.8.

38.8 Constituent or Structural Identification

Most specimens of materials when observed orthoscopically under normal bright field illumination manifest weak shades on a usually bright white background (unless etched). The object is to enhance these shades so that they can be observed more clearly, and if necessary exaggerate any surface deviation to study the microstructure.

Some specimens reflect the incident ray so weakly, say 5% (coke and ceramics), that the internal reflections have to be reduced to achieve any contrast. The use of the polarizer and reduced aperture iris is not on this occasion comparable with using an oil immersion objective (even for low power). The oil reduces light scattering from the specimen top surface and the objective front lens.

Constituent information relating to bireflection is easy to identify if the specimen is birefringent (different colours). Take, for example, an alumina sprayed coating. This subject in bright field looks relatively featureless and between crossed polars it also looks featureless but brighter; however, unmelted particles are extinguished (look black). When observing these materials exhibiting porosity the black features have to be identified and grouped. The unmelted particles are identified as above, the remainder by the first-order grey DIC technique.

Some materials are extremely difficult to prepare; none more so than the two examples in Figure 38.9 (see Plate 17), a soft bearing metal, and Figure 38.10 (Plate 17), a glass matrix material (both micrographs photographed in DIC). The 'in-depth' look *into* the bearing metal clearly shows all the constituents and the condition at the steel bearing interface.

The glass matrix material has silicon carbide fibres distributed transversely and longitudinally. The information revealed has to show the matrix/fibre interface; poor preparation techniques would make this impossible.

The judges when comparing entries such as these will allow extra marks for the preparation achievements. Additionally there is the use of DIC in colour to show structural conditions as well as constituent differences. Note the rosette effect within the heat-resistant glass matrix in Figure 38.10.

38.9 Looking Within

Figure 38.11 (Plate 18) embraces the best of preparation techniques for an extremely difficult material (large silicon carbide particles in a nickel aluminide/aluminium matrix). This is associated with the appropriate optical technique required to check the integrity of the preparation. Without DIC one would be unable to look *within* the subject. High pressure polishing would obscure any porosity level. Additionally without preparation on special formulated metal composite surfaces the sample would be in gross relief. Incorrect choice of abrasive size would result in chipped particles. This photograph is a winner in my book.

The 'in-depth' look at the slag micrograph of Figure 38.12 (Plate 18) is interesting not on this occasion for its preparation technique so much as the careful manner in which the dual dendrites manifest themselves at different levels. This subject is also a candidate for dark field illumination. The results would be as shown in Figures 38.5 and 38.6.

Figure 38.13 (see Plate 18), shot in grey DIC with first-order red tint, on first appearance could be compared with other examples of similar MMC materials (alumina fibres in aluminium). A closer inspection, however, will reveal the sharpness of every detail. This quality of integrity and resolution will withstand optical observations using the ×100 objective without losing its sharpness. This test emphasizes the high level of preparation.

38.10 Combined Techniques

The object of using a micrograph in a technical report or for publication is to convey information by visual impact rather than resort to many words. Words can be interpreted in different ways, just as statistics can be made to suit a particular slant. Micrographs, because they do not use words or numbers, should portray a clear and unequivocal description of all aspects of interest. This is not always possible when confining activities to one optical technique. Because of this the chosen optical technique can be a compromise, allowing the observer the chance of misinterpretation. To illustrate this point a coated ceramic (aluminium nitride coated with molybdenum, nickel, gold) has been photographed in bright field, dark ground and differential interference contrast (Figures 38.14 to 38.16; Plate 18).

(a) *Bright field* (Figure 38.14)
    Substrate   Evidence of grains
    Diffusion layer   The substrate grains are still apparent within the diffusion layer
    Molybdenum   Showing porosity though it is not clear where the nickel interface takes place
    Nickel   Not in evidence
    Gold   This is at the resin interface but is not revealed in bright field

(b) *Dark ground* (Figure 38.15)
    Substrate   Still evidence of grain boundaries; had any subsurface damage occurred it would have manifested itself as dense white areas
    Diffusion layer   Still clearly defined
    Molybdenum   More structural information appearing
    Nickel   Now clearly defined by its dark black appearance
    Gold   A perfect technique for illustrating this submicrometre layer

(c) *DIC* (Figure 38.16)

Substrate	The grains are so well resolved, this is the ideal technique
Diffusion layer	The penetration could not be clearer
Molybdenum	Information now confused
Nickel	It all looks like gold
Gold	Confused with nickel

The three examples adequately demonstrate the need to explore different optical techniques when displaying visual information and to define the particular element of interest applicable to that technique.

38.11 Selective Extinction of Colours

The geologist makes use of the technique of selective extinction when compensating for different minerals. When image colour is the result of wave interference it is possible to change that resultant colour by varying the optical path difference for different wavelengths and thereby extinguishing selective colours. The first two examples are a result of interference by inverted DIC techniques, the third example by classical reflected light polarizing methods. The first example 'The Christmas Tree' (Figure 38.17; see Plate 19) shows the effect when most of the columnar silica crystals have been extinguished. Notice how the central growth line has 'end-on' needle-like structures, the remaining features being totally extinguished. Without selective extinction the whole micrograph would be a mass of crystal colours. To capture this picture using a 100 ASA film on a projection microscope on a 5 in × 4 in format required an exposure time of 50 minutes. Although every effort was made to avoid reciprocity failure, the familiar green appearance is starting to show. This same picture if taken using a smaller format film (35 mm) devoid of the necessity for projection lenses and mirrors would only require approximately 2 minutes of exposure.

'Cleavage' (Figure 38.18; Plate 19) is used to illustrate extreme depth. Remember that we are looking at a flat subject; the appearance of depth is an optical creation. Unlike the previous example where the crystal growth line was surrounded by similarly coloured crystals, on this occasion only one side of the growth line is coloured.

The last example 'The Cavern' (Figure 38.19; see Plate 20) has used a combination of extinction, colour and magnification in creating this three-dimensional effect from a mineral that has been ground and polished perfectly flat and planar.

These three micrographs were produced for their aesthetic qualities. It nevertheless illustrates another facet of highlighting features.

38.12 Capitalizing on Reciprocity Failure

To create a balanced coloured micrograph it always helps if the three primary colours are present. Since green is present in our everyday landscape views, we take it for granted. Optical staining, on the other hand, is virtually exclusive of green. We can, however, induce green by ignoring the required colour compensating requirements of the film. 'The Flying Moth' (Figure 38.20; see Plate 20) is an illustration of colours introduced by avoiding compensation.

38.13 Constructing the Picture

Composition is very important. It is not good enough to make all the necessary adjustments to the microscope and illumination without carefully scanning the specimen for the most suitable FOV. Having selected the FOV the picture then requires composing; the choice of filters, optical technique and magnification orientation are all to be considered. Often as composition progresses some undesirable feature manifests itself. The ability to wash out unwanted information must be pursued. Orientation of the specimen can sometimes 'wash out' scratches. The polarizer, analyser, filter, illumination focus, etc., can help to preferentially dampen, allowing dominance of the important feature. The example illustrated, 'Horoscope' (Figures 38.21 and 38.22) is a case where the bireflecting features of the subject also enhance unwanted information.

On this occasion the field diaphragm must be sharply focused and stopped down as shown. Every micrograph has a natural

Photomicrography in Practice 335

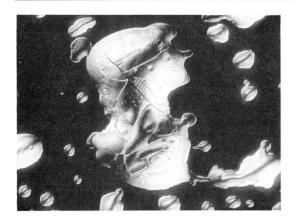

Figure 38.21 Horoscope. Oxide slurry—open field diaphragm (×100)

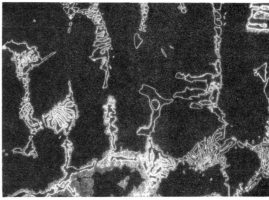

Figure 38.24 Animal. Cast alloy steel—DG (×400)

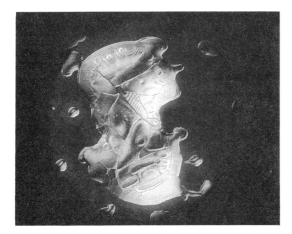

Figure 38.22 Horoscope. Oxide slurry—reduced field diaphragm (×100)

orientation. Before presenting any micrograph for others to see keep turning it around until this position is realized. 'Horoscope' is an excellent example of this point; rotate the image 90° and the subject is destroyed.

38.14 Novel Photomicrographs

The world of microscopy is a fascinating world, full of different shapes and colours. The microscopist looking for the novel shot must be imaginative as the sections are scanned. Take, for example, the 'Fourth Horseman of the Apocalypse' in Figure 38.23 (see Plate 21). Look at the sky, the hills, the sea, the horserider, the mushroom cloud from the atom bomb—yes, you need imagination.

Can you make out the animal figure in Figure 38.24? It is on the left-hand side of the picture. What about 'Blackpool Illuminations' ('The 4th of July' for the USA) in Figure 38.25 (Plate 21) or 'The Sky at Night' in Figure 38.26 (Plate 21).

Figure 38.27 (see Plate 21) shows how you can superimpose two images together to further enhance your ideas in 'Attack on Planet Earth'.

What about 'The Mountain Range' in Figure 38.28 (Plate 22) or 'The Isle of Skye' in Figure 38.29 (see Plate 22) or even 'The Sea Bed' in Figure 38.30 (see Plate 22)? Thinking of 'Going for Colour' in Figure 38.31 (Plate 22) or better still, 'Double Interference' in Figure 38.32 (Plate 22), 'Amoeba' in Figure 38.33 (see Plate 22) is actually a prepared steel sample. 'Diamonds Are Forever' (Figure 38.34, Plate 23), unfortunately, is only fractured silicon—still, it is better to be imaginative. Can you imagine the joy of finding 'The Easter Chick' as shown in Figure 38.35 (see Plate 23). See if you can find it in Figure 14.13 (Plate 9), 'The Easter Chick' is also being closely watched by a 'Wild Cat'.

38.14.1 Cleanliness

Cleanliness has always been important in materials surface preparation; the transfer of abrasives from one stage to the next has necessitated good housekeeping. With the advent of oxide and silica polishing it has become more prudent to ensure that the specimen is clean before making a microstructural analysis. The example shown in Figure 38.36 (see Plate 23) called 'The Stormy Plain' is in fact what greeted the microscopist when trying to make an analysis of a prepared titanium surface. This stain/crystal growth is the residue left after inadequate cleaning, using colloidal silica.

38.14.2 Morphology by contrast

Although this book has been, in general, concerned with the microscopy of mechanically worked surfaces, there are cases in the electronics industry where it is necessary to make an analysis from unworked surfaces. One such example is shown in Figures 38.37 and 38.38 (see Plate 23), where both darkground and crossed polars are used to highlight morphology. This subject when observed in bright field would not be very revealing.

38.15 Poster Presentation

It has become popular at many conferences to offer a poster display facility. This allows additional material applicable to the conference to be displayed at strategic points (refreshment area, etc.). In order to attract attention to individual presentation, there must be some visual impact.

The two illustrations, Figures 38.39 and 38.40 (see Plate 24), are examples of poster work presented by the author in the past. The object of these posters was to attract the attention of the delegate to the beautiful micrographs and hope that the commercial impact would result from reading the simple advertising slogan. Note how both examples have some kind of subject title and how the chosen card complements the colours in the micrographs; the hand writing is neat. Had it proved necessary to use a typed text, this would be first photographed to give maximum contrast and then mounted on card. The card would then be positioned behind the framed window, which would have been cut in advance.

38.16 Facing up to the Competition

Far be it for me to suggest how to enter (or win) photographic competitions, but one can through experience in judging give some useful guide-lines:

(i) If the maximum size is quoted in the rules it usually pays to use that size.
(ii) Present your entry on strong card.
(iii) Avoid writing by hand unless you are very neat.
(iv) Do everything in accordance with the rules.
(v) Pay as much attention to the general presentation as you do the photographs.
(vi) When using typed words, they can be photographed to give extra contrast before mounting on presentation cards.
(vii) Presentation card is available in many different colours. Select one that complements the photographs.

38.17 That Once in a Lifetime Shot

Over the last four decades many different creatures have looked back at me from within the object field of view. The dragon-like creatures illustrated at the front of this book are examples of how they can be captured and used in a marketing context. Some others have been illustrated in this last chapter. In order to share the experience with others, creatures of imagination have to be enhanced by changing the FOV or optical technique, to display the percepted image. However, once in my lifetime this fish came swimming along. Experiments being carried out on crystallization habits of certain polishing mediums necessitated a microstructural analysis of crystal shapes and forms. These tests were usually carried out after the crystals had formed. The rate of crystallization, however, necessitated staged interval analysis. It was during one of those interval tests that 'Jaws' came swimming into view.

Out with the camera, literally shaking with excitement, 'Jaws' was captured with the most wonderful colours imaginable—blue, green and even red seas. The colour scales on the fish were unbelievable. Yes, this truly was not imagination, with its eye looking back at me it was difficult to stop shooting off all the film. Eventually my thoughts went to sharing this experience with others, but in my excitement

I had not realized that everyone had gone home. *Then* I realized that all my shots had been exposed at 400 ASA, which was correct for the 'framing' Polaroid shot, but totally incorrect for 50 ASA film being used. To my horror I found I had only one more shot left in the camera; without that last shot this story would have been just another fisherman's tale about the one that got away. Within minutes of this last shot the crystals regrouped and 'Jaws' swam away into the unknown, never to return. 'The one that didn't get away' is shown in Figure 38.41 (see Plate 24).

Bibliography

References

Although this book is a personal résumé of the author's experience in this specialized field, we all at some time are indebted to knowledge passed on by others, be it in person or by published material. An attempt to acknowledge some of these people has been included under 'Further Reading'. I extend my sincere apologies to those I have inadvertently omitted.

Weiler, W. W., *A New Microhardness Test Method*, The American Society for Testing and Materials.
The European Society for Microstructural Traceability, Physics Department, University of Warwick, UK.
Education Training Authority, 54 Clarendon Road, Watford, UK.

Further Reading

Alexander, W. and Street, A., *Metals in the Service of Man*, Pelican, London, 1989.
Hallimond, A. F., *The Polarizing Microscope*, Vickers Instruments, York, 1970.
Kerr, P. F., *Optical Mineralogy*, McGraw-Hill, New York, 1977.
Marshall, C. E., *Crystal Optics*, Cooke, Troughton & Sims, York, 1953.
Martin & Johnson, *Practical Microscopy*, Blackie, London, 1931.
Metals and Materials, Back issues to 1985, The Institute of Metals, London.
Oliver, C. W., *The Intelligent Use of the Microscope*, Chapman & Hall, London, 1951.
Payne, B. O., *Microscope Design and Construction*, Cooke, Troughton & Sims, York, 1957.
RMS, *Dictionary of Light Microscopy*, OUP, Oxford, 1990.
Samuels, L. E., *Metallographic Polishing by Mechanical Methods*, Pitman, London, 1971.
Tolansky, S., *Surface Microtopography*, Longmans, London, 1960.
Van der Voort, George F., *Metallography Principles and Practice*, McGraw-Hill, New York, 1984.

Acknowledgements

Ian English	For his support, encouragement and review
Margaret Walters	For her unstinting work in the preparation of the manuscript
Laboratory staff at Buehler UK Limited	For their assistance in the preparation of microsections

Thanks must also go to Peter Strong for making this book possible.

Index

Abbe 272
Abradable wheels 12, 14
Abrasive, charged 33, 61, 71, 120, 123
  critical 26
  degrading 35, 63
  fixed 33, 61, 120, 126
  function 34, 118
  impressed 101, 117
  rate 18, 20
  shape 28, 64, 118
  size 28, 62, 97, 120, 123
  types 12, 27, 33
Achromatic 246
Aery 262
Amosite 299
Amplitude 275
Analyser 291, 324
Anisotropic 277, 293
Anodizing 292, 300
Aplanatic 244, 272
Apochromatic 246
Artefact 16
Aspherized 245
Astigmatic 250
Auditing 74, 82, 88, 196

Bacteriologist 247
Beam-splitter 282
Becke line 276
Bellows 238
Bertrand lens 291
Biaxial 299
Biodegradation 137
Bireflectance 293

Birefringent 283, 292, 324
Blooming 250
Bond, chip 150
Bond, cutting wheels 13
Bone 151
Brewster angle 253
Bright field 72, 105, 285, 326
Brittle 117, 165

Camera 311
Catoptric 286
CCTV 259, 270, 287
Chip, primary 149
  secondary 150
Chromosomes 247
Chrysotile 299
Classification, platen surface 33, 74
  polishing cloths 67, 79
Cleaning, lens 320
Coating, specification 98
Colour temperature 316
Coma 249
Comet tail 250
Compensators 295
Competence 210
Concept, new 30, 58, 67
Conjugates 313
Conoscopic 298
Contrast 293
Conventions, operating 44
Cosmetic finishing 65
Crocidolite 299
Crossed polars 105, 110
Cutting 25, 143

Damage, structural 27, 85, 119
Dark ground 72, 257, 272, 285, 326
Data imprinting 319
Definition, sample/specimen 24
    abrasive function 37
    platen surface 38
Deformation 7, 10, 35, 64, 72, 76
Delamination 100
Delicate components 164
Depth of field 239, 249, 256
Depth of focus 249
Desmear 167
Diaphragm 247, 264, 322
Differential interference contrast (DIC) 72, 106, 108, 112, 134, 142, 327
Diffraction 254, 273
Dioptre 312
Dissimilar constituents 165
Dispersion 276
Distortion 250
Ductility 25, 71, 78
Dyes 102

Edge retention 17
EDS analysis 139
Electrolytic polishing 68, 302
Encapsulation, *see* Mounting
Etching 168, 200, 301
Exposure 235, 322
Extinction 284, 334
Eyepiece 233, 259, 314
    micrometer 260
    filar 260, 270
    Huygenian 270
    Compensating 270

Faithful reproduction 1, 3, 99, 113
Family classification 47
Field curvature 249
Field of view 259, 269
Filament 267
Film 314, 322, 329
Filter 316, 323
First order red excitation 104, 323, 332
Fluorescence 102, 104, 247
Fluorite 246
Focus, precise 261
Friable 117
Friction 204

Ghost images 268
Globular-particles 99
Graticule 177, 288
Gratings 273
Grey, first-order 284
Grinding 37, 63, 89, 102, 117, 119, 138, 211
Grit size 58

Halo 244
Hard 165
Hardness, micro 206
    diagonals 208
Huygenian eyepiece 270

Igneous 294
Illumination 235, 285
Image, real 241
    primary 242
    secondary 242
    splitting 260
    virtual 241
Image analysis 98, 197
Indentation 203
Integrity, curve 31
    thin sections 146
Intensity 257
Isotropic 292

Kelner 270
Kerf 10
Kodak 289
Kohler 263, 320

Lamp, tungsten 304
Lapping 37, 63
Lens aperture 239
Lidding 101
Light, transmitted 145
    reflected 152, 192, 235
Load 204, 209
Lubricant 11, 27, 63, 66, 71, 117

Magnification 208, 241, 242, 253, 332
Magnification changer 233
Maltese cross 129
Material, removal differential (MRD) 29
Materialographer 2
Measuring 176, 261
Mercury vapour 304
Metallized film 275
Metallography 3, 89
Metamorphic 295
Methods, empirical 74, 83, 84
    see also Family classification
Michelson 278
Microscaler 260
Microscope, compound 232, 242
    inverted 236
    macro 238
    stereoscopic 237
Microstructure 3
Mirau 278
Morphology 204, 293
Mounting, cold 17, 20, 144, 167, 192, 211
    hot 17, 211

Nomarski 282
Nosepiece 234
Numerical aperture 247, 250, 252, 272, 286

Objective 234
  characteristics 248
  finite 243
  infinite 243
  statistics 255
Optical disk 289
Optical sign 299
Optimization 44, 69, 72
Orthoscopic 243–283
Osseointegration 137
Oxide 100

Parfocal 243
PCB 166, 261
pH 116, 120
Photomicrography 98, 285, 310
Photomultiplier 312
Planar grinding 60, 144, 193
Platen surface, affinity 71
  efficiency 42
  function 38
  selection 43, 61
  threshold 81
  types 39
Plating 166
Pleochroism 294
Ploughing 25
Polarizer 291, 323
Polaroid 289, 315
Polars 283
Polishing 37, 90, 103, 168, 194, 212
  cloths 64, 73
  electrolytic 68
  stage 33, 145
  vibratory 67
Porosity 99, 117
Preparation, examples format 47
  see also Family classification
Pressure 27, 67, 117

Quarter wave plate 296
Quartz wedge 296

Rake angles 25, 27
Ramsden eyepiece 270
Rate 204
Reciprocity 315, 318, 334
Red excitation, see First order red excitation
Reflectivity 98, 140, 196, 250
Refractive index 244, 274, 252
Relief 17, 19, 65, 126
Residual damage 4
Resolution 247, 254
Resultant damage 7
Retardation 275

Roughness 204
Round robin 97
Royal Microscopical Society thread 243
Rubbing 25/65

Sample, integrity 33, 60, 144, 168, 193
  material 98
Scratch pattern 63, 72, 100
Screen 233, 257
Sectioning 7, 167, 192, 210
Sedimentary 295
Sensitive tint plate 285
Sine waves 279
Single lens reflex 313
Smearing 101
Snell's law 253
Soft 165
Specimen 24, 164
Spectra 273
Speed 12, 29, 66
Standards 82, 85, 87, 209
Structural damage, see Damage, structural
Substage 235

Technician training 1
Thermal cycling 167
Thickness 166
Time 77, 120, 204
Tolansky 278
Topography 278, 284
Tough 166
Traceability 6, 88, 205
Transmission lamps 304
Trielectrode 288
Tube length 243, 256, 268

Ultraviolet 303
Uniaxial 298
Universal language format 87
Unrefracted 277

Vacuum 102, 111, 138
Vibration 205, 312
Vibratory polishing 67
Video printers 289
Violet 303
Viscosity 102, 129

Wicking 167
Working clearance 274

Xenon arc 304

Young's modulus 137

Z axis 35, 97, 108, 129, 140, 194
  curves 4, 77, 81
  data bank 79, 86
  depth 78, 261
Zero order 275